Systems Genetics

Linking Genotypes and Phenotypes

Whereas genetic studies have traditionally focused on explaining heritance of single traits and their phenotypes, recent technological advances have made it possible to comprehensively dissect the genetic architecture of complex traits and to quantify how genes interact to shape phenotypes. This exciting new area has been termed systems genetics and is born out of a synthesis of multiple fields, integrating a range of approaches and exploiting our increased ability to obtain quantitative and detailed measurements on a broad spectrum of phenotypes.

Gathering the contributions of leading scientists, both computational and experimental, this book shows how experimental perturbations can help us to understand the link between genotype and phenotype. A snapshot of current research activity and state-of-the-art approaches to systems genetics is provided, including work from model organisms such as *Saccharomices cerevisiae*, *Drosophila melanogaster*, as well as from human studies.

Researchers and graduate students in genetics, functional genomics, bioinformatics, computational biology, systems biology, and biotechnology will find this a valuable and timely resource.

Florian Markowetz is a Group Leader at Cancer Research UK's Cambridge Research Institute. His research is concerned with developing statistical and mathematical models of complex biological systems and analysing large-scale molecular data. His research interests range from the analysis of molecular clinical data to inference of cellular networks from high-throughput gene perturbation screens and integration of heterogeneous data sources using machines learning techniques and probabilistic graphic models.

Michael Boutros is a Group Leader at the German Cancer Research Centre (DKFZ) in Heidelberg, where he heads the Division of Signalling and Functional Genomics. He also holds a Professorship at the University of Heidelberg. His research focuses on the systematic dissection signalling pathways in *Drosophila* and mammalian cells, which are important during development and cancer. He attempts to define key components of signalling pathways, discovering interaction between pathways, and characterisation of signalling networks under normal and perturbed conditions.

CAMBRIDGE SERIES IN SYSTEMS GENETICS

Series Editors:

Jason A. Moore, PhD
University of Pennsylvania

Scott M. Williams, PhD
Geisel School of Medicine at Dartmouth College

Systems Genetics is the study of DNA sequence variation and biological traits as a complex system characterised by spatial and temporal interactions. This includes phenomena such as epigenetics, epistasis, plastic reaction norms, and locus heterogeneity. The Cambridge Series in Systems Genetics covers all areas of genetics approached from a complex systems point of view. This series is of interest to researchers across several areas of the life sciences including bioinformatics, evolution, genomics, human genetics, molecular genetics, precision medicine, and systems biology.

Systems Genetics

Linking Genotypes and Phenotypes

Edited by

FLORIAN MARKOWETZ

Cancer Research UK Cambridge Institute

MICHAEL BOUTROS

German Cancer Research Center, Heidelberg

CAMBRIDGE
UNIVERSITY PRESS

University Printing House, Cambridge CB2 8BS, United Kingdom

One Liberty Plaza, 20th Floor, New York, NY 10006, USA

477 Williamstown Road, Port Melbourne, VIC 3207, Australia

314-321, 3rd Floor, Plot 3, Splendor Forum, Jasola District Centre, New Delhi - 110025, India

79 Anson Road, #06-04/06, Singapore 079906

Cambridge University Press is part of the University of Cambridge.

It furthers the University's mission by disseminating knowledge in the pursuit of
education, learning and research at the highest international levels of excellence.

www.cambridge.org
Information on this title: www.cambridge.org/9781108794596

© Cambridge University Press 2015

First published 2015
First paperback edition 2020

A catalogue record for this publication is available from the British Library

Library of Congress Cataloging in Publication data
Systems genetics : linking genotypes and phenotypes / edited by
Florian Markowetz, Cancer Research UK Cambridge Institute,
Michael Boutros, German Cancer Research Center, Heidelberg.
    pages   cm
Includes index.
ISBN 978-1-107-01384-1
1. Phenotype.   2. Genetics.   3. Functional genomics.
I. Markowetz, Florian.   II. Boutros, Michael.
QH438.5.S97   2015
576.5–dc23

2014050295

ISBN 978-1-107-01384-1 Hardback
ISBN 978-1-108-79459-6 Paperback

Additional resources for this publication at www.cambridge.org/SystemsGenetics

Contents

The colour plate section can be found between pages 148 and 149.

Contributors

Joel S. Bader
Johns Hopkins University, Baltimore, MD 21218, USA

Niko Beerenwinkel
Department of Biosystems Science and Engineering, ETH Zurich, CH-4058 Basel, Switzerland

Maximilian Billmann
German Cancer Research Center (DKFZ) and Heidelberg University, Division of Signaling and Functional Genomics, D-69120 Heidelberg, Germany

Charles Boone
Banting and Best Department of Medical Research and Department of Molecular Genetics, Terrence Donnelly Center for Cellular and Biomolecular Research, University of Toronto, Toronto, ON M5S 3E1, Canada

Michael Boutros
German Cancer Research Center (DKFZ) and Heidelberg University, Division of Signaling and Functional Genomics, D-69120 Heidelberg, Germany

Peter R. Braun
Steinbeis-Innovationszentrum, Center for Systems Biomedicine, D-14612, and Max Planck-Institüt für Infektionsbiologie, D-10117 Berlin

André E. X. Brown
MRC Clinical Sciences Centre, Faculty of Medicine, Imperial College, Du Cane Road, London, W12 0NN, UK

Ben Calderhead
Centre for Mathematics and Physics in the Life Sciences and Experimental Biology (CoMPLEX), University College London, London, WC1E 6BT, UK

Michael Costanzo
Banting and Best Department of Medical Research and Department of Molecular Genetics, Terrence Donnelly Center for Cellular and Biomolecular Research, University of Toronto, Toronto, ON M5S 3E1, Canada

Edgar Delgado-Eckert
Department of Biosystems Science and Engineering, ETH Zurich, CH-4058 Basel, Switzerland

Paolo Gallipoli
Paul O'Gorman Leukaemia Research Centre, College of Medical, Veterinary and Life Sciences, Institute of Cancer Sciences, University of Glasgow, Glasgow, G12 0ZD, UK

Mark A. Girolami
Department of Statistical Science, University College London, London, WC1E 6BT, UK

Andreas Hadjiprocopis
The Institute of Cancer Research (ICR), London, SW3 6JB, UK

Kate Holden-Dye
Steinbeis-Innovationszentrum, Center for Systems Biomedicine, D-14612 Berlin, Germany

Tessa L. Holyoake
Paul O'Gorman Leukaemia Research Centre, College of Medical, Veterinary and Life Sciences, Institute of Cancer Sciences, University of Glasgow, Glasgow, G12 0ZD, UK

Pengyu Hong
Department of Computer Science, Volen Center for Complex Systems, Brandeis University, Waltham, MA 02454, USA

Lisa E. M. Hopcroft
School of Computing Science, College of Science and Engineering, University of Glasgow, Glasgow, G12 8QQ, and Stem Cell and Leukaemia Proteomics Laboratory, Institute of Cancer Sciences, University of Manchester, Manchester, M20 3LJ, UK

Christina Laufer
German Cancer Research Center (DKFZ) and Heidelberg University, Division of Signaling and Functional Genomics, D-69120 Heidelberg, Germany

Rune Linding
Cellular Signal Integration Group (C-SIG), Center for Biological Sequence Analysis (CBS), Department of Systems Biology, Technical University of Denmark (DTU), DK-2800 Lyngby, Denmark

Florian Markowetz
Cancer Research UK Cambridge Institute, Cambridge, CB2 0RE, UK

André P. Mäurer
Steinbeis-Innovationszentrum, Center for Systems Biomedicine, Haydnallee 21, D-14612, and Max Planck-Institüt für Infektionsbiologie, Charitéplatz 1, D-10117 Berlin, Germany

Kimberly Maxfield
University of North Carolina – Chapel Hill School of Medicine, Chapel Hill, NC 27599, USA

Thomas F. Meyer
Max Planck-Institüt für Infektionsbiologie, Charitéplatz 1, D-10117 Berlin, Germany

Stephanie Mohr
Department of Genetics, Harvard Medical School, Boston, MA 02115, USA

Chad L. Myers
Department of Computer Science and Engineering, University of Minnesota-Twin Cities, Minneapolis, MN 55455, USA

Yongjin Park
Johns Hopkins University, Baltimore, MD 21218, USA

Norbert Perrimon
Department of Genetics, Harvard Medical School, Boston, MA, USA, and Howard Hughes Medical Institute, Harvard Medical School, Boston, MA 02115, USA

Carles Pons
Department of Computer Science and Engineering, University of Minnesota-Twin Cities, Minneapolis, MN 55455, USA

William R. Schafer
Medical Research Council Laboratory of Molecular Biology, Cambridge, CB2 0QH, UK

Xiaoyun Sun
Department of Computer Science, Volen Center for Complex Systems, Brandeis University, Waltham, MA 02454, USA

Arunachalam Vinayagam
Department of Genetics, Harvard Medical School, Boston, MA 02115, USA

Xin Wang
Cancer Research UK Cambridge Institute, Cambridge, CB2 0RE, UK

Angelique Whitehurst
University of North Caroline – Chapel Hill School of Medicine, Chapel Hill, NC 27599, USA

Ke Yuan
Cancer Research UK Cambridge Institute, Cambridge, CB2 0RE, UK

1 An introduction to systems genetics

Florian Markowetz and Michael Boutros

Systems genetics is an emerging field based on old approaches going back to the genetic studies performed by Gregor Mendel (Mendel 1866). Mendel's experiments primarily focused on explaining inheritance of single traits and their phenotypes – for example how specific genetic alleles influence colour or size of peas – but recently developed technologies can comprehensively dissect the genetic architecture of complex traits and quantify how genes interact to shape phenotypes by using natural variation or experimental perturbations as a basis to understand links from genotypes to phenotypes. This exciting new area has recently been termed 'systems genetics' (Civelek & Lusis 2014).

While the basic, underlying questions are not new, systems genetics builds upon major methodological advances that facilitate the measurement of genotypes and phenotypes in a previously unforeseen and comprehensive manner. With this arsenal at hand, one of the major aims of systems genetics is to understand "how genetic information is integrated, coordinated and ultimately transmitted through molecular, cellular and physiological networks to enable the higher-order functions and emergent properties of biological systems" (Nadeau & Dudley 2011).

1.1 Definition of systems genetics

Systems genetics is born out of a synthesis of multiple fields: it integrates approaches of genetics, genomics, systems biology and 'phenomics', that is, our increased ability to obtain quantitative and detailed measurements on a broad spectrum of phenotypes. One of the first papers using the term 'systems genetics' defines it as "the integration and anchoring of multi-dimensional data-types to underlying genetic variation" (Threadgill 2006). Since then, many studies have aimed at integrating genome-wide data across many different levels, and possibly different environments, in approaches that are closely related to quantitative genetics.

In our view, a systems genetic approach should bring together three dimensions: it should combine (i) a genome-wide analysis with (ii) many quantitative phenotypes, both at the molecular and organismal level, (iii) in many different conditions or environments (Fig. 1.1). As such, the integration that needs to be achieved by systems genetics goes

Systems Genetics: Linking Genotypes and Phenotypes, ed. F. Markowetz and M. Boutros. Published by Cambridge University Press. © Cambridge University Press 2015.

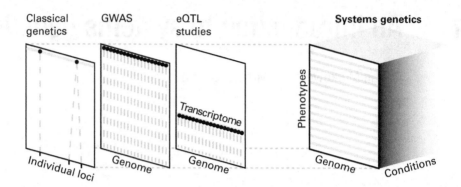

Figure 1.1   Systems genetics ideally combines genome-wide analysis with many quantitative phenotypes, both at the molecular and organismal level, in many different conditions or environments. This subsumes previous approaches that were focused on linking individual loci to a single phenotype (in classical genetical and epistatic analysis) or comprehensively linking many genomic loci to a single phenotype in a single condition (in GWAS or eQTL studies; see text).

well beyond a 'single' dimension of genotypes or phenotypes. For example, genome-wide association studies (GWAS) link natural variation to a single phenotype in a single condition. While systems biology often focuses on the dynamic behaviour of systems, systems genetics uses genomic techniques to link genotypes with large-scale, quantitative measurements of phenotypes.

While in general this might sound like a lofty goal which is hard to achieve, studies that come close already exist in model systems like plants (Sozzani & Benfey 2011, Topp et al. 2013), *Drosophila* (Ayroles et al. 2009) or human cell lines (ENCODE Consortium et al. 2012). For example, single-nucleotide polymorphisms (eQTL) analysis uses expression quantitative trait loci (SNPs) and transcripts to establish links between genotypes and phenotypes. The strength of such approaches is most easily demonstrated in model organisms where large amounts of phenotype and genotype data at different levels already exist. For example Skelly et al. (2013) measured transcript, protein, metabolite and morphological traits in 22 genetically diverse strains of *Saccharomyces cerevisiae* to gain insights into the spectrum and structure of phenotypic diversity and the characteristics influencing the ability to accurately predict phenotypes. Similarly in *Drosophila*, Ayroles et al. (2009) integrate gene expression and phenotypes across multiple conditions, leading to the identification of overlapping groups of transcripts that influence different organismal phenotypes (see Fig. 1.2). These studies address many fundamental questions, such as how many different SNPs influence the same phenotype (complex phenotypes) or how many different phenotypes are influenced by the same SNP (pleiotropy). Because such studies look globally across many phenotypes, they can shed light on interactions between genes. Genetic interactions, also called epistatic interactions, can then be used to dissect the genetic architecture of particular processes (Avery & Wasserman 1992, Phillips 1998, Phillips 2008).

Why is this an important area of research? Firstly, many traits are complex and not determined by a single gene. Instead, combinations of alleles with low abundance or

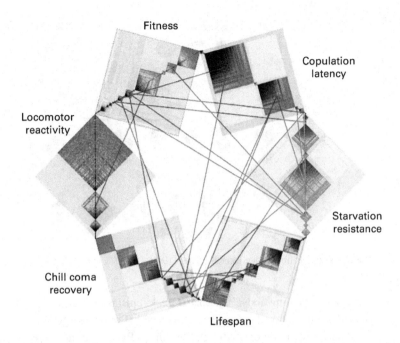

Fitness

Copulation
latency

Locomotor
reactivity

Starvation
resistance

Chill coma
recovery

Lifespan

Figure 1.2   Pleiotropy between phenotypic modules in *Drosophila*. Grey lines connect modules with a significant overlap of greater than four genes. Adapted by permission from Macmillan Publishers Ltd (Ayroles et al. 2009). A black and white version of this figure will appear in some formats. For the colour version, please refer to the plate section.

context-dependent effects (environmental or genetic background) are responsible for the bulk of phenotypic variation. Secondly, genetic interactions have also been proposed to account for missing heritability (Zuk et al. 2012) but this issue is still contentious and widely discussed (Eichler et al. 2010, Hemani et al. 2013). Thirdly, comprehensive network models built on principles of correlation and causation might be able to predict phenotypic outcome in response to different genetic backgrounds and environmental or therapeutic perturbations. These models could also identify targets for modulating phenotypic outcome which would be an important contribution to treatment and prevention of disease. These technological advances have been argued to entirely transform healthcare in the future (Friend & Ideker 2011).

1.2   History of systems genetics

The simplest example of a systems genetic question is how two genes influence a single phenotype. This question and the concept of epistasis have a long tradition in genetics, with theoretical underpinnings that go back more than a hundred years.

Epistasis: from one gene to two genes

Different definitions of epistasis emerged that vary between sub-disciplines (Phillips 1998, Cordell 2002, Phillips 2008). Bateson's original definition is based on the idea of

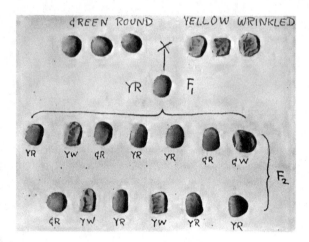

Figure 1.3   Seed-character inheritance after Bateson (1909).

one gene 'standing above' another (Greek: epistasis) (Bateson 1909) and is sometimes called 'compositional epistasis' or 'masking' (Phillips 2008).

Bateson wrote on epistatic relationships: "Pending a more precise knowledge of the nature of this relationship it will be enough to regard those factors which prevent others from manifesting their effects as higher, and the concealed factors as lower. In accordance with this suggestion the terms epistatic and hypostatic may conveniently be introduced" (Bateson 1909, p79) (see Fig. 1.3).

Fisher's statistical definition of epistasis (originally called 'epistacy') compares the phenotype of a two-loci genotype with the phenotype predicted by a model simply adding up the effects of the two loci (Fisher 1918). If the prediction of the additive model differs from the observed phenotype, the two loci are called epistatic (Phillips 2008). These concepts were established a hundred years ago, decades before genetic information was understood to be encoded by DNA.

Genomics: from one gene to genome-wide
In the last few decades the amount of genomic information has dramatically increased. The first eukaryotic genome, *S. cerevisiae*, was sequenced in 1995 and since the early 2000s the human genome sequence has been determined and the collection of human genes is known. Microarray and later sequencing technologies have led to the identification of genetic variants and the ability to measure gene activity genome-wide. New experimental tools like gene deletion libraries, RNA interference (RNAi) and transposon mutagenesis allow manipulation of all the genes in a genome (Boutros & Ahringer 2008). In observational studies, epistasis can be measured in genome-wide association studies (GWAS) (Cordell 2009), which allows modelling of genetic traits on a genome-wide level. Perturbation studies, combining loss-of-function alleles to assess their combined phenotypes, have been spurred by genome-wide deletion libraries. For example in yeast, a large part of the 18 million possible genetic interactions has been discovered by systematic double-knockout experiments (Costanzo et al. 2010)

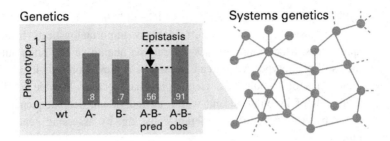

Figure 1.4   Systems genetics comprehensively uses concepts from classical genetics. One example is epistasis: where the phenotype of a double perturbation deviates from the prediction based on the phenotypes of the single perturbations. Automated, large-scale approaches extend this analysis to a genome-wide level (Costanzo et al. 2010). Each edge in a large genetic interaction network corresponds to one classical genetics experiment.

(see Fig. 1.4). Similarly, genetic epistasis has been discussed as a major driver underlying genotype-to-phenotype relationships of complex traits and of genome evolution (Mackay 2014).

Phenomics: from one phenotype to many phenotypes

In the past years, there has been much progress in technologies to quantitatively measure phenotypic data on a genome-wide scale. Molecular phenotypes, such as expression levels or transcription factor binding sites have been mapped in many organisms and across many conditions. For example, to better understand how genetic variation influences gene expression levels, genetic linkage and association mapping have identified cis- and trans-acting DNA variants in so-called expression quantitative trait loci (eQTL) studies (Cheung & Spielman 2009). The ENCODE (and modENCODE) projects generated many of such phenotypes that can be correlated to underlying genetic or environmental changes, including regions of transcription, transcription factor association, chromatin structure and histone modification in the human genome sequence (Celniker et al. 2009, ENCODE Consortium et al. 2012). At one level above, phenotypes can be measured at the level of the cell. For example, large-scale studies have interrogated the whole genome for genes that are required for cell proliferation (Berns et al. 2004, Boutros et al. 2004, Cheung et al. 2011, Marcotte et al. 2012, Vizeacoumar et al. 2013). Similarly, cells can be interrogated for more complex phenotypes, such as changes in cell shape, cell division, cell migration or to the response of specific signalling pathways (Kiger et al. 2003, Snijder et al. 2009, Fuchs et al. 2010, Snijder et al. 2012, Yin et al. 2013). Whole-organism phenotyping, e.g. in plants, has become feasible through novel technologies that can comprehensively characterize phenotypes in an automated manner (Sozzani & Benfey 2011). Phenome-wide association studies can help to understand the genetic basis of diseases: "There is great interest in making use of the large amounts of phenotypic data that are stored in electronic medical records (EMRs) in the quest to understand the genetic basis of disease. One way to do this is to carry out phenome-wide association studies (PheWASs), in which genetic variants are tested for association with a wide range of phenotypes" (Flintoft 2014).

Systems genetics: a synthesis of genetics, genomics and phenomics

Modern systems genetics differs in many important aspects from historic genetic approaches. Modern approaches can rely on quantitive and often automated measurements of phenotypes, which can be now be achieved on individuals (cells, organisms) rather than on populations. Genotyping has become feasible for hundreds or even thousands of individuals, and within an individual, deep sequencing allows to identify subsets of genetically different cells, e.g. by assessing genetic heterogeneity in tumours.

Advances in biological and medical imaging by high-resolution microscopy or functional MRI allow to measure phenotypes in a much more quantitative manner. Also, technological advances allow measurements of a variety of intermediate phenotypes in a high-throughput manner, including chromatin state, histone modifications and gene and soon protein expression. The ENCODE is a good example of a large-scale phenotyping project that characterised large parts of the genome (ENCODE Consortium et al. 2012). The analysis of these comprehensive data sets is enabled by new computational approaches to represent and interpret biological information. Examples are the network and graph theory approaches that have permeated wide parts of biological research in the last few years (Barabási & Oltvai 2004, Ideker & Krogan 2012). Network models can help to understand and interpret the roles of genetics and epigenetics in disease predisposition and etiology. By providing the backbone of molecular interactions through which signals are transduced and gene expression is regulated, networks have been proposed to limit the search space of allele variants and alterations that can be causally linked to the presentation of a phenotype (Califano et al. 2012).

The combination of new high-throughput experimental technologies with new computational analysis methods allows a comprehensive view on complex interactions between genes, beyond measuring epistasis on individual gene pairs. The focus of research has shifted from analysing individual genetic interactions to comparing detailed profiles of genetic interactions, which are much more informative of cellular genetic architecture (Costanzo et al. 2010, Baryshnikova et al. 2013). These technologies are far beyond what we were able to envision in the year 2000 and promise to revolutionise our understanding of the genetics of complex traits.

In cancer, genomic efforts to characterise tumour genomes on many molecular levels (The Cancer Genome Atlas Network 2012) have started to be complemented with quantitative assessment of cellular and tissue morphology (Beck et al. 2011, Yuan et al. 2012). Cell line studies have led to a better mechanistic understanding of how cancer genes influence complex cellular phenotypes, such as cell morphology and invasion (Yin et al. 2013). In pathology, computational methods reach back half a century (Smith & Melton 1964) and modern methods are mainly being used for quantifying immunohistochemical stainings in tissue microarrays (Schüffler et al. 2010, Fuchs & Buhmann 2011, Schüffler et al. 2013). Computational methods have been very successful in making standard pathological analyses more objective and reproducible, but they have generally not quantified the global spatial organisation of the tumour tissue and were generally not linked closely to genomics. With large multi-level data collections, e.g in breast cancer (Curtis et al. 2012, The Cancer Genome Atlas Network 2012), this situation is beginning to change (Ali et al. 2013) and we expect it to lead to a systems genetics understanding of cancer that links tumour genotypes to intermediate molecular and tissue phenotypes as well as organismal phenotypes like progression, survival and treatment response (Yuan et al. 2012).

1.3 Future challenges

We see a number of areas where we believe systems genetics will have an immediate impact. One such area is systems genetics of disease, where some work has already been done in metabolic syndromes (Schadt et al. 2005), but less so in cancer research. Looking at cancer from a systems genetics perspective offers new opportunities for understanding basic concepts, finding ways to apply combination treatments and for the development of new therapeutic strategies.

Tumours are ideal objects for systems genetic analyses, because they contain many different genotypes, many different cellular phenotypes, ongoing clonal evolution, interactions with the organismal environment and between the cells within the tumour microenvironment. These interactions influence disease progression and outcome as well as the development of metastases and are of key importance for treatment decisions. Often treatments fail because of outgrowth of a resistant cancer subpopulation. Systems genetics approaches will have great impact on understanding how heterogeneity in cancer genotypes influences different outcomes. Systems genetics approaches might also lead to rational design on combinatorial drug treatment, identifying key intervention points that might be masked by buffering through other pathways.

Cancer is also an ideal application for systems genetics because many genomes and intermediate phenotypes have already been collected, on a population scale in projects like The Cancer Genome Atlas (TCGA) and the International Cancer Genome Consortium (ICGC) and also by multiple sampling within individual patients. However, the question of how to connect genotype to phenotype on a single-cell level in a complex tissue is still an experimental and theoretical open problem. An experiment one could envision to tackle this question is to take one metastasis, determine the genome of every single cell and as phenotypes the spatial organisation of the tumour tissue and molecular phenotypes like gene expression. This would allow to associate individual genotypes with the environment the cell lives in and link them to intermediate molecular phenotypes as well as morphology. This will depend on further advances in single-cell genomics techniques, which is currently a rapidly emerging research area (Shapiro et al. 2013).

Another area where we predict a large impact of sytems genetics is in complementing observational studies with large-scale perturbation studies that build on foundations in yeast and extend them to more complex organisms. First examples exist in flies, worms and human cells (Ayroles et al. 2009, Muellner et al. 2011). Understanding the genetic architecture of organisms more complex than yeast is an important ongoing area of research. Additionally, how drug treatments interact with the genotypes of individual cells to influence subpopulations and lead to resistance is an important field.

The success of these research programmes will depend on theoretical and computational advances. Standards for the basic data analysis of single-cell phenotyping have yet to be established. In addition, how to integrate data from different molecular and phenotypic levels in populations of cells is currently poorly understood.

1.4   What is covered in the book

Systems genetics is a broad field with many aspects ranging from population genetics to molecular networks, which cannot all be covered in a single volume. In this book we focus on how experimental perturbations can help us to understand the link between genotype and phenotype. We provide a snapshot of current research activity and state-of-the-art approaches to systems genetics.

The book chapters cover both experimental and theoretical approaches: on the experimental side the topics range from large-scale RNA interference studies and mutant analysis to combinatorial perturbations and epistatic interactions; on the theoretical side from network reconstruction and reliability analysis to data integration and conceptual discussion of the nature of phenotypes. To show the wide applicability of systems genetics methods, we chose work from a variety of model organisms, including *S. cerevisiae*, *Drosophila melanogaster* and human.

Myers and colleagues (Chapter 2) introduce the definition of genetic interactions and describe how comprehensive genetic interaction analysis has been performed in yeast.

Boutros and colleagues (Chapter 3) introduce genetic interaction analysis and how it has been applied to map genetic networks in different organisms. They introduce how genetic interactions can be experimentally measured and how quantitative genetic interaction profiles can be deduced in order to make conclusions about the genetic architecture of processes.

Beerenwinkel and Delgado-Eckert (Chapter 4) concentrate on the mathematical analysis of genetic interaction networks. They introduce a mathematical framework to link epistasis to the redundancy and reliability of biological networks. They present statistical methodology for analysing epistatic relationships.

Whitehurst and Maxfield (Chapter 5) extend the discussion of genetic interactions and epistasis by describing how to use functional screening approaches to identify genetic vulnerabilities in cancer cells. Overcoming chemoresistancy and exploiting chemosensitivity is an important area of cancer research and this chapter presents how synthetic lethal analysis can make its way from the bench to the bedside.

Markowetz and colleagues (Chapter 6) give an overview of statistical analysis strategies for genetic screens ranging from genome-wide screens with single reporters to targeted screens with rich molecular phenotypes. They describe statistical methods for functional annotation and network reconstruction.

Meyer and colleagues (Chapter 7) describe the application of genetic perturbation studies in infectious diseases, and how RNAi screens identify novel host cell targets with previously unknown roles in infection and its pathology.

Perrimon and colleagues (Chapter 8) expand on the description of network inference methods by focusing on the integration of gene perturbation studies with mass spectrometry data, which can play an important role in generating hypotheses, driving further experimentation and providing novel insights.

Girolami and colleagues (Chapter 9) describe an application of Bayesian model selection to signalling pathways in chronic myeloid leukaemia. They describe how gene perturbations can be integrated with a dynamic phenotype to infer pathway structure.

Bader and Park (Chapter 10) describe statistical modelling of dynamic protein complexes and apply them to networks in *S. cerevisiae* and *Arabidopsis thaliana*.

Linding and Hadjiprocopis (Chapter 11) take a broader view on phenotypes and different phenotype states and describe theoretical concepts to formalise cellular decision-making.

In the last chapter **Schafer and Brown** (Chapter 12) go beyond cells to organisms and behavioural phenotypes, how they can be quantitatively measured in *Caenorhabditis elegans* and linked to phenotypes.

References

Ali, H. R., Irwin, M., Morris, L., Dawson, S.-J., Blows, F. M. et al. (2013), 'Astronomical algorithms for automated analysis of tissue protein expression in breast cancer.' *Br J Cancer* **108**(3), 602–612.

Avery, L. & Wasserman, S. (1992), 'Ordering gene function: the interpretation of epistasis in regulatory hierarchies.' *Trends Genet* **8**(9), 312–316.

Ayroles, J. F., Carbone, M. A., Stone, E. A., Jordan, K. W., Lyman, R. F. et al. (2009), 'Systems genetics of complex traits in *Drosophila melanogaster*.' *Nat Genet* **41**(3), 299–307.

Barabási, A.-L. & Oltvai, Z. N. (2004), 'Network biology: understanding the cell's functional organization.' *Nat Rev Genet* **5**(2), 101–113.

Baryshnikova, A., Costanzo, M., Myers, C. L., Andrews, B. & Boone, C. (2013), 'Genetic interaction networks: toward an understanding of heritability.' *Annu Rev Genomics Hum Genet* **14**, 111–133.

Bateson, W. (1909), *Mendel's principles of heredity*, Cambridge University Press.

Beck, A. H., Sangoi, A. R., Leung, S., Marinelli, R. J., Nielsen, T. O. et al. (2011), 'Systematic analysis of breast cancer morphology uncovers stromal features associated with survival.' *Sci Transl Med* **3**(108), 108–113.

Berns, K., Hijmans, E. M., Mullenders, J., Brummelkamp, T. R., Velds, A. et al. (2004), 'A large-scale RNAi screen in human cells identifies new components of the p53 pathway.' *Nature* **428**(6981), 431–437.

Boutros, M. & Ahringer, J. (2008), 'The art and design of genetic screens: RNA interference.' *Nat Rev Genet* **9**(7), 554–566.

Boutros, M., Kiger, A. A., Armknecht, S., Kerr, K., Hild, M. et al. (2004), 'Genome-wide RNAi analysis of growth and viability in *Drosophila* cells.' *Science* **303**(5659), 832–835.

Califano, A., Butte, A. J., Friend, S., Ideker, T. & Schadt, E. (2012), 'Leveraging models of cell regulation and GWAS data in integrative network-based association studies.' *Nat Genet* **44**(8), 841–847.

Celniker, S. E., Dillon, L. A. L., Gerstein, M. B., Gunsalus, K. C., Henikoff, S. et al. (2009), 'Unlocking the secrets of the genome.' *Nature* **459**(7249), 927–930.

Cheung, H. W., Cowley, G. S., Weir, B. A., Boehm, J. S., Rusin, S. et al. (2011), 'Systematic investigation of genetic vulnerabilities across cancer cell lines reveals lineage-specific dependencies in ovarian cancer.' *Proc Natl Acad Sci USA* **108**(30), 12 372–12 377.

Cheung, V. G. & Spielman, R. S. (2009), 'Genetics of human gene expression: mapping DNA variants that influence gene expression.' *Nat Rev Genet* **10**(9), 595–604.

Civelek, M. & Lusis, A. J. (2014), 'Systems genetics approaches to understand complex traits.' *Nat Rev Genet* **15**(1), 34–48.

Cordell, H. J. (2002), 'Epistasis: what it means, what it doesn't mean, and statistical methods to detect it in humans.' *Hum Mol Genet* **11**(20), 2463–2468.

Cordell, H. J. (2009), 'Detecting gene–gene interactions that underlie human diseases.' *Nat Rev Genet* **10**(6), 392–404.

Costanzo, M., Baryshnikova, A., Bellay, J., Kim, Y., Spear, E. et al. (2010), 'The genetic landscape of a cell.' *Science* **327**(5964), 425.

Curtis, C., Shah, S. P., Chin, S.-F., Turashvili, G., Rueda, O. M. et al. (2012), 'The genomic and transcriptomic architecture of 2,000 breast tumours reveals novel subgroups.' *Nature* **486**(7403), 346–352.

Eichler, E. E., Flint, J., Gibson, G., Kong, A., Leal, S. M. et al. (2010), 'Missing heritability and strategies for finding the underlying causes of complex disease.' *Nat Rev Genet* **11**(6), 446–450.

ENCODE Consortium, Bernstein, B. E., Birney, E., Dunham, I., Green, E. D. et al. (2012), 'An integrated encyclopedia of DNA elements in the human genome.' *Nature* **489**(7414), 57–74.

Fisher, R. A. (1918), 'The correlations between relatives on the supposition of Mendelian inheritance.' *Trans R Soc Edinburgh* **52**, 399–433.

Flintoft, L. (2014), 'Disease genetics: phenome-wide association studies go large.' *Nat Rev Genet* **15**(1), 2.

Friend, S. H. & Ideker, T. (2011), 'Point: are we prepared for the future doctor visit?' *Nat Biotechnol* **29**(3), 215–218.

Fuchs, F., Pau, G., Kranz, D., Sklyar, O., Budjan, C. et al. (2010), 'Clustering phenotype populations by genome-wide RNAi and multiparametric imaging.' *Mol Syst Biol* **6**, 370.

Fuchs, T. J. & Buhmann, J. M. (2011), 'Computational pathology: challenges and promises for tissue analysis.' *Comput Med Imaging Graph* **35**(7–8), 515–530.

Hemani, G., Knott, S. & Haley, C. (2013), 'An evolutionary perspective on epistasis and the missing heritability.' *PLoS Genet* **9**(2), e1003295.

Ideker, T. & Krogan, N. J. (2012), 'Differential network biology.' *Mol Syst Biol* **8**, 565.

Kiger, A. A., Baum, B., Jones, S., Jones, M. R., Coulson, A. et al. (2003), 'A functional genomic analysis of cell morphology using RNA interference.' *J Biol* **2**(4), 27.

Mackay, T. F. C. (2014), 'Epistasis and quantitative traits: using model organisms to study gene–gene interactions.' *Nat Rev Genet* **15**(1), 22–33.

Marcotte, R., Brown, K. R., Suarez, F., Sayad, A., Karamboulas, K. et al. (2012), 'Essential gene profiles in breast, pancreatic, and ovarian cancer cells.' *Cancer Discov* **2**(2), 172–189.

Mendel, G. (1866), 'Versuche über Pflanzen-Hybriden.' *Verhandl Naturforsch Vereines Brünn* **4**, 3–47.

Muellner, M. K., Uras, I. Z., Gapp, B. V., Kerzendorfer, C., Smida, M. et al. (2011), 'A chemical–genetic screen reveals a mechanism of resistance to PI3K inhibitors in cancer.' *Nat Chem Biol* **7**(11), 787–793.

Nadeau, J. H. & Dudley, A. M. (2011), 'Genetics: systems genetics.' *Science* **331**(6020), 1015–1016.

Phillips, P. C. (1998), 'The language of gene interaction.' *Genetics* **149**(3), 1167–1171.

Phillips, P. C. (2008), 'Epistasis: the essential role of gene interactions in the structure and evolution of genetic systems.' *Nat Rev Genet* **9**(11), 855–867.

Schadt, E. E., Lamb, J., Yang, X., Zhu, J., Edwards, S. et al. (2005), 'An integrative genomics approach to infer causal associations between gene expression and disease.' *Nat Genet* **37**(7), 710–717.

Schüffler, P. J., Fuchs, T. J., Ong, C. S., Roth, V. & Buhmann, J. M. (2010), 'Computational TMA analysis and cell nucleus classification of renal cell carcinoma.' *Pattern Recognition: Proceedings of 32nd DAGM Symposium*, Darmstadt, Germany, 22–24 September 2010, pp. 202–211.

Schüffler, P. J., Fuchs, T. J., Ong, C. S., Wild, P. J., Rupp, N. J. et al. (2013), 'TMARKER: a free software toolkit for histopathological cell counting and staining estimation.' *J Pathol Inform* **4**(Suppl), S2.

Shapiro, E., Biezuner, T. & Linnarsson, S. (2013), 'Single-cell sequencing-based technologies will revolutionize whole-organism science.' *Nat Rev Genet* **14**(9), 618–630.

Skelly, D. A., Merrihew, G. E., Riffle, M., Connelly, C. F., Kerr, E. O. et al. (2013), 'Integrative phenomics reveals insight into the structure of phenotypic diversity in budding yeast.' *Genome Res* **23**(9), 1496–1504.

Smith, J. C. & Melton, J. (1964), 'Manipulation of autopsy diagnoses by computer technique.' *JAMA* **188**, 958–962.

Snijder, B., Sacher, R., Rämö, P., Damm, E., Liberali, P. et al. (2009), 'Population context determines cell-to-cell variability in endocytosis and virus infection.' *Nature* **461**(7263), 520–523.

Snijder, B., Sacher, R., Rämö, P., Liberali, P., Mench, K. et al. (2012), 'Single-cell analysis of population context advances RNAi screening at multiple levels.' *Mol Syst Biol* **8**(1), 579.

Sozzani, R. & Benfey, P. N. (2011), 'High-throughput phenotyping of multicellular organisms: finding the link between genotype and phenotype.' *Genome Biol* **12**(3), 219.

The Cancer Genome Atlas Network (2012), 'Comprehensive molecular portraits of human breast tumours.' *Nature* **490**(7418), 61–70.

Threadgill, D. W. (2006), 'Meeting report for the 4th Annual Complex Trait Consortium meeting: from QTLs to systems genetics.' *Mamm Genome* **17**(1), 2–4.

Topp, C. N., Iyer-Pascuzzi, A. S., Anderson, J. T., Lee, C.-R., Zurek, P. R. et al. (2013), '3D phenotyping and quantitative trait locus mapping identify core regions of the rice genome controlling root architecture.' *Proc Natl Acad Sci USA* **110**(18), E1695–E1704.

Vizeacoumar, F. J., Arnold, R., Vizeacoumar, F. S., Chandrashekhar, M., Buzina, A. et al. (2013), 'A negative genetic interaction map in isogenic cancer cell lines reveals cancer cell vulnerabilities.' *Mol Syst Biol* **9**, 696.

Yin, Z., Sadok, A., Sailem, H., McCarthy, A., Xia, X. et al. (2013), 'A screen for morphological complexity identifies regulators of switch-like transitions between discrete cell shapes.' *Nat Cell Biol* **15**(7), 860–871.

Yuan, Y., Failmezger, H., Rueda, O. M., Ali, H. R., Gräf, S. et al. (2012), 'Quantitative image analysis of cellular heterogeneity in breast tumors complements genomic profiling.' *Sci Transl Med* **4**(157), 143–157.

Zuk, O., Hechter, E., Sunyaev, S. R. & Lander, E. S. (2012), 'The mystery of missing heritability: genetic interactions create phantom heritability.' *Proc Natl Acad Sci USA* **109**(4), 1193–1198.

2 Computational paradigms for analyzing genetic interaction networks

Carles Pons, Michael Costanzo, Charles Boone, and Chad L. Myers

The advent of sequencing technologies has revolutionized our understanding and approach to studying biological systems. Indeed, whole-genome sequencing projects have already targeted many different species, enabling the identification of most genes in those organisms. However, observed phenotypes cannot be explained by genes alone, but rather by the interactions that their products establish under some environmental conditions (Waddington 1957). Thus, it is through the analysis of these interaction networks (e.g. regulatory, metabolic, molecular, or genetic) that we can better understand the genotype-to-phenotype relationship, the complexity and evolution of organisms, or the differences among individuals of the same species. The topology and dynamics of these biological networks can be unveiled by systematic perturbation of their nodes (i.e. genes). For instance, upon single-gene deletions in *Saccharomyces cerevisiae* under standard laboratory conditions, most genes ($\sim$80%) were not found to be essential for cell viability (Giaever et al. 2002). Though many of these genes may be required for growth in other environments (Hillenmeyer et al. 2008), this result suggests extensive functional redundancy among genes. Such functional buffering confers robustness to biological networks and shields the cellular machinery from genetic perturbations (Hartman et al. 2001). Additionally, the small effect on phenotype that many gene deletions exhibit (see Figure 2.1) evidences that single perturbations alone cannot capture the complexity of the genotype-to-phenotype relationship. Therefore, a combinatorial approach to gene perturbations is best suited to elucidate biological systems and can enable a better characterization of genes and cellular functioning.

2.1 Definition of genetic interaction

Genetic interactions reveal functional relations between genes that contribute to a phenotypic trait. William Bateson first introduced the term, formerly known as epistasis (see Phillips [1998] for a description on the origin and evolution of the definition), to refer to an allele at one locus preventing a variant at another from manifesting its effect (Bateson 1909). Later, Fisher described it as a general deviation from additivity in the

Systems Genetics: Linking Genotypes and Phenotypes, ed. F. Markowetz and M. Boutros. Published by Cambridge University Press. © Cambridge University Press 2015.

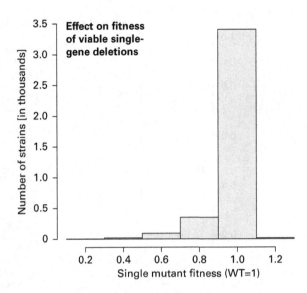

Figure 2.1   Effect on fitness of viable single-gene deletions. Single mutant fitness for a collection of viable gene-deletion mutants in *S. cerevisiae*. Wild-type fitness is 1. Most genes can be deleted without a strong effect on cell growth. Data from Costanzo et al. (2010).

effect of alleles at different loci with respect to a quantitative phenotype (Fisher 1918). The current interpretation extends that of Fisher's, and states that interacting genes are identified by an unexpected phenotype when mutated simultaneously, which cannot be explained from the combined phenotypes of the individual mutants (Mani et al. 2008). Genetic interactions can be classified as negative if the phenotype of the combined mutant is more severe than expected, and as positive if the phenotype is less compromised than expected (see Figure 2.2). Synthetic lethal interactions are a particular case of negative genetic interactions, where combinations of mutations result in an unexpected lethal phenotype (Lucchesi 1968).

The detection of genetic interactions requires the evaluation of a phenotypic trait, and the definition of a null model to determine the expected phenotype under the assumption of independent mutations. For instance, in *S. cerevisiae*, the most widely studied phenotype is cell fitness using colony growth as a proxy, and a multiplicative model is used to calculate the expected fitness of the combined mutations based on the fitnesses of the individual mutants (see Figure 2.2) (St Onge et al. 2007). Importantly, different null models can often diverge extensively in the identification of genetic interactions (Mani et al. 2008, Segre et al. 2005). Phenotypic traits other than fitness have also been evaluated for the detection of genetic interactions (Drees et al. 2005, Jonikas et al. 2009, Van Driessche et al. 2005), but their integration in high-throughput experiments is challenging. Still, the evaluaton of these phenotypes can complement the information obtained from the unidimensional cell fitness. For instance, phenotypic evaluation of single-gene deletion mutants by an automated digital image process revealed that half of the viable single mutant yeast strains were in fact showing strong morphological defects (Ohya et al. 2005).

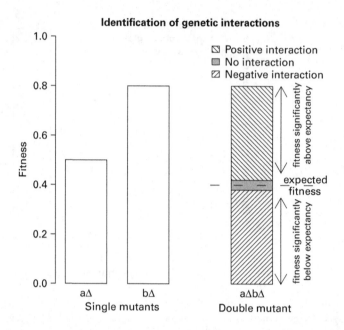

Figure 2.2   Identification of genetic interactions. In this example, cell fitness is the evaluated phenotype (wild-type fitness is 1) and the expected double mutant fitness is calculated by a multiplicative model (horizontal dashed line). Double mutants with fitness higher and lower than expected identify positive and negative genetic interactions, respectively. Not interacting genes result in double mutants with fitness similar to the calculated expectancy.

A complete characterization of cellular behavior depends upon understanding the functional roles of all genes within a species, which requires the use of combinatorial genetic perturbations. Still, even in *S. cerevisiae*, one of the simplest eukaryotes, there are almost 20 million possible double and 10^{10} possible triple mutants. The exhaustive exploration of such combinatorial space is a daunting task and requires the use of automated approaches. Indeed, this promoted the development of high-throughput technologies, such as Synthetic Genetic Array (SGA) analysis, the first technology that enabled systematic large-scale genetic interaction screenings (Tong et al. 2001) (see next section). Later, two complementary approaches were also developed: diploid synthetic lethal analysis by microarray (dSLAM) (Pan et al. 2004) and genetic interaction mapping (GIM) (Decourty et al. 2008).

All these technologies were designed for *S. cerevisiae*, which has also pioneered the development of other omics technologies (Suter et al. 2006), given that it is highly amenable to genetic manipulation. Indeed, it was the first eukaryote with a sequenced genome (Goffeau et al. 1996) and a gene-deletion collection available (Giaever et al. 2002). Initial systematic genetic interaction screens using SGA identified several thousand synthetic lethal negative genetic interactions, which allowed the detection of genes involved in parallel pathways of the same essential process (Tong et al. 2004). The improvement of this protocol with the quantitative evaluation of genetic interaction data facilitated the detection of more subtle functional relations, i.e. viable double mutants with unexpected fitness, and enabled the identification of positive genetic interactions (Collins et al. 2006, Schuldiner et al. 2005). However, SGA screenings, like other high-throughput techniques (Leek et al. 2010), are prone to systematic experimental effects that introduce bias into the data, and can potentially account for a large fraction of the observed variance (Baryshnikova et al. 2010*a*), which is especially problematic given the low density of genetic interaction networks (Tong et al. 2001). Thus, computational tools for the removal of systematic effects and data normalization are required to enable the unbiased analysis of genetic networks (Baryshnikova et al. 2010*a*, Baryshnikova et al. 2010*b*). Development of such tools and advances in the experimental SGA pipeline enabled the first high-quality genome-scale screening of quantitative genetic interactions, which depicted the first functional map of a eukaryotic cell (Costanzo et al. 2010). Recently, genome-scale screenings were also performed in *Schizosaccharomyces pombe* (Frost et al. 2012, Ryan et al. 2012) using the deletion collection in that species (Kim et al. 2010), which allowed for the first time the unbiased cross-species comparison of genetic networks. Previous studies in *Caenorhabditis elegans* (Byrne et al. 2007, Lehner et al. 2006), *Escherichia coli* (Babu et al. 2011), *Drosophila melanogaster* (Horn et al. 2011), and *S. pombe* (Dixon et al. 2008, Roguev et al. 2008) also systematically screened for genetic interactions, but not at such scale, making the cross-species comparisons in those cases possible only for specific pathways or modules.

Screening technology is developing at a very fast pace, which is expected to enable soon the completion of the genetic networks in *S. cerevisiae* and *S. pombe* but also the targeting of other species currently unapproachable at a genome scale. Certainly, the amount of genetic data available is huge, and growing steadily, making analysis increasingly challenging. Network-based analysis can characterize specific features of the genetic networks. However, understanding the biological relevance of genetic networks requires the integration of heterogeneous data and cross-species comparisons, and such strategies can only be addressed by computational methods.

This review focuses on the computational analysis of large-scale genetic interaction data. We first describe Synthetic Genetic Array (SGA) technology and the computational tools applied for the unbiased quantitative detection of genetic interactions. Next, we present different computational approaches used to derive biological insights from genetic interaction networks.

2.2 Toward the first reference global genetic interaction network: Synthetic Genetic Array analysis in yeast

Saccharomyces cerevisiae serves as a powerful model system for understanding the fundamental properties of eukaryotic cells at a molecular level. Indeed, it shares many of the basic cell division and growth functions of human cells and numerous genes are conserved from yeast to humans. Due to its facile genetics, *S. cerevisiae* has catalyzed the development of numerous genomic technologies, including methods for large-scale mapping of genetic interactions, and has played a primary role in deciphering the basic functional wiring diagram of the eukaryotic cell (Botstein & Fink 2011). Specifically, genome-scale mapping of genetic interactions requires three fundamental tools. These include: large collections of mutant strains designed to systematically perturb gene activity; high-throughput methodologies for combining mutations; and finally, a phenotypic assay that can be easily and accurately measured on a genome-wide scale in order to identify genetic interactions quantitatively. Below, we describe experimental and analytical methodologies as they pertain to mapping genetic interactions in budding yeast.

Synthetic Genetic Array (SGA) was the first approach to automate classical yeast genetics and enable large-scale construction of double mutants from ordered arrays of single mutants (Tong et al. 2001). SGA combines arrays of non-essential gene deletion mutants (Winzeler et al. 1999) or conditional alleles of essential genes (Ben-Aroya et al. 2008, Li et al. 2011) with robotic manipulations for high-throughput construction of haploid yeast double mutants (Tong et al. 2001). In a typical SGA screen, a 'query' mutant strain, carrying haploid-specific reporter genes and selectable markers, is crossed to an array of ~4800 viable deletion mutants (Winzeler et al. 1999) or conditional alleles of essential genes (Ben-Aroya et al. 2008, Li et al. 2011), carrying a different selectable marker. The resulting heterozygous diploids are sporulated and robotically replica-pinned onto a series of selective media enabling precise stepwise selection of double mutant cells (Baryshnikova et al. 2010a). As a result, an SGA screen produces an ordered array of double mutants, which can be scored for fitness through measurement of their colony size, or assessed for a variety of other quantitative phenotypes.

In one of its first applications, SGA was used to cross 132 query strains to the complete array of ~4800 haploid deletion mutants, resulting in a large-scale genetic interaction network consisting of ~1000 genes and ~4000 synthetic sick/lethal interactions (Tong et al. 2004). This network showed that synthetic lethality, despite being generally rare, tends to occur between genes that share similar biological functions and thus can be used to uncover novel functional relationships on a global scale (Tong et al. 2004). More recently, SGA was used to generate a large-scale genetic interaction map of a cell, revealing fundamental properties of genetic networks and illustrating the effectiveness of genetic interactions for organizing genes into specific biological pathways and complexes (Costanzo et al. 2010). The global yeast genetic network revealed both the organization and the complexity of the constellation of genetic interactions. Estimates, based on the current genetic interaction network, indicate that a complete *S. cerevisiae* genetic interaction network may consist of as many as ~200 000 synthetic sick/lethal gene pairs (Davierwala et al. 2005, Tong et al. 2004) and a smaller but comparable number of positive genetic interactions (Costanzo et al. 2010). These numbers clearly indicate that genetic interactions may significantly complicate the task of mapping genotypes to phenotypes in natural populations (Hartman et al. 2001). For example, while in yeast there are ~1000 single-gene perturbations that result in a lethal phenotype (i.e. ~1000 essential genes), there seem to be at least ~200-fold more digenic mutant combinations that result in the same phenotypic outcome.

2.2.1   Mapping quantitative genetic interaction networks

Early genetic interaction studies were based on the binary assessment of cellular fitness (sick/lethal vs. no fitness defect) (Tong et al. 2001, Tong et al. 2004). Although synthetic genetic relationships of this kind are informative, quantitative analysis enables identification of more subtle interactions and construction of higher-resolution genetic networks encompassing both negative and positive interactions. For example, liquid growth profiling was used to accurately measure genetic interactions between a subset of genes involved in DNA replication and repair (St Onge et al. 2007). In addition to identifying positive and negative genetic interactions, positive interactions were differentiated into distinct subclasses associated with different biological interpretations (St Onge et al. 2007). In another example, fitness was measured from fluorescence-labeled populations of wild-type cells mixed with either single or double mutant yeast strains to map a quantitative genetic interaction network for genes encoding components of the 26S proteasome (Breslow et al. 2008). These various assays have also been applied to quantify genetic interactions between duplicated genes (Dean et al. 2008, DeLuna et al. 2008, Jasnos & Korona 2007, Musso et al. 2008, VanderSluis et al. 2010).

Another study combined SGA with a new genome-scale quantitative scoring methodology to examine ∼5.4 million gene pairs covering ∼30% of the *S. cerevisiae* genome (Baryshnikova et al. 2010*a*, Costanzo et al. 2010). This large-scale endeavor measured single and double mutant yeast fitness to uncover ∼170 000 genetic interactions (both negative and positive) and provided the first glimpse of a quantitative, genome-scale genetic interaction network for a eukaryotic cell. Consistent with the degree distribution of other biological networks (Barabási & Oltvai 2004) the majority of genes are sparsely connected in the genetic interaction network while a small number have many interactions and serve as network 'hubs.' While most genetic interactions occur between genes involved in the same biological process (Costanzo et al. 2010, Tong et al. 2004), network hubs tend to be pleiotropic and interact with many functionally diverse sets of genes (Costanzo et al. 2010). Importantly, genes annotated to chromatin/transcription, secretion, and membrane trafficking showed a significant number of genetic interactions with numerous different processes indicating that genes involved in these functions are important for mediating cross-process connections in the genetic network (Costanzo et al. 2010).

2.3   Computational paradigms for genetic interaction networks

2.3.1   Function prediction

Complete understanding of the genotype to phenotype relationship requires the functional characterization of all genes within a species. Publicly available databases like the Gene Ontology (GO) (Ashburner et al. 2000) provide functional annotations of genes for most organisms including *S. cerevisiae* where the majority of genes are associated with at least one GO annotation. Nonetheless, a significant number of genes remain completely uncharacterized, and annotations for other genes may not reflect their precise or multiple functional roles in the cell. The functional and unbiased nature of large-scale genetic interactions networks is well suited to complete and re-evaluate gene annotations. For instance, phenotypic analysis of double mutants can be a powerful means for uncovering gene function (Hartman et al. 2001). However, interpretation of even simple phenotypes, such as cell fitness, is not straightforward and the underlying function relating the interacting genes remains elusive in many cases. An alternative is the use of guilt-by-association approaches, which assume that genes sharing a certain network property are more likely to share function. However, in order to predict function in such a way it should be first established the extent to which this principle applies in genetic networks (how a shared property relates to functional similarity), and what is the best strategy to implement it (what network property should be evaluated). With this purpose, functional similarity between genes within the genetic network has been widely studied in *S. cerevisiae*, where large-scale unbiased genetic interaction data and functional annotations for most genes are available.

Initial studies found that functional similarity between direct, genetically interacting genes is significant but rather weak (Tong et al. 2004), probably reflecting that interactions can link genes belonging to distant pathways or the pleiotropy of specific genes (Costanzo et al. 2010). For this reason, a single genetic interaction may often be difficult to interpret. Additionally, genetic interactions are known to be more informative when analyzed within the context of the larger genetic network, where experimental bias is less likely to affect global properties than individual interactions. Therefore, the function of a gene can be better characterized by considering all its interacting partners within the genetic network, i.e. the genetic interaction profile. Certainly, genes that perform a similar biological function are expected to disrupt the cell in a similar way upon deletion (see Figure 2.3). For instance, genes belonging to the same pathway are expected to interact with similar partners. Thus, the more similar the interaction profiles of two genes, the more likely they perform the same function (Tong et al. 2004). Profile similarity can be calculated by many different means. For instance, the overlap of interactions has been used to evaluate similarity between binary interaction profiles, which describe synthetic lethal interactions (Tong et al. 2004). An alternative approach proposed a genetic congruence score for each pair of genes, where the number of shared interactions was normalized by the number of interactions of both genes and the significance of the overlap between two profiles was assessed using the hypergeometric distribution (Ye et al. 2005). This statistic was found to be more suited for the identification of functional associations between genes than the absolute number of shared interactions. In quantitative interaction data, a more common choice to evaluate similarity between profiles is the Pearson's correlation coefficient (PCC) (Costanzo et al. 2010), which can capture subtle differences in interaction strength. However, more recent work found that PCC is less informative when applied on thresholded genetic interaction data (Deshpande et al. 2013). In that study, a diverse set of profile similarity measures were computed using two large-scale data sets and the ability of each method to identify functionally associated genes from either continuous or thresholded genetic interaction data was evaluated. In general, linear similarity measures such as dot product, Pearson's correlation, or cosine similarity outperformed overlapping measures such as the Jaccard

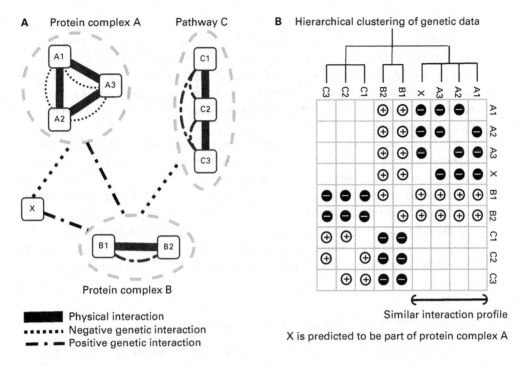

A   Protein complex A   Pathway C

B   Hierarchical clustering of genetic data

Protein complex B

█████ Physical interaction
▪▪▪▪▪▪▪ Negative genetic interaction
▬ ▪ ▬ Positive genetic interaction

Similar interaction profile

X is predicted to be part of protein complex A

Figure 2.3   Functional prediction by genetic interaction profiles. (A) Members of complex A and the uncharacterized gene interact in the same way with the rest of the genes. (B) This profile similarity can be identified by a clustering algorithm. In the resulting clustered data, functionally similar genes are in close proximity.

coefficient. Surprisingly, dot product had the most consistent performance across the different conditions tested but the study highlights that the optimal choice of similarity measure is conditioned by the data type and the biological problem in question.

In a large-scale experiment, the resulting genetic interaction data can be represented as a matrix that contains the different genetic interaction profiles. For analysis purposes, this matrix is often rearranged by a two-dimensional hierarchical clustering algorithm (Eisen et al. 1998), which independently clusters rows and columns based on a selected profile similarity measure: the more similar the interactions between two rows (or two columns), the closer they will be located. In the resulting matrix, similar interaction profiles are grouped together, identifying genes likely to share a similar function (see Figure 2.3), and genetic interactions are clustered in comprehensive blocks, identifying functional modules such as protein complexes and pathways (Collins et al. 2006, Tong et al. 2004). This reflects the intrinsic modularity of the genetic network (Hartwell et al. 1999, Segre et al. 2005) where isolated genetic interactions are expected to be rare. Indeed, this modularity was evident in the global similarity network built by Costanzo et al. The authors systematically calculated the PCC between all pairs of genetic interaction profiles in a genome-scale data set. Using these similarity measures, they defined a network where nodes represented genes, edges connected genes sharing similar genetic profiles (i.e. high PCC), and the distance between genes was inversely proportional to their profile similarity. In this global similarity network, nodes clustered together tended to form coherent groups of genes sharing the same GO annotation (Costanzo et al. 2010). Importantly, profile similarity can identify genes with a tight functional relationship, such as protein complexes or biological pathways, but also connections between functional neighborhoods which describe the higher-level structure of the genetic network.

Altogether, these studies show that profile similarity measures are powerful tools to capture functional similarity between genes and can unveil the functions of uncharacterized genes (see Figure 2.3). For instance, PAR32, ECM30, and UBP15 were associated to the Gap1-sorting module in *S. cerevisiae* (Costanzo et al. 2010), and SPAC323.03c and SPCC188.10c to the peroxisome regulation and G protein signaling machinery, respectively, in *S. pombe* (Ryan et al. 2012). Additionally, processing of the results derived from their systematic application (e.g. hierarchical clustering or network representation) can capture and delineate the inherent functional organization of the cell.

2.3.2 Elucidating the modular organization of the genetic network

Explaining the cellular mechanisms underlying genetic interactions is not straightforward: they reveal functional and not necessarily physical relationships. Certainly, the overlap between the genetic and protein–protein interactions is low (Costanzo et al. 2010), and genetic networks can connect distant genes, i.e. not necessarily belonging to the same molecular pathway or complex. Still, genetic interactions are expected to cluster together in comprehensive modules and seldomly appear in isolation (Hartwell et al. 1999). For instance, a theoretical study focused on the yeast metabolic network showed that genes belonging to the same biological modules were mostly connected by entirely positive or negative genetic interactions. This property was defined as monochromaticity (Segre et al. 2005) and can be used to drive the clustering of genetic interaction profiles for the identification of modules and their members. Indeed, analyses of experimental genetic interaction data found that groups of interacting genes tended also to be connected mostly by either negative or positive interactions (Costanzo et al. 2010, Schuldiner et al. 2005). The identification of such monochromatic modules in the genetic network allowed the functional prediction of uncharacterized genes (Costanzo et al. 2010). Additionally, genes belonging to the same GO biological process (identifying mostly protein complexes) showed a significant tendency to interact monochromatically (Michaut et al. 2011). These approaches highlight the relevance of the context of a genetic interaction: it is usually within a specific neighborhood that genetic interactions make biological sense, and seldomly when isolated.

Two complementary models have been proposed for the mechanistic interpretation of genetic interactions: the between-pathway and the within-pathway models (see Figure 2.4 below) (Guarente 1993, Tucker & Fields 2003). In the between-pathway model, two physically independent pathways are interconnected by genetic interactions. On the other hand, the within-pathway model proposes that genes belonging to the same

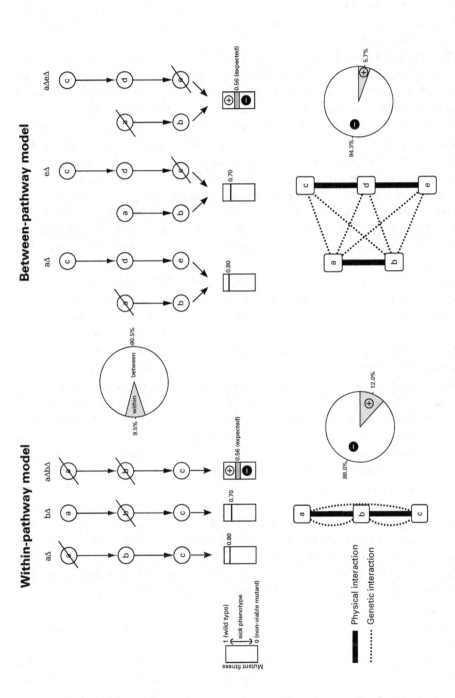

Figure 2.4 Modular organization of the genetic network. Representation of the within- and between-pathway models for genetic interactions. Data in the pie charts result from applying a modular decomposition algorithm to the negative and positive genetic interaction data separately (Bellay et al. 2011). Up to 90.5% of the modules found in that study connect two different pathways. Most modules are connected by negative genetic interactions: 88.0% in the within-pathway model, and 94.3% in the between-pathway model.

pathway are intraconnected by genetic interactions. In both cases pathway refers to a group of genes that physically interact. Indeed, empirical evidence supports the modular organization proposed by both models (Collins et al. 2007, Kelley & Ideker 2005, St Onge et al. 2007).

Several studies have evaluated the extent to which these two models can account for the molecular mechanisms behind genetic networks. A successful approach has been to complement genetic with physical interaction data. For instance, despite the limited overlap between the physical and genetic networks, members of the same protein complex often exhibit the same pattern of genetic interactions, which can be detected by clustering of genetic interaction profiles (Tong et al. 2004). Building on this, Kelley and Ideker proposed the systematic interpretation of genetic data by integrating physical interactions. They defined a probabilistic model that assessed the likelihood of a group of genes to be more densely connected (genetically or physically) than expected, where density was determined by an arbitrary threshold. A pathway was defined as a set of proteins densely connected by protein interactions. Individual pathways also densely connected by genetic interactions were interpreted as within-pathway structures, whereas pairs of pathways densely interconnected by genetic interactions were interpreted as between-pathway structures (Kelley & Ideker 2005). Using this probabilistic approach, 40% of the interactions were assigned to one of the models, with between-pathway interactions 3.5 times more common than within-pathway interactions, and where more than half of the pathways detected were significantly enriched in GO annotations. In contrast to this definition of pathways, which required the density of physical interactions between a group of genes to be above a certain arbitrary threshold, Bandyopadhyay et al. proposed a supervised approach where the optimal pattern of interactions was learned from a set of known complexes. The detection of protein complexes and their interactions was based on the genetic interaction score and the likelihood of physical interaction between pairs of genes (Bandyopadhyay et al. 2008). Strong genetic and physical signals were used as evidence of within-pathway modularity, and strong genetic but weak physical signal were used as evidence of between-pathway relationships. Interestingly, the sign of the genetic interaction and the correlation of genetic profiles did not improve the detection of complexes in their gold standard. Most detected complexes were enriched in positive genetic interactions, and were composed basically of non-essential genes. Conversely, the few complexes enriched in negative interactions were more likely to be essential, which was later confirmed on a larger study (Baryshnikova et al. 2010a). Their supervised and integrative approach proved better than hierarchical clustering or that of Kelley and Ideker for protein complex identification, and identified 91 complexes involved in the chromosome organization, suggesting 10 new complexes and completing the subunits of other known complexes. In both previous studies, pathways were identified by groups of genes densely connected by physical interactions, which favors the detection of protein complexes but misses other biological pathways less tightly connected, such as signaling cascades. Taking this into account, Ulitsky and Shamir proposed a less stringent definition of pathway, which they considered as any connected subnetwork within the physical network, regardless of its density of interactions (Ulitsky & Shamir 2007). Physical interactions were used for the detection of pathways, but not for the identification of between-pathway modules. Altogether, this approach allowed the identification of more modules than in the work by Kelley and Ideker. Additionally, by analysis of experimentally determined mRNA half-lives and computationally predicted phosphorylation sites, they noted that genes belonging to between-pathway modules tended to be more tightly regulated than the rest of genes, probably reflecting the more complex interplay between redundant mechanisms. Interestingly, they identified characteristic proteins which were densely connected by both constituent pathways of the between-pathway modules. These pivot proteins were found to be more essential and more conserved across species than the average gene, which highlights their importance for cross-species analyses. A different study noted that genes with the most highly correlated profiles tended to lack a negative interaction, which the authors interpreted as pairs of genes that belonged to the same linear pathway or protein complex: deleting one gene could disable the module, making the second deletion inconsequential. This observation was formalized in the COP (complex or linear pathway) score for the identification of pairs of genes with high profile similarity and absence of negative interaction. Remarkably, most pairs identified by high COP scores were part of known protein complexes or linear pathways, and also unknown complexes that were experimentally verified (Collins et al. 2006, Collins et al. 2007).

Altogether, these studies depicted a compartmentalized and structured organization of genetic networks, with an intrinsic difference between the modules described by positive and negative interactions between non-essential genes. Negative genetic interactions would mostly belong to between-pathway modules, identifying two parallel pathways responsible for the same function. If one pathway was compromised, the redundant pathway could still perform the function, acting as a functional buffer. However, if both pathways were disrupted, the function could not be fulfilled and the damage to the cell would be higher than suggested by the individual perturbation of the pathways. Positive genetic interactions involving non-essential genes would belong to the same linear pathway or protein complex: if the pathway was already compromised by one of the mutations, then a second mutation affecting the same pathway would not add the expected extra damage to the cell (St Onge et al. 2007). As shown above, this mindset was key in the characterization of many gene functions and the identification of several protein complexes. However, unbiased genome-scale genetic interaction data have revealed a more complex interplay between pathway models and positive and negative genetic interactions, evidencing that previous interpretations only partially captured the modular principles of genetic networks, and highlighting the relevance of unbiased data in theoretical studies (see Figure 2.4). For instance, Costanzo et al. showed that both negative and positive interactions had a similar overlap with protein–protein interactions (Costanzo et al. 2010).

Further work described the complete modular decomposition of the genome-scale genetic network using a data-mining approach, without any use of physical interaction data (Bellay et al. 2011). An exhaustive algorithm searched the genome-scale genetic network for all possible biclusters, disjoint or overlapping sets of genes where all genes interact among themselves, which should contain all the within- and between-pathway modules. For instance, negative interactions connecting different pathway modules (between-pathway interactions) result in bicliques (biclusters with disjoint genes) and positive interactions connecting members of the same pathway module (within-pathway interactions) in cliques (biclusters with overlapping genes) (Bellay et al. 2011). More than 10 000 negative biclusters were found, containing 58% of all negative genetic interactions. As expected, most negative biclusters (91%) corresponded to bicliques, identifying between-pathway modules, but a significant number (9%) also organized into within-pathway modules. Interestingly, genetic interactions between duplicate gene pairs were mostly isolated and did not belong to any modular structure. On the other hand, the algorithm detected ~600 positive biclusters, which could only explain 19% of the positive genetic interactions. The rest of the interactions may be part of biclusters too small to be detected by the algorithm, but evidence also suggested a higher false positive rate for these non-structured positive interactions. Surprisingly, the within-pathway model (identified by clique structures) could only explain 18% of the positive interactions found in modules, and in fact positive interactions tended to connect different modules (identified by bicliques) more than members of the same pathway or complex, which is in agreement with recent work showing that positive genetic interactions connect relatively functionally distant modules in metabolic networks (He et al. 2010) and with the large amount of suppression interactions found between protein complexes (Baryshnikova et al. 2010a). Still, structured positive interactions were more likely to organize in within-pathway modules than structured negative interactions. However, given the prevalence of negative biclusters, the within-pathway modules were more likely to be linked by negative than positive interactions (see Figure 2.4). Interestingly, the number of modules populated by a gene correlated with the number of GO annotations and with its chemical degree, which indicates that the detected modules capture the functional role of genes, and reveal the different, sometimes multiple, contexts in which genes participate. Indeed, most modules identified functionally coherent genes. Using module membership to assign functional contexts to genes evidenced a surprisingly high degree of multifunctionality within the yeast genome. In fact, 74 genes were associated with more than 20 biological processes.

In summary, analysis on large-scale data proves that genetic interactions are largely modular. However, recent studies present a more complex organization than what was previously suggested. Negative interactions generally organize in comprehensive modules, mostly following the between-pathway model. Positive interactions are much less structured, which points to a larger fraction of noisy data, but also to the need of new models that can explain the molecular mechanisms behind them. In contradiction with previous studies, genetic interactions within protein complexes tended to be negative, and structured positive interactions fell more often into between-pathway models than into within-pathway models.

2.3.3   Defining pathway order from genetic interactions

Positive genetic interactions have long been used by geneticists to define gene order within pathways, given that they result from a mutation alleviating the effects of another. Masking of a phenotypic effect can be detected by comparing the phenotypes of single mutants to that of the double mutant. Under certain assumptions, one may identify the gene acting downstream in the pathway (Avery & Wasserman 1992). For instance, given two single mutants (A and B) and their combined double mutant (AB), if the phenotype of AB is closer to A, this may suggest that A acts upstream of B within a given pathway. This idea was adapted for quantitative genetic interaction data, where it was used to correctly predict the order of genes involved on several DNA recombination repair pathways based solely on the single and double mutant fitnesses (St Onge et al. 2007). Using a novel approach (Battle et al. 2010), large-scale individual genetic interaction data were used to estimate the preferred pathway organization between each pair of genes. Specifically, a statistical score was used to assess how likely were each pair of genes to be organized across the different types of functional relationships considered: linear, parallel, independent, or a combined linear–parallel pathway for partially alleviating interactions. Each functional relationship had an expected double mutant phenotype, and the statistical score assessed how consistent was the actual double mutant phenotype to that expectancy. In a second phase, optimal global activity pathway networks were determined based on the pairwise statistical scores. Importantly, unlike previous approaches that analyzed each pair of genes independently, here all genes were considered together to build optimal pathways that best satisfied the individual pairwise relations, which allowed explicit ordering of multi-gene pathways. The use of evidence from multiple pairs of genes enabled a global optimization that handled noise, missing measurements, and conflicting data. The method was tested on data of genes involved in protein folding in the endoplasmic reticulum and accurately predicted gene order in the N-linked glycosylation and ER-associated protein degradation pathways.

2.3.4   Prediction of genetic interactions

Screening for genetic interactions is costly and time consuming. Additionally, interactions in higher eukaryotes are difficult to measure, where high-throughput techniques are under development, but still not available, at least at the scale used in lower model organisms like yeast. Importantly, a huge number of experimentally detected genetic interactions is stored in databases such as BioGRID (Stark et al. 2006). Though most of them belong to yeast species, screenings in other eukaryotes also contribute a significant fraction.

Machine learning has been used for the prediction of genetic interactions using heterogeneous sources of biological information. An early study applied decision trees for the prediction of genetic interactions in *S. cerevisiae* by integration of genomic and proteomic data, and local topology features of the genetic network (Wong et al. 2004). Local topology around a gene pair was found as the best predictor for interactions. For instance, reflecting the between-pathway modularity, a genetic interaction between A and B may be predicted if A physically interacts with C, and B genetically interacts with C. Their method reached 80% true positive rate and up to 18% of non-interacting pairs were misclassified as interacting when tested on a large data set (Tong et al. 2004). More impressively, a 20% true positive rate was achieved with only 0.2% false positive rate, resulting in 31% accuracy (Wong et al. 2004). Later, more than 18 000 genetic interactions were predicted for ~2200 genes in *C. elegans* with the integration of functional annotations, genetic and physical interactions, gene expression, and phenotype data from *S. cerevisiae*, *Drosophila melanogaster*, and *C. elegans* (Zhong & Sternberg 2006). The different parameters were calibrated with almost 2000 previously reported genetic interactions. Experimental validation of the predictions in two gene sets showed a precision of at least 44% and 60%.

Interactions can also be predicted solely on partial genetic interaction data. Qi et al. first showed that within an incompletely sampled interaction network, interactions between non-screened pairs could be inferred just by the properties of the subset of screened interactions. The authors observed that genetically interacting pairs were connected by odd-length paths in the genetic network, reflecting the properties of two redundant pathways, and devised a graph diffusion kernel that captured specific local structures in the network that could predict novel interactions with a precision of 50% (Qi et al. 2008). On a more technical note, other methods compensate for the incompleteness of high-throughput techniques. For instance, high-throughput screens can fail to provide interaction data for one-third of the screened pairs, either for technical reasons or because of unreliable measurements. These missing values for specific gene pairs can be inferred by identifying the most similar profiles within the screens (the nearest neighbors) and combining their measurements into quantitative predictions (Ryan et al. 2010, Ulitsky et al. 2009). This same approach can be extended to map interactions in non-tested pairs of genes using two independent but partially overlapping genetic networks, resulting in similar precision as replicate screens (Ryan et al. 2011).

Additionally, genetic interactions between metabolic genes can also be predicted by flux balance analysis (FBA). FBA models the flow of all known metabolites within the metabolic network of a species, and is able to predict growth rate (Orth et al. 2010), including for gene deletion mutants (Förster et al. 2003). Thus, upon single and double deletions, it can identify genetic interactions as has been shown in yeast (Deutscher et al. 2006, Segre et al. 2005, Snitkin & Segre 2011). Such theoretical studies provided insights into the organization of genetic networks, for instance introducing the concept of monochromaticity (Segre et al. 2005). Later work identified genetic interactions by evaluating different phenotypes, and found eight times more interactions than when using growth rate alone (Snitkin & Segre 2011). Intriguingly, many individual gene pairs showed positive interactions by some phenotypes and negative by others. Szappanos et al. used metabolic network models to predict genetic interactions in *S. cerevisiae*. In spite of their moderate success on a genome-scale data set, their approach remarkably identified a biosynthesis pathway erroneously inferred from *E. coli* based on inconsistencies between the observed interactions and those predicted by the metabolic network model (Szappanos et al. 2011).

Genetic interactions can be accurately predicted by a combination of different sources of biological information, or, to a certain extent, using only genetic data from partial screens. This enables the rational design of future genetic interaction screens that could dramatically reduce the number of genes needed to be experimentally tested within a species to unveil its functional network, given that interactions between untested pairs of genes could be inferred. This strategy may prove especially useful as screening technology in higher eukaryotes continues to develop, where experiments may be more costly, and the combinatorial space much larger than in model organisms like *S. cerevisiae*.

2.3.5   Cross-species studies

Saccharomyces cerevisiae has pioneered the large-scale study of genetic interactions, but techniques to systematically explore genetic networks in other organisms are being developed, including *S. pombe* (Dixon et al. 2008, Frost et al. 2012, Roguev et al. 2007, Ryan et al. 2012), *C. elegans* (Byrne et al. 2007, Lehner et al. 2006), *E. coli* (Butland et al. 2008, Typas et al. 2008), and *D. melanogaster* (Horn et al. 2011). However, genome-scale application is challenging, and in fact in most model organisms, large-scale unbiased data are still not available. In those cases, an alternative is to infer interactions by establishing gene orthology relationships with the screened species, which proved fruitful in other biological networks. For instance, in the case of protein–protein interactions, the conservation of interacting pairs across species is significant, the so-called interologs (Walhout et al. 2000), and in fact 90% of co-complex proteins in human are also found in the same complex in yeast, when orthology relationships exist (van Dam & Snel 2008). Therefore, determining the extent to which the individual genetic interactions and the properties of the genetic networks are conserved across species is crucial for the reliable mapping of genetic interactions in uncharted organisms. Additionally, if global network properties were conserved, a rational screen design could maximize the amount of information obtained by each tested pair, minimizing the screening cost in new species.

In recent years, large-scale genetic interaction screens have been performed in *C. elegans* and *S. pombe*, though only in the latter at a genome scale, and the resulting networks have been compared to that of *S. cerevisiae*. In the case of *C. elegans*, non-viable phenotypes are conserved for 61% of the essential genes in *S. cerevisiae* (Tischler et al. 2006), and 31% of the protein–protein interactions are found in both species (Matthews et al. 2001). However, the genetic interaction network is much less conserved, and in fact, the synthetic lethal interactions in *C. elegans* showed a very low conservation rate (less than 5%) in *S. cerevisiae* (Byrne et al. 2007, Tischler et al. 2008). However, it is not clear to what extent the technical differences in the assays could account for part of the observed divergence. Additionally, in *C. elegans*, gene deletions were performed by RNAi-based methods, and the phenotype evaluated was different than colony size. Still, genes involved in chromatin and transcription tend to be more connected in both genetic networks (Costanzo et al. 2010), and a higher rate of conservation was found for a set of chromosomal instability (CIN) genes (McLellan et al. 2009). These findings suggest that specific parts of the genetic network (e.g. some biological processes) may be more conserved than others. Additionally, the structure and the general principles of the genetic network appear to be conserved (Byrne et al. 2007, Lehner et al. 2006). For instance, in both species the general degree distribution is similar to scale-free networks (most nodes with very few connections, and few nodes highly connected, i.e. the hubs), hubs tend to be more essential (have stronger single mutant phenotypes), and both networks have comparable clustering coefficient and average density.

Other studies have focused on the comparison between *S. cerevisiae* and *S. pombe*, which is closer in evolutionary terms than *C. elegans*. Both yeast species diverged around 400 million years ago, but maintained a high number of one-to-one orthology relationships (Wood 2006), and as many as 75% of genes in *S. pombe* have an ortholog in *S. cerevisiae*. Additionally, 83% of the single-copy orthologs are essential in both species (Kim et al. 2010). However, both species have important biological differences, and *S. pombe* is in some cases closer to metazoans (Sunnerhagen 2002). The initial large-scale screenings of genetic interactions in *S. pombe* reported 18% (Roguev et al. 2008) and 29% (Dixon et al. 2008) overlap with *S. cerevisiae*. Overlap was higher (more than 50%) for positive interactions within complexes, which suggested a higher conservation for functional modules than individual genetic interactions (Roguev et al. 2008). As a first step toward the definition of a conserved eukaryotic genetic interaction network, Dixon et al. compiled the conserved orthologous genes reporting an interaction in both species. They noticed that some of the nodes had human orthologs involved in cancer, which highlights the ability of genetic interactions to map core processes conserved across eukaryotic cells, and evidences the potential benefit of genetic networks for the rational selection of targets of therapeutic interest (Dixon et al. 2008). Very recently, two genome-scale screens were performed in *S. pombe* (Frost et al. 2012, Ryan et al. 2012). The size of these two screenings offered an unprecedented opportunity for the unbiased comparison with the genetic network of *S. cerevisiae*. Importantly, the general properties of genetic networks found in previous studies were confirmed. The overall conservation of genetic interactions was ~25%, and higher conservation was found within protein complexes (~70%), followed by interactions within biological processes, and being lower between different biological processes (Ryan et al. 2012). Additionally, ortho-essential genes (genes essential in only one of the species) showed 2.5 times more interactions than non-essential genes.

In genetic networks, the most connected genes tend to show specific features like evidence of a single mutant phenotype, a high level of structural disorder, or a high protein–protein interaction degree. Based on these observations, Koch et al. built a model using 16 different gene features to predict interaction degree in the *S. cerevisiae* genetic network (Koch et al. 2012). The model was trained on a large-scale data set applying a regression tree approach and accurately predicted degree for genes excluded from the training set ($r = 0.80$). Interestingly, while gene features correlated only weakly between *S. cerevisiae* and *S. pombe* (e.g. correlation of single mutant fitness between both species was 0.20), the correlation of each feature to interaction degree was very similar across species suggesting that the learned model in *S. cerevisiae* could be used to predict degree in *S. pombe*. Indeed, using that very same model showed good performance also in *S. pombe* ($r = 0.51$). Importantly, training the model on *S. pombe* data yielded very similar predictive capabilities, evidencing that the principles defining the topology of genetic networks are conserved over large evolutionary distances.

The general picture arising from these studies suggests that, while the extent of conservation of individual genetic interactions between orthologous pairs remains unclear, structural and topological properties of genetic networks appear to be more generally conserved suggesting that a complete genetic interaction network for a simple eukaryote like budding yeast should serve as a powerful reference network to guide genetic interaction studies in other species and organisms.

2.4   Perspectives

The analysis of large-scale genetic interaction data requires the use of computational methods. These tools improve our understanding of genetic networks, and evolve as the quantity and quality of data available increases. Certainly, new approaches must be devised to tackle the developments, challenges, and opportunities that will emerge in the field in forthcoming years. For instance, fewer than $\sim20\%$ of yeast genes are essential for cell viability under standard laboratory conditions (Giaever et al. 2002), and most gene deletions do not show any effect on cell growth (see Figure 2.1). However, fitness of single-gene deletions assessed in more than 1000 chemical and environmental conditions revealed a growth defect for 97% of genes in at least one condition (Hillenmeyer et al. 2008). Therefore, mapping of genetic interactions under a set of diverse conditions is necessary in order to unveil functional roles and modular organizations which remain inactive and elusive under standard laboratory conditions. Differential analysis on these data is best suited to discern the parts of the network most affected by each condition, rather than performing an independent analysis in each case (Bean & Ideker 2012, Ideker & Krogan 2012). This differential network approach is also valid to elucidate the genetic interactions specific to developmental stages or tissues.

So far, genetic interaction screens have focused on deletion mutants, unveiling functional roles of non-essential genes. Interactions of essential genes have been explored to a lesser extent, by temperature-sensitive or hypomorphic alleles (Collins et al. 2007, Costanzo et al. 2010). Interestingly, studies on these partial data revealed different properties with respect to the non-essential network (Bandyopadhyay et al. 2008, Bellay et al. 2011, Costanzo et al. 2010). Therefore, screens on essential genes are required to unveil the properties of a specific part of the genetic network that now remains hidden. Besides, essential genes tend to form more protein complexes and are more conserved across species than non-essential genes, which could lead to higher conservation also for their individual genetic interactions. The digenic space of genetic interactions is being thoroughly explored in different species (Costanzo et al. 2010, Frost et al. 2012, Ryan et al. 2012). However, functional buffering may occur between three or even more genes, which would not be identified by current experimental setups. For instance, this would be the case for duplicate genes, where their genetic interactions would remain hidden given that they functionally buffer each other. Unveiling the genetic interactions on these redundant duplicates would require the screening of triple (or higher-order) mutants (VanderSluis et al. 2010).

New methodology will be also needed to benefit from the use of complex phenotypes in high-throughput screens. So far, the only realistic choice in genome-scale experiments has been the evaluation of cell fitness as phenotype. However, recent advances in imaging techniques allow the measurement of more complex phenotypic traits in a systematic manner (Horn et al. 2011, Vizeacoumar et al. 2009). Very promisingly, the combination of this approach with SGA detected four times more interactions than using cell fitness as reference phenotype (Vizeacoumar et al. 2010). Therefore, it is expected that the evaluation of complex phenotypes will soon be part of the genetic interactions high-throughput screenings, which will increase the quality of the functional predictions, and will facilitate the mapping of genetic interactions in other organisms sharing specific phenotypic traits. This could for instance allow a better understanding of morphological defects that appear in higher eukaryotes, such as human. Further, as screening in different species progresses, cross-species analysis will improve our understanding about the general rules governing genetic networks which will enhance predictions in other organisms. Importantly, predictions of genetic interactions in human and analysis of interactions of human orthologs in other organisms can bring specific insights about complex diseases, which are more often caused by the combined effect of multiple mutations than just single perturbations (Barabási et al. 2011).

References

Ashburner, M., Ball, C., Blake, J., Botstein, D., Butler, H. et al. (2000), 'Gene Ontology: tool for the unification of biology', *Nature Genetics* **25**(1), 25–29.

Avery, L. & Wasserman, S. (1992), 'Ordering gene function: the interpretation of epistasis in regulatory hierarchies', *Trends in Genetics* **8**(9), 312–316.

Babu, M., Daz-Meja, J. J., Vlasblom, J., Gagarinova, A., Phanse, S. et al. (2011), 'Genetic interaction maps in *Escherichia coli* reveal functional crosstalk among cell envelope biogenesis pathways', *PLoS Genetics* **7**(11), e1002377.

Bandyopadhyay, S., Kelley, R., Krogan, N. J. & Ideker, T. (2008), 'Functional maps of protein complexes from quantitative genetic interaction data', *PLoS Computational Biology* **4**(4).

Barabási, A.-L. & Oltvai, Z. N. (2004), 'Network biology: understanding the cell's functional organization', *Nature Review Genetics* **5**(2), 101–113.

Barabási, A.-L., Gulbahce, N. & Loscalzo, J. (2011), 'Network medicine: a network-based approach to human disease', *Nature Review Genetics* **12**(1), 56–68.

Baryshnikova, A., Costanzo, M., Dixon, S., Vizeacoumar, F. J., Myers, C. L. et al. (2010*a*), 'Synthetic genetic array (SGA) analysis in *Saccharomyces cerevisiae* and *Schizosaccharomyces pombe*', *Methods in Enzymology* **470**, 145–179.

Baryshnikova, A., Costanzo, M., Kim, Y., Ding, H., Koh, J. et al. (2010*b*), 'Quantitative analysis of fitness and genetic interactions in yeast on a genome scale', *Nature Methods* **7**(12), 1017–1024.

Bateson, W. (1909), *Mendel's principles of heredity*, Cambridge University Press.

Battle, A., Jonikas, M. C., Walter, P., Weissman, J. S. & Koller, D. (2010), 'Automated identification of pathways from quantitative genetic interaction data', *Molecular Systems Biology* **6**, 379.

Bean, G. J. & Ideker, T. (2012), 'Differential analysis of high-throughput quantitative genetic interaction data', *Genome Biology* **13**(12), R123.

Bellay, J., Atluri, G., Sing, T. L., Toufighi, K., Costanzo, M. et al. (2011), 'Putting genetic interactions in context through a global modular decomposition', *Genome Research* **21**(8), 1375–1387.

Ben-Aroya, S., Coombes, C., Kwok, T., O'Donnell, K. A., Boeke, J. D. et al. (2008), 'Toward a comprehensive temperature-sensitive mutant repository of the essential genes of *Saccharomyces cerevisiae*', *Molecular Cell* **30**(2), 248–258.

Botstein, D. & Fink, G. R. (2011), 'Yeast: an experimental organism for 21st century biology', *Genetics* **189**(3), 695–704.

Breslow, D. K., Cameron, D. M., Collins, S. R., Schuldiner, M., Stewart-Ornstein, J. et al. (2008), 'A comprehensive strategy enabling high-resolution functional analysis of the yeast genome', *Nature Methods* **5**(8), 711–718.

Butland, G., Babu, M., Daz-Meja, J. J., Bohdana, F., Phanse, S. et al. (2008), 'eSGA: *E. coli* synthetic genetic array analysis', *Nature Methods* **5**(9), 789–795.

Byrne, A. B., Weirauch, M. T., Wong, V., Koeva, M., Dixon, S. J. et al. (2007), 'A global analysis of genetic interactions in *Caenorhabditis elegans*', *Journal of Biology* **6**(3), 8.

Collins, S., Miller, K., Maas, N., Roguev, A., Fillingham, J. et al. (2007), 'Functional dissection of protein complexes involved in yeast chromosome biology using a genetic interaction map', *Nature* **446**(7137), 806–810.

Collins, S. R., Schuldiner, M., Krogan, N. J. & Weissman, J. S. (2006), 'A strategy for extracting and analyzing large-scale quantitative epistatic interaction data', *Genome Biology* **7**(7), R63.

Costanzo, M., Baryshnikova, A., Bellay, J., Kim, Y., Spear, E. et al. (2010), 'The genetic landscape of a cell', *Science* **327**(5964), 425.

Davierwala, A. P., Haynes, J., Li, Z., Brost, R. L., Robinson, M. D. et al. (2005), 'The synthetic genetic interaction spectrum of essential genes', *Nature Genetics* **37**(10), 1147–1152.

Dean, E. J., Davis, J. C., Davis, R. W. & Petrov, D. A. (2008), 'Pervasive and persistent redundancy among duplicated genes in yeast', *PLoS Genetics* **4**(7), e1000113.

Decourty, L., Saveanu, C., Zemam, K., Hantraye, F., Frachon, E. et al. (2008), 'Linking functionally related genes by sensitive and quantitative characterization of genetic interaction profiles', *Proceedings of the National Academy of Sciences of the USA* **105**(15), 5821–5826.

DeLuna, A., Vetsigian, K., Shoresh, N., Hegreness, M., Coln-Gonzlez, M. et al. (2008), 'Exposing the fitness contribution of duplicated genes', *Nature Genetics* **40**(5), 676–681.

Deshpande, R., Vandersluis, B. & Myers, C. L. (2013), 'Comparison of profile similarity measures for genetic interaction networks', *PLoS One* **8**(7), e68664.

Deutscher, D., Meilijson, I., Kupiec, M. & Ruppin, E. (2006), 'Multiple knockout analysis of genetic robustness in the yeast metabolic network', *Nature Genetics* **38**(9), 993–998.

Dixon, S. J., Fedyshyn, Y., Koh, J. L. Y., Prasad, T. S. K., Chahwan, C. et al. (2008), 'Significant conservation of synthetic lethal genetic interaction networks between distantly related eukaryotes', *Proceedings of the National Academy of Sciences of the USA* **105**(43), 16 653–16 658.

Drees, B. L., Thorsson, V., Carter, G. W., Rives, A. W., Raymond, M. Z. et al. (2005), 'Derivation of genetic interaction networks from quantitative phenotype data', *Genome Biology* **6**(4), R38.

Eisen, M. B., Spellman, P. T., Brown, P. O. & Botstein, D. (1998), 'Cluster analysis and display of genome-wide expression patterns', *Proceedings of the National Academy of Sciences of the USA* **95**(25), 14 863–14 868.

Fisher, R. A. (1918), 'The correlations between relatives on the supposition of Mendelian inheritance', *Transactions of the Royal Society Edinburgh* **52**, 399–433.

Förster, J., Famili, I., Palsson, B. O. & Nielsen, J. (2003), 'Large-scale evaluation of in silico gene deletions in *Saccharomyces cerevisiae*', *Omics* **7**(2), 193–202.

Frost, A., Elgort, M. G., Brandman, O., Ives, C., Collins, S. R. et al. (2012), 'Functional repurposing revealed by comparing *S. pombe* and *S. cerevisiae* genetic interactions', *Cell* **149**(6), 1339–1352.

Giaever, G., Chu, A. M., Ni, L., Connelly, C., Riles, L. et al. (2002), 'Functional profiling of the *Saccharomyces cerevisiae* genome', *Nature* **418**(6896), 387–391.

Goffeau, A., Barrell, B. G., Bussey, H., Davis, R. W., Dujon, B. et al. (1996), 'Life with 6000 genes', *Science* **274**(5287), 546, 563–567.

Guarente, L. (1993), 'Synthetic enhancement in gene interaction: a genetic tool come of age', *Trends in Genetics* **9**(10), 362–366.

Hartman, J. L. t., Garvik, B. & Hartwell, L. (2001), 'Principles for the buffering of genetic variation', *Science* **291**(5506), 1001–1004.

Hartwell, L. H., Hopfield, J. J., Leibler, S. & Murray, A. W. (1999), 'From molecular to modular cell biology', *Nature* **402**(6761 Suppl), C47–52.

He, X., Qian, W., Wang, Z., Li, Y. & Zhang, J. (2010), 'Prevalent positive epistasis in *Escherichia coli* and *Saccharomyces cerevisiae* metabolic networks', *Nature Genetics* **42**(3), 272–276.

Hillenmeyer, M. E., Fung, E., Wildenhain, J., Pierce, S. E., Hoon, S. et al. (2008), 'The chemical genomic portrait of yeast: uncovering a phenotype for all genes', *Science* **320**(5874), 362–365.

Horn, T., Sandmann, T., Fischer, B., Axelsson, E., Huber, W. et al. (2011), 'Mapping of signaling networks through synthetic genetic interaction analysis by RNAi', *Nature Methods* **8**(4), 341–346.

Ideker, T. & Krogan, N. J. (2012), 'Differential network biology', *Molecular Systems Biology* **8**, 565.

Jasnos, L. & Korona, R. (2007), 'Epistatic buffering of fitness loss in yeast double deletion strains', *Nature Genetics* **39**(4), 550–554.

Jonikas, M. C., Collins, S. R., Denic, V., Oh, E., Quan, E. M. et al. (2009), 'Comprehensive characterization of genes required for protein folding in the endoplasmic reticulum', *Science* **323**(5922), 1693–1697.

Kelley, R. & Ideker, T. (2005), 'Systematic interpretation of genetic interactions using protein networks', *Nature Biotechnology* **23**(5), 561–566.

Kim, D.-U., Hayles, J., Kim, D., Wood, V., Park, H.-O. et al. (2010), 'Analysis of a genome-wide set of gene deletions in the fission yeast *Schizosaccharomyces pombe*', *Nature Biotechnology* **28**(6), 617–623.

Koch, E. N., Costanzo, M., Bellay, J., Deshpande, R., Chatfield-Reed, K. et al. (2012), 'Conserved rules govern genetic interaction degree across species', *Genome Biology* **13**(7), R57.

Leek, J. T., Scharpf, R. B., Bravo, H. C., Simcha, D., Langmead, B. et al. (2010), 'Tackling the widespread and critical impact of batch effects in high-throughput data', *Nature Reviews Genetics* **11**(10), 733–739.

Lehner, B., Crombie, C., Tischler, J., Fortunato, A. & Fraser, A. G. (2006), 'Systematic mapping of genetic interactions in *Caenorhabditis elegans* identifies common modifiers of diverse signaling pathways', *Nature Genetics* **38**(8), 896–903.

Li, Z., Vizeacoumar, F. J., Bahr, S., Li, J., Warringer, J. et al. (2011), 'Systematic exploration of essential yeast gene function with temperature-sensitive mutants', *Nature Biotechnology* **29**(4), 361–367.

Lucchesi, J. C. (1968), 'Synthetic lethality and semi-lethality among functionally related mutants of *Drosophila melanogaster*', *Genetics* **59**(1), 37–44.

Mani, R., St Onge, R., Hartman, J., Giaever, G. & Roth, F. (2008), 'Defining genetic interaction', *Proceedings of the National Academy of Sciences of the USA* **105**(9), 3461–3466.

Matthews, L. R., Vaglio, P., Reboul, J., Ge, H., Davis, B. P. et al. (2001), 'Identification of potential interaction networks using sequence-based searches for conserved protein–protein interactions or "interologs"', *Genome Research* **11**(12), 2120–2126.

McLellan, J., O'Neil, N., Tarailo, S., Stoepel, J., Bryan, J. et al. (2009), 'Synthetic lethal genetic interactions that decrease somatic cell proliferation in *Caenorhabditis elegans* identify the alternative RFC CTF18 as a candidate cancer drug target', *Molecular Biology of the Cell* **20**(24), 5306–5313.

Michaut, M., Baryshnikova, A., Costanzo, M., Myers, C. L., Andrews, B. J. et al. (2011), 'Protein complexes are central in the yeast genetic landscape', *PLoS Computational Biology* **7**(2), e1001092.

Musso, G., Costanzo, M., Huangfu, M., Smith, A. M., Paw, J. et al. (2008), 'The extensive and condition-dependent nature of epistasis among whole-genome duplicates in yeast', *Genome Research* **18**(7), 1092–1099.

Ohya, Y., Sese, J., Yukawa, M., Sano, F., Nakatani, Y. et al. (2005), 'High-dimensional and large-scale phenotyping of yeast mutants', *Proceedings of the National Academy of Sciences of the USA* **102**(52), 19 015–19 020.

Orth, J. D., Thiele, I. & Palsson, B. (2010), 'What is flux balance analysis?', *Nature Biotechnology* **28**(3), 245–248.

Pan, X., Yuan, D. S., Xiang, D., Wang, X., Sookhai-Mahadeo, S. et al. (2004), 'A robust toolkit for functional profiling of the yeast genome', *Molecular Cell* **16**(3), 487–496.

Phillips, P. C. (1998), 'The language of gene interaction', *Genetics* **149**(3), 1167–1171.

Qi, Y., Suhail, Y., Lin, Y., Boeke, J. D. & Bader, J. S. (2008), 'Finding friends and enemies in an enemies-only network: a graph diffusion kernel for predicting novel genetic interactions and co-complex membership from yeast genetic interactions', *Genome Research* **18**(12), 1991–2004.

Roguev, A., Bandyopadhyay, S., Zofall, M., Zhang, K., Fischer, T. et al. (2008), 'Conservation and rewiring of functional modules revealed by an epistasis map in fission yeast', *Science* **322**(5900), 405–410.

Roguev, A., Wiren, M., Weissman, J. S. & Krogan, N. J. (2007), 'High-throughput genetic interaction mapping in the fission yeast *Schizosaccharomyces pombe*', *Nature Methods* **4**(10), 861–866.

Ryan, C., Greene, D., Cagney, G. & Cunningham, P. (2010), 'Missing value imputation for epistatic MAPs', *BMC Bioinformatics* **11**, 197.

Ryan, C., Greene, D., Gunol, A., van Attikum, H., Krogan, N. J. et al. (2011), 'Improved functional overview of protein complexes using inferred epistatic relationships', *BMC Systems Biology* **5**, 80.

Ryan, C. J., Roguev, A., Patrick, K., Xu, J., Jahari, H. et al. (2012), 'Hierarchical modularity and the evolution of genetic interactomes across species', *Molecular Cell* **46**(5), 691–704.

Schuldiner, M., Collins, S. R., Thompson, N. J., Denic, V., Bhamidipati, A. et al. (2005), 'Exploration of the function and organization of the yeast early secretory pathway through an epistatic miniarray profile', *Cell* **123**(3), 507–519.

Segre, D., Deluna, A., Church, G. M. & Kishony, R. (2005), 'Modular epistasis in yeast metabolism', *Nature Genetics* **37**(1), 77–83.

Snitkin, E. S. & Segre, D. (2011), 'Epistatic interaction maps relative to multiple metabolic phenotypes', *PLoS Genetics* **7**(2), e1001294.

St Onge, R. P., Mani, R., Oh, J., Proctor, M., Fung, E. et al. (2007), 'Systematic pathway analysis using high-resolution fitness profiling of combinatorial gene deletions', *Nature Genetics* **39**(2), 199–206.

Stark, C., Breitkreutz, B.-J., Reguly, T., Boucher, L., Breitkreutz, A. et al. (2006), 'BioGRID: a general repository for interaction datasets', *Nucleic Acids Research* **34**(Database issue), D535–539.

Sunnerhagen, P. (2002), 'Prospects for functional genomics in *Schizosaccaromyces pombe*', *Current Genetics* **42**, 73–84.

Suter, B., Auerbach, D. & Stagljar, I. (2006), 'Yeast-based functional genomics and proteomics technologies: the first 15 years and beyond', *BioTechniques* **40**(5), 625–644.

Szappanos, B., Kovacs, K., Szamecz, B., Honti, F., Costanzo, M. et al. (2011), 'An integrated approach to characterize genetic interaction networks in yeast metabolism', *Nature Genetics* **43**(7), 656–662.

Tischler, J., Lehner, B., Chen, N. & Fraser, A. G. (2006), 'Combinatorial RNA interference in *Caenorhabditis elegans* reveals that redundancy between gene duplicates can be maintained for more than 80 million years of evolution', *Genome Biology* **7**(8), R69.

Tischler, J., Lehner, B. & Fraser, A. G. (2008), 'Evolutionary plasticity of genetic interaction networks', *Nature Genetics* **40**(4), 390–391.

Tong, A. H., Evangelista, M., Parsons, A. B., Xu, H., Bader, G. D. et al. (2001), 'Systematic genetic analysis with ordered arrays of yeast deletion mutants', *Science* **294**(5550), 2364–2368.

Tong, A. H., Lesage, G., Bader, G. D., Ding, H., Xu, H. et al. (2004), 'Global mapping of the yeast genetic interaction network', *Science* **303**(5659), 808–813.

Tucker, C. L. & Fields, S. (2003), 'Lethal combinations', *Nature Genetics* **35**(3), 204–205.

Typas, A., Nichols, R. J., Siegele, D. A., Shales, M., Collins, S. R. et al. (2008), 'High-throughput, quantitative analyses of genetic interactions in *E. coli*', *Nature Methods* **5**(9), 781–787.

Ulitsky, I. & Shamir, R. (2007), 'Pathway redundancy and protein essentiality revealed in the *Saccharomyces cerevisiae* interaction networks', *Molecular Systems Biology* **3**, 104.

Ulitsky, I., Krogan, N. J. & Shamir, R. (2009), 'Towards accurate imputation of quantitative genetic interactions', *Genome Biology* **10**(12), R140.

van Dam, T. J. P. & Snel, B. (2008), 'Protein complex evolution does not involve extensive network rewiring', *PLoS Computational Biology* **4**(7), e1000132.

Van Driessche, N., Demsar, J., Booth, E. O., Hill, P., Juvan, P. et al. (2005), 'Epistasis analysis with global transcriptional phenotypes', *Nature Genetics* **37**(5), 471–477.

VanderSluis, B., Bellay, J., Musso, G., Costanzo, M., Papp, B. et al. (2010), 'Genetic interactions reveal the evolutionary trajectories of duplicate genes', *Molecular Systems Biology* **6**, 429.

Vizeacoumar, F. J., Chong, Y., Boone, C. & Andrews, B. J. (2009), 'A picture is worth a thousand words: genomics to phenomics in the yeast *Saccharomyces cerevisiae*', *FEBS Letters* **583**(11), 1656–1661.

Vizeacoumar, F. J., van Dyk, N., Vizeacoumar, F. S., Cheung, V., Li, J. et al. (2010), 'Integrating high-throughput genetic interaction mapping and high-content screening to explore yeast spindle morphogenesis', *Journal of Cell Biology* **188**(1), 69–81.

Waddington, C. H. (1957), *The strategy of the genes*, Allen & Unwin, London.

Walhout, A. J., Sordella, R., Lu, X., Hartley, J. L., Temple, G. F. et al. (2000), 'Protein interaction mapping in *C. elegans* using proteins involved in vulval development', *Science* **287**(5450), 116–122.

Winzeler, E. A., Shoemaker, D. D., Astromoff, A., Liang, H., Anderson, K. et al. (1999), 'Functional characterization of the *S. cerevisiae* genome by gene deletion and parallel analysis', *Science* **285**(5429), 901–906.

Wong, S. L., Zhang, L. V., Tong, A. H. Y., Li, Z., Goldberg, D. S. et al. (2004), 'Combining biological networks to predict genetic interactions', *Proceedings of the National Academy of Sciences of the USA* **101**(44), 15 682–15 687.

Wood, V. (2006), '*Schizosaccharomyces pombe* comparative genomics: from sequence to systems', *Topics in Current Genetics* **15**, 233–285.

Ye, P., Peyser, B. D., Pan, X., Boeke, J. D., Spencer, F. A. et al. (2005), 'Gene function prediction from congruent synthetic lethal interactions in yeast', *Molecular Systems Biology* **1**, 2005.0026.

Zhong, W. & Sternberg, P. W. (2006), 'Genome-wide prediction of *C. elegans* genetic interactions', *Science* **311**(5766), 1481–1484.

3   Mapping genetic interactions across many phenotypes in metazoan cells

Christina Laufer, Maximilian Billmann, and Michael Boutros

Interactions between genes can be experimentally determined by combining multiple mutations and identifying combinations where the resulting phenotype differs from the expected one. Such genetic interactions, for example measured in yeast for cell proliferation and growth phenotypes, provided intricate insights into the genetic architecture and interplay of pathways. Due to the lack of comprehensive deletion libraries, similar experiments in higher eukaryotic cells have been challenging. Recently, we and others described methods to perform systematic, comprehensive double-perturbation analyses in *Drosophila* and human cells using RNA interference. We also introduced methods to use multiple phenotypes to map genetic interactions across a broad spectrum of processes.

This chapter focuses on the systematic mapping of genetic interactions and the use of image-based phenotypes to improve genetic interaction calling. It also describes experimental approaches for the analysis of genetic interactions in human cells and discusses concepts to expand genetic interaction mapping towards a genomic scale.

3.1   A short history of genetic interaction analysis

Using quantitative traits to map genetic interactions has a long tradition in *Drosophila*. In the 1960s and 1970s, Dobzhansky, Rendel, and others used externally visible phenotypes or overall fitness to study non-mendelian inheritance and dissect the heritability of complex traits (Fig. 3.1a). One of the underlying assumptions was that genetic loci in the *Drosophila* genome interact to shape complex phenotypes or buffer detrimental alleles.

In 1965, Dobzhansky and colleagues analyzed epistatic interactions between the components of genetic variants in *Drosophila*. They crossed flies carrying mutant alleles into a wild-type background obtained from a natural habitat and found that the combination of particular chromosomes showed synthetic sick phenotypes, whereas both chromosomes alone did not. This for the first time demonstrated the presence of bichromosomal synthetic interactions in *Drosophila* populations. Similarly, Rendel and colleagues demonstrated the existence of epistatic modifiers in *Drosophila* by analysis of scute alleles, which reduce the number of scutellar bristles on the dorsal thorax from four to an average of one. Artificial selection of flies for high bristle number lead to

Systems Genetics: Linking Genotypes and Phenotypes, ed. F. Markowetz and M. Boutros. Published by Cambridge University Press. © Cambridge University Press 2015.

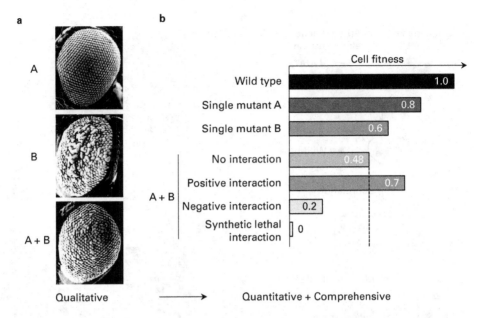

Figure 3.1   Qualitative and quantitative analysis of genetic epistasis. (a) Eye phenotypes in *Drosophila melanogaster* by scanning electron microscopy. Whereas allele A shows a wild-type eye phenotype with regularly arranged ommatidia, allele B displays a severe rough eye. The combination (A + B) shows an intermediate phenotype. (b) Quantitative evaluation of genetic interactions based on the cell fitness phenotype. The phenotype of a double mutant (A + B) is expected to be the product of both single mutant phenotypes. Deviation from this model in either direction is defined as a positive or a negative interaction, respectively. The extreme of the latter is a synthetic lethal interaction, where the combination of both mutations is lethal, while the single mutations have no or only a mild effect on cell fitness.

seven to eight bristles in flies carrying at least one scute+ allele and three to four bristles in scute mutant flies, providing evidence for epistatic modifiers suppressing the scute phenotype.

These experiments laid the foundation for today's genetic interaction studies, yet they involved labor-intensive crossing schemes and manual phenotyping while mapping only few interaction loci. Experiments in *Drosophila* populations can be scaled to analyze tens of thousands of crosses; however, one of the major limitations of the approaches mentioned was the constraint to easily observable phenotypes as those could not be scored in an automated fashion. This only became feasible recently by employing automated imaging (for visual phenotypes) and genomic experiments (for molecular phenotypes).

3.2   Perturbation-based genetic interaction studies in yeast

Alternative approaches to observational studies have been pioneered in yeast. They largely focused on the identification of synthetic sick/lethal interactions, in which the

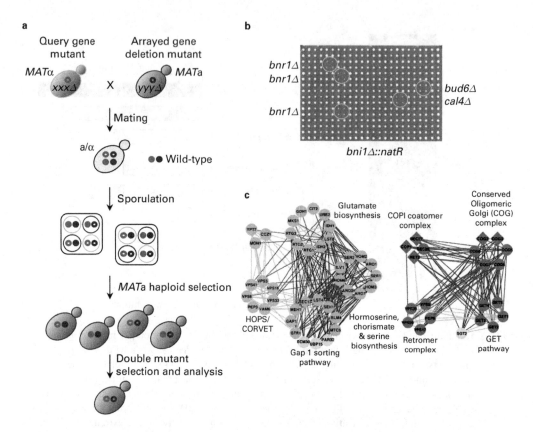

Figure 3.2 Genetic interaction studies in yeast. (a) Synthetic genetic arrays are used for the generation and evaluation of double mutant strains in *Saccharomyces cerevisiae*. A strain carrying a query mutation is crossed to an arrayed library of deletion mutants. All mutations are linked to dominant selectable markers. After inducing sporulation, double mutant haploids are selected and their growth rate is analyzed in an arrayed format. (b) Result of an SGA analysis with the query deletion *bni1*Δ crossed to an array of 96 deletion mutants. (Reprinted from Tong et al. [2001] with permission from AAAS.) The yellow circles mark the formation of residual colonies caused by synthetic sick/lethal interactions, which were scored with *bnr1*Δ, *cla4*Δ, and *bud6*Δ. *Bni1*Δ double mutants could not form in haploid progeny. (c) Genetic interaction networks of gene subsets belonging to amino acid biosynthesis and uptake (green) and ER-Golgi (violet). Uncharacterized genes are colored yellow. (Reprinted from Costanzo et al. [2010] with permission from AAAS.) Nodes represent non-essential (circles) and essential (diamonds) genes, which are grouped according to the similarity of their genetic interaction profiles obtained in a large-scale genetic interaction screen. The edges represent direct negative (red) and positive (green) genetic interactions. A black and white version of this figure will appear in some formats. For the color version, please refer to the plate section.

combination of two mutations unexpectedly leads to cell death or reduced fitness (Fig. 3.1b). Synthetic genetic array (SGA) analysis was developed as an automated approach for the systematic generation and evaluation of double mutant strains in the budding yeast *Saccharomyces cerevisiae* (Tong et al. 2001) (Fig. 3.2a). This technique takes advantage of deletion libraries available in yeast which comprise ~4700 viable mutants and additional conditional or hypomorphic alleles for the ∼1000 essential yeast genes (Winzeler et al. 1999, Schuldiner et al. 2005, Ben-Aroya et al. 2008). Typically, a strain carrying a query mutation linked to a dominant selectable marker is crossed to an arrayed mutant library. Robotics is used for the generation, selection, and analysis of haploid double mutants and colony size is scored as an estimate for double mutant fitness (Fig. 3.2b). Using this method, Tong and colleagues generated a genetic interaction network of ∼1000 genes and ∼4000 interactions. They found that similar interaction patterns of genes can point to functional connections between them (Tong et al. 2004).

In 2005, Schuldiner and colleagues described a further development of SGA analysis termed epistatic miniarray profile (E-MAP). They aimed at maximizing the output of genetic interaction analyses by integrating two strategies that cope with the technical challenges of such studies (Schuldiner et al. 2005). In this approach, genetic interactions are measured quantitatively, with the purpose of analyzing the complete interaction spectrum including positive interactions. Second, screening for genetic interactions within a subset of logically connected genes increases the signal-to-noise ratio, as interactions are more frequent between functionally associated genes. Furthermore, this reduces experimental size to typically 400 to 800 genes, which can in turn be screened in pairwise fashion. Selection of gene subsets can for example be based on protein–protein interactions, functional annotations, or domain architecture. This method may miss factors that indirectly influence the process of interest, however, Collins and colleagues could demonstrate that analyzing a gene subset using the E-MAP approach provides substantial power for high-resolution mapping of genetic interactions. They chose a set of 743 genes involved in chromosome biology and obtained a detailed map of pairwise genetic interactions, which they used to group genes into functional clusters and to systematically dissect multi-protein complexes into modules (Collins et al. 2007).

A major result of the analyses in yeast was that these methods can in principle be taken to a very large or even genome-wide scale – at least using fitness as readout. Detailed genetic interaction maps provide insight into the genomic architecture of a cell and help to place components into pathways and to identify buffering between biological processes (Costanzo et al. 2010) (Fig. 3.2c). This highlights the power of genetic interaction mapping for the understanding of complex genetic networks. However, while some studies used more complex phenotypes to identify genetic interactions (Schuldiner et al. 2005, Kuzmin et al. 2014), there is a limitation to these experimental approaches in yeast due to the rather small size of the (visual) phenotypic spectrum.

3.3   Genetic interaction analysis in *Drosophila*

The fruit fly *Drosophila* has been an important tool for the dissection of epistatic relationships between genes. For example, many core RAS pathway members, which are central to human diseases such as cancer, have been discovered in *Drosophila* and their epistatic relationship clarified their role in RAS/RTK signal transduction (Wassarman et al. 1995). As a metazoan model system, *Drosophila* allows to study cell-to-cell signaling, yet has a compact genome without many gene duplications. Moreover, in comparison to other model systems such as *Caenorhabditis elegans*, *Drosophila* shows a high conservation of its key pathway components when compared to humans (St Johnston 2002). This makes *Drosophila* a valuable tool for the analysis of interactions of genes and pathways functionally relevant in human biology and disease.

The discovery of RNAi has also enabled large-scale genetic screens for loss-of-function phenotypes in *Drosophila*. In cultured *Drosophila* cells, introduction of 100–800 base-pair-long dsRNAs reliably induces RNAi and very efficient gene knockdown. The availability of genome-wide RNAi libraries enabled the first large-scale screens in *Drosophila* to deplete almost every gene in the genome (Boutros et al. 2004, Müller et al. 2005, Bartscherer et al. 2006, Björklund et al. 2006). In genome-wide screens, epistatic one-by-one follow-up of candidate genes was used to place them into existing pathways (Müller et al. 2005, Bartscherer et al. 2006). In 2006, Friedman et al. reasoned that the specificity to detect regulators of receptor tyrosine kinase (RTK) signaling would be increased by RNAi screening in the presence of different RTK signaling stimuli (Friedman & Perrimon 2006). In a two-step process, they screened a genome-wide RNAi library with and without insulin stimulation and validated a subset of 362 genes in five distinct RTK stimulating backgrounds. Genes such as GC31302 and CG30387 that had a condition-independent effect on RTK signaling were considered potential novel core signaling members.

In contrast to alternating growth conditions, Bakal and colleagues used RNAi to introduce distinct signaling backgrounds (Bakal et al. 2008). Specifically, they depleted 12 genes likely to alter or interconnect JNK signaling and screened all kinases and phosphatases for their effect on a dJUN reporter in those genetic backgrounds. Primarily, they used those backgrounds to expand their hit list to 79 genes, of which 70% only influenced JNK signaling in a context-dependent fashion. To functionally subdivide the tested genes, they clustered the profiles of their dJUN-specific phenotype in the different backgrounds (Bakal et al. 2008). In summary, those studies showed that depletion of genes in a context-specific background suggests gene function more precisely.

3.3.1 Mapping genetic interactions in *Drosophila* using a square double perturbation matrix

Previously, we conducted a study that aimed at systematically mapping gene interactions and function in *Drosophila*. We depleted 93 genes involved in kinase signaling in a pairwise fashion. We used high-content microscopy and automated image analysis to extract quantitative cellular phenotypes like cell or nuclear size or intensity of cytoskeletal staining (Horn et al. 2011) (Fig. 3.3a). Based on these data, we estimated genetic interactions for each gene pair by using a so-called multiplicative model. This model assumes that a combinatorial depletion phenotype equals the product of both single depletion phenotypes if the two genes do not interact genetically. A combinatorial phenotype that deviates from this model is considered a genetic interaction (Dixon et al. 2009, Axelsson et al. 2011). We generated genetic interaction profiles for each of the 93

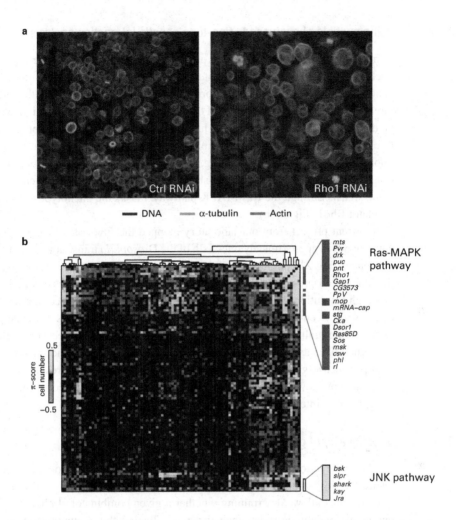

Figure 3.3   Genetic interaction mapping in *Drosophila*. (a) For multiparametric phenotyping, nuclei and cytoskeleton are visualized using DAPI, phalloidin, and a tubulin antibody. The left panel shows *Drosophila* D.mel-2 cells treated with control RNAi. Knockdown of *Rho1* (right panel) causes defects in cytokinesis, leading to large, multinucleated cells. (b) Square genetic interaction matrix obtained by pairwise knockdown of 93 genes involved in signal transduction (Horn et al. 2011). Genetic interactions were calculated based on observed cell number. Hierarchical clustering based on the genes' interaction profiles revealed several groups of functionally connected genes, such as members of the Ras-MAPK or JNK pathway. A black and white version of this figure will appear in some formats. For the color version, please refer to the plate section.

genes and used those for hierarchical clustering and to train a classifier. This approach reconstructed the RAS-MAPK pathway and identified Cka as a novel regulator, downstream of Ras (Horn et al. 2011) (Fig. 3.3b). Our work illustrated that in *Drosophila*, similar to studies in yeast (Tong et al. 2004, Schuldiner et al. 2005, Collins et al. 2007), epistatic or genetic interaction profiles can be used to systematically map the functional relationship of genes.

3.3.2   Using multiple phenotypes to fine-tune genetic interaction analysis

Using an image-based readout, we observed that distinct phenotypic features such as nuclear count, nuclear area, and intensity provide complementary information about genetic interactions upon combinatorial RNAi (Horn et al. 2011). With regard to those three quantitative features, we found that as few as 21% of all genetic interactions could be detected in all features, whereas half of the genetic interactions appeared to be feature specific (Fig. 3.4a). In fact, multiparametric image analysis not only allows for improved genetic interaction calling, but also highlights how functional relationships can be pinpointed. For instance, we showed that depletion of drk, which provides mitogenic signals through the Ras-MAPK pathway, masks the phenotype of the cytokinesis regulator Rho1 (Fig. 3.4b).

A current project from our laboratory mapped the function of >1300 genes, which corresponds to approximately one-sixth of the *Drosophila* transcriptome in this cell type. By considering 21 computationally selected phenotypic features, we reconstructed and interconnected processes regulating transcription, translation, or the mitotic cell cycle at a high resolution. Interestingly, we observed that mapping of certain biological processes was driven by features that describe the very same process: for example, mitotic features essentially pinpointed the functional relationship between the gamma-TuRC, the Dynein/Dynactin complex and axillary regulators of the mitotic spindle. In prospect, multiparametric genetic interaction analysis might enable high resolution mapping of gene function as well as process-specific reconstruction of functional relationships and assignment of interaction directionality.

3.3.3   Directionality of genetic interactions

Genetic interaction profiles more accurately group the function of genes than the respective single depletion phenotypes alone. Similarly, the profile of multiple phenotypic features allows determining whether a given combinatorial phenotype resembles one single phenotype rather than the other. Testing this assumption, Carter and colleagues analyzed the multiparametric data of the 93 × 93 matrix generated in our study (Horn et al. 2011, Carter 2013). They found that, when measuring a genetic interaction, the combinatorial effect of two genes often resembles one phenotype. In this constellation, the respective other gene aggravates the phenotype. This suggests a directionality of the genetic interaction from the gene that aggravates the phenotype towards the gene whose phenotype is aggravated. In conclusion, by assigning directionality to gene pairs, multiparametric genetic interaction mapping allows to integrate linear pathways into networks.

3.4   Expanding genetic interaction mapping towards the genomic scale

To date, genome-level studies of genetic interactions have been restricted to yeast. Due to the larger size of their genomes, current techniques limit the systematic exploration of genetic interactions in higher eukaryotes. Here, we discuss several approaches to

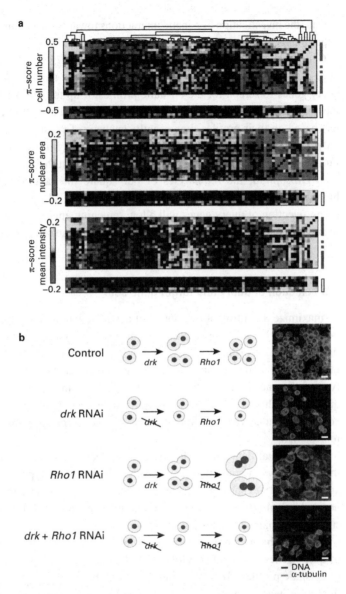

Figure 3.4 Genetic interaction analysis based on multiparametric phenotypes. (a) Genetic interaction profiles of genes clustering to the Ras-MAPK or JNK pathway based on cell number, nuclear area, and mean intensity. The order of genes is always based on the clustering obtained for cell number. The additional phenotypes contribute new and specific genetic interactions (Horn et al. 2011). (b) Scheme depicting the effects of *drk* and *Rho1* RNAi, which both lead to reduced cell numbers, but have opposite effects on cell size (Horn et al. 2011). The combined knockdown of both genes phenocopies the single knockdown of *drk*, revealing the epistatic effect of *Rho1* to *drk*. A black and white version of this figure will appear in some formats. For the color version, please refer to the plate section.

expand the investigation of genetic interactions towards the genomic scale while keeping experiments at a feasible scale.

Genetic interaction studies have investigated up to 75% of all yeast genes (Costanzo et al. 2010). To reduce the experimental scale, those studies used only a small set of query genes to screen for the epistatic profiles of up to 6000 target genes and the data they obtained was detailed enough to allow functional analysis of genes and gene clusters based on their profiles. Importantly, analysis of yeast E-MAP data has shown that only a fraction of query genes provides most of the information (Casey et al. 2008). Furthermore, the authors suggest an approach to select the most suitable query genes.

In *Drosophila*, the re-analyses of our square interaction map of 93 × 93 genes highlights a similar trend (Horn et al. 2011, Carter 2013). This suggests that an appropriate selection of query genes will allow screening epistatic profiles for a substantially increased number of genes as compared to a square matrix and highlights the importance of proper selection methods.

3.4.1   Pre-selection of query and target gene sets

To maximize the information content of a genetic interaction experiment, careful selection of query but also target genes is required. The selection of query genes comprises literature and experimental research. For experimental selection, transcriptome data of the respective model system can be used as a first filter. Next, the single depletion phenotype of all genes in the genome or a set of genes of interest can be assessed. The set of query genes should cover the entire spectrum of phenotypic diversity: each information-containing phenotype at different levels of phenotypic penetrance should be equally represented.

Experimental data can further be used to reduce the size of the target gene set. In this context, we assume that (i) genes that show a phenotype upon single perturbation are more likely to show genetic interactions and that (ii) those interactions are richer in non-redundant information. The latter largely depends on the biological focus of a given study and remains to be tested carefully.

Literature-based curation further refines the set of selected genes. For the selection of query genes, an equal coverage of biological processes assures unbiased generation of epistatic profiles. This is essential for a global view on gene function. However, in the future, we might focus more on process-specific genetic interaction networks and a process-focused set of query genes might be one way to perform those experiments. Finally, literature-based curation can maximize the number of investigated biological processes while keeping a feasible experimental workload. Here, the observation that allows us to assign genes to the same complex also increases the experimental coverage of processes: since members of the same complex have similar epistatic profiles, large complexes such as the proteasome or the ribosomal machinery can be represented by a selected number of their members. This approach enables a global reconstruction of functional relationships but might miss out a functional subdivision of processes.

In conclusion, several options support a systematic investigation of genetic interactions in the large genomes of higher eukaryotes. A combination of process-focused and more global approaches as well as further technical improvements might facilitate reconstruction of the genetic interaction landscape in a metazoan cell.

3.5 Towards genetic interaction mapping in human cells

While genetic interaction studies have given promising results in model organisms like yeast and *Drosophila*; similar studies in higher-order organisms are more challenging due to genomic complexity and due to the fact that it is far more difficult to create specific genetic perturbations. In yeast, deletion strains or hypomorphic alleles are available for all genes of the genome. In mammalian or even human cells it is not trivial to create true loss of function of genes, especially on a large scale. The current method of choice for genetic perturbations in mammalian cells is RNAi, induced by the presence of small interfering RNA (siRNA), small hairpin RNA (shRNA), or endonuclease-prepared siRNA (esiRNA). However, there are two issues when working with RNAi in mammalian cells as compared to mutations. First, the levels of target knockdown can vary substantially, so that only partial phenotypes may be evaluated. Second, every RNAi reagent can bear unspecific off-target effects. The main strategy to overcome those drawbacks in large-scale RNAi experiment is to employ multiple reagents targeting the same gene and to require that they give the same result.

Three independent studies have recently opened the field of genetic interaction analysis in mouse and human cells (Fig. 3.5a). They all employ pairwise genetic perturbations with subsequent evaluation of the combinatorial phenotypes, but the three methods vary in the RNAi methodology, the readouts used to detect genetic interactions and the approach of controlling unspecific effects exhibited by the RNAi reagents.

Roguev et al. developed a pipeline for the generation of E-MAPs in mouse fibroblast cells that is based on the methods developed in yeast (Collins et al. 2010, Roguev et al. 2013). This method applies esiRNA, a mixture of siRNAs targeting the same gene, which is believed to exhibit fewer off-target effects because each single sequence is present at only low concentrations (Kittler et al. 2007). One hundred and thirty genes with a function in chromatin regulation were targeted in pairwise combination in an arrayed format. Cell number was determined as a measure of cell growth and fitness using an Acumen plate cytometer. Following this procedure, about 11 000 phenotype measurements were performed to create genetic interaction profiles for the genes of interest. The resulting chromatin E-MAP of mouse fibroblasts revealed several functional modules of genes, in which known complex partners showed similar interaction profiles and positive genetic interactions with each other.

A different approach was taken by Bassik et al., who adopted the pooled shRNA screening design to analyze genetic interactions in the human myelogenous leukemia cell line K562 (Bassik et al. 2013, Kampmann et al. 2013). Their study followed a two-stage approach. In a first step, a genome-wide screen was conducted to identify genes involved in a cell's susceptibility to ricin as well as efficient shRNA constructs to target those genes. The second step was to analyze genetic interactions between 60 hit genes identified from the primary screen. For the pairwise gene knockdown, efficient

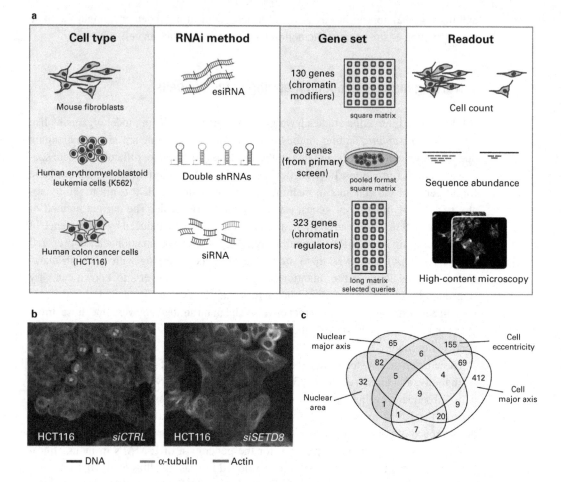

Figure 3.5 Genetic interaction studies in mammalian cells. (a) An overview about three different approaches to analyze genetic interactions in mouse and human cells. The methods differ in the RNAi reagents themselves, experimental design, and the cellular phenotypes measured for genetic interaction calling. (b) Multiparametric imaging in human HCT116 cells (Laufer et al. 2013). Staining DNA, tubulin, and actin allows the quantification of morphological features such as cell or nuclear size or intensity of cytoskeletal staining. Compared to control treated cells in the left panel, knockdown of *SETD8* (right panel) leads to enlarged cells and nuclei. (c) Venn diagram displaying the genetic interactions identified in a combinatorial RNAi screen, exemplarily shown for four non-redundant phenotypes (Laufer et al. 2013). A substantial amount of interactions is found only for selected phenotypes. A black and white version of this figure will appear in some formats. For the color version, please refer to the plate section.

shRNAs identified in the genome-wide screen were used to create a double-shRNA library, which was then again screened in a pooled format. With a deep-sequencing readout the presence of shRNA constructs was determined as a measure for cell fitness. Again, hierarchical clustering of genes according to the correlation of their interaction profiles recovered a number of known complexes. Furthermore, the genetic interaction map allowed functional predictions for several genes.

In our laboratory we developed a third experimental approach based on our recent experience in *Drosophila* studies. We employed siRNAs for pairwise gene knockdown in the human colorectal carcinoma cell line HCT116 (Laufer et al. 2013, Laufer et al. 2014). In contrast to the other studies, we used high-content microscopy for deep phenotype profiling, covering changes in nuclear and cytoskeletal morphology (Fig. 3.5b). Like the approach of Bassik and colleagues, our method involved two steps. In the first step, we analyzed the single knockdowns of 323 genes with a function in human chromatin biology using three individual siRNA designs per gene. We generated phenotype profiles and analyzed the knockdown efficiency of every single siRNA. Based on knockdown performance, we chose two out of three siRNA designs for the interaction screen. Out of our 323 target genes, we selected 20 genes that represented the full phenotypic space as queries for the double knockdowns. Finally, in the second step, double knockdowns were performed using two individual siRNA designs per target gene and per query gene. In this way we screened the interaction between two genes using four siRNA combinations, allowing for thorough quality control regarding design coherence. We generated genetic interaction profiles based on non-redundant phenotypes, such as cell number, cell major axis, or nuclear eccentricity. Just as in the *Drosophila* study, the different phenotypes enabled the detection of more and seemingly specific genetic interactions (Fig. 3.5c). Clustering of highly interactive genes recapitulated several known complex partners and also appears to be useful for functional predictions.

3.5.1   Scaling up human genetic interaction studies

The aforementioned studies clearly illustrate the value of genetic interaction studies in mammalian and especially human cells for the understanding of gene networks and functional connections between genes. However, each method has advantages over the others, but also drawbacks that have to be taken into account when planning large-scale interaction studies in the future. For example, esiRNAs can be produced inexpensively in large amounts. Also, the esiRNA approach relies widely on the fact that esiRNAs are believed to have a strong, combined effect on the target gene, but exhibit weak off-target effects. This saves time, as reagents are not evaluated prior to the experiment. However, in order to truly control reagent effects one would need to assay at least two independent reagents per gene. In the studies using shRNAs and siRNAs the reagents were controlled for unspecific effects. Bassik and colleagues pre-evaluated the reagents with much effort and only effective shRNAs were chosen. The correlation of interaction profiles of shRNAs targeting the same gene was then compared to the correlation of shRNAs targeting different genes and was observed to be higher, indicating similar effects of reagents targeting the same gene. In our study we also invested significant time in the selection and evaluation of siRNAs. After testing three individual RNAi reagents for each gene, pairs of siRNAs were chosen based on similarity of knockdown performance. In the final screen the correlation between two siRNAs targeting the same gene was measured and genes with incoherent results of both reagents were taken from the analysis.

Assaying RNAi reagent prior to screening increases the confidence in the results. However, this approach substantially increases experimental size and is cost- and time-intensive which might limit scale. Future studies will even more face the challenge of balancing the specificity and reliability of results against the scalability of the approach, since future experiments will be of growing complexity as a consequence of larger gene sets or combinations of different conditions or cell lines. To generate information about reagent reliability, data from large-scale screens could be integrated to identify coherent reagents. On a similar basis, screening individual reagents on a genome-wide scale in a cell line of interest by analyzing multiple broad phenotypes can provide a rich source for the selection of reliable reagents in the future.

Another factor influencing experimental complexity is the choice of phenotype used to detect genetic interactions. Many studies are based on cell fitness readout, because cell growth and survival integrate the outcome of many biological processes. In yeast, colony size as a proxy for cell fitness has revealed great insight into the genetic interaction network of cells (Costanzo et al. 2010). However, we demonstrated in our *Drosophila* and human studies that deep phenotyping greatly improves the calling of genetic interactions. Here again the time and effort of scoring multiple phenotypes and using them for genetic interaction analysis has to be weighed against time and cost.

3.6   Conclusions

Genetic interaction analysis across multiple phenotypes is a powerful method to dissect the underlying molecular and functional networks. With the advance in image analysis and high-throughput screening, automated phenotyping using many readouts on a cellular – and in the future also organismal – basis will provide a rich resource to map epistatic interactions. Genetic interaction networks can also be analyzed under multiple environmental conditions, for example using drug treatment or cancer mutations to create differential interaction networks. Disease-associated pathways can be dissected in detail, and the results can be integrated with protein–protein interaction networks or cancer mutation data. Genetic interaction networks should provide a detailed understanding of a gene's action under normal and pathophysiological conditions.

References

Axelsson, E., Sandmann, T., Horn, T., Boutros, M., Huber, W. et al. (2011), 'Extracting quantitative genetic interaction phenotypes from matrix combinatorial RNAI', *BMC Bioinformatics* **12**, 342.

Bakal, C., Linding, R., Llense, F., Heffern, E., Martin-Blanco, E. et al. (2008), 'Phosphorylation networks regulating jnk activity in diverse genetic backgrounds', *Science* **322**(5900), 453–6.

Bartscherer, K., Pelte, N., Ingelfinger, D. & Boutros, M. (2006), 'Secretion of wnt ligands requires evi, a conserved transmembrane protein', *Cell* **125**(3), 523–33.

Bassik, M. C., Kampmann, M., Lebbink, R. J., Wang, S., Hein, M. Y. et al. (2013), 'A systematic mammalian genetic interaction map reveals pathways underlying ricin susceptibility', *Cell* **152**(4), 909–22.

Ben-Aroya, S., Coombes, C., Kwok, T., O'Donnell, K. A., Boeke, J. D. et al. (2008), 'Toward a comprehensive temperature-sensitive mutant repository of the essential genes of *Saccharomyces cerevisiae*', *Mol Cell* **30**(2), 248–58.

Björklund, M., Taipale, M., Varjosalo, M., Saharinen, J., Lahdenperä, J. et al. (2006), 'Identification of pathways regulating cell size and cell-cycle progression by RNAI', *Nature* **439**(7079), 1009–13.

Boutros, M., Kiger, A. A., Armknecht, S., Kerr, K., Hild, M. et al. (2004), 'Genome-wide rnai analysis of growth and viability in *Drosophila* cells', *Science* **303**(5659), 832–5.

Carter, G. W. (2013), 'Inferring gene function and network organization in *Drosophila* signaling by combined analysis of pleiotropy and epistasis', *G3 (Bethesda)* **3**(5), 807–14.

Casey, F. P., Cagney, G., Krogan, N. J. & Shields, D. C. (2008), 'Optimal stepwise experimental design for pairwise functional interaction studies', *Bioinformatics* **24**(23), 2733–9.

Collins, S. R., Miller, K. M., Maas, N. L., Roguev, A., Fillingham, J. et al. (2007), 'Functional dissection of protein complexes involved in yeast chromosome biology using a genetic interaction map', *Nature* **446**(7137), 806–10.

Collins, S. R., Roguev, A. & Krogan, N. J. (2010), 'Quantitative genetic interaction mapping using the E-MAP approach', *Methods Enzymol* **470**, 205–31.

Costanzo, M., Baryshnikova, A., Bellay, J., Kim, Y., Spear, E. D. et al. (2010), 'The genetic landscape of a cell', *Science* **327**(5964), 425–31.

Dixon, S. J., Costanzo, M., Baryshnikova, A., Andrews, B. & Boone, C. (2009), 'Systematic mapping of genetic interaction networks', *Annu Rev Genet* **43**, 601–25.

Friedman, A. & Perrimon, N. (2006), 'A functional RNAI screen for regulators of receptor tyrosine kinase and ERK signalling', *Nature* **444**(7116), 230–4.

Horn, T., Sandmann, T., Fischer, B., Axelsson, E., Huber, W. et al. (2011), 'Mapping of signaling networks through synthetic genetic interaction analysis by RNAI', *Nat Methods* **8**(4), 341–6.

Kampmann, M., Bassik, M. C. & Weissman, J. S. (2013), 'Integrated platform for genome-wide screening and construction of high-density genetic interaction maps in mammalian cells', *Proc Natl Acad Sci U S A* **110**(25), E2317–26.

Kittler, R., Surendranath, V., Heninger, A.-K., Slabicki, M., Theis, M. et al. (2007), 'Genome-wide resources of endoribonuclease-prepared short interfering RNAS for specific loss-of-function studies', *Nat Methods* **4**(4), 337–44.

Kuzmin, E., Sharifpoor, S., Baryshnikova, A., Costanzo, M., Myers, C. L. et al. (2014), 'Synthetic genetic array analysis for global mapping of genetic networks in yeast', *Methods Mol Biol* **1205**, 143–68.

Laufer, C., Fischer, B., Billmann, M., Huber, W. & Boutros, M. (2013), 'Mapping genetic interactions in human cancer cells with RNAI and multiparametric phenotyping', *Nat Methods* **10**(5), 427–31.

Laufer, C., Fischer, B., Huber, W. & Boutros, M. (2014), 'Measuring genetic interactions in human cells by RNAI and imaging', *Nat Protoc* **9**(10), 2341–53.

Müller, P., Kuttenkeuler, D., Gesellchen, V., Zeidler, M. P. & Boutros, M. (2005), 'Identification of JAK/STAT signalling components by genome-wide RNA interference', *Nature* **436**(7052), 871–5.

Roguev, A., Talbot, D., Negri, G. L., Shales, M., Cagney, G. et al. (2013), 'Quantitative genetic-interaction mapping in mammalian cells', *Nat Methods* **10**(5), 432–7.

Schuldiner, M., Collins, S. R., Thompson, N. J., Denic, V., Bhamidipati, A. et al. (2005), 'Exploration of the function and organization of the yeast early secretory pathway through an epistatic miniarray profile', *Cell* **123**(3), 507–19.

St Johnston, D. (2002), 'The art and design of genetic screens: *Drosophila melanogaster*', *Nat Rev Genet* **3**(3), 176–88.

Tong, A. H., Evangelista, M., Parsons, A. B., Xu, H., Bader, G. D. et al. (2001), 'Systematic genetic analysis with ordered arrays of yeast deletion mutants', *Science* **294**(5550), 2364–8.

Tong, A. H. Y., Lesage, G., Bader, G. D., Ding, H., Xu, H. et al. (2004), 'Global mapping of the yeast genetic interaction network', *Science* **303**(5659), 808–13.

Wassarman, D. A., Therrien, M. & Rubin, G. M. (1995), 'The Ras signaling pathway in *Drosophila*', *Curr Opin Genet Dev* **5**(1), 44–50.

Winzeler, E. A., Shoemaker, D. D., Astromoff, A., Liang, H., Anderson, K. et al. (1999), 'Functional characterization of the *S. cerevisiae* genome by gene deletion and parallel analysis', *Science* **285**(5429), 901–6.

4   Genetic interactions and network reliability

Edgar Delgado-Eckert and Niko Beerenwinkel

The biochemical and molecular mechanisms underlying epistatic gene interactions observed in various living organisms are poorly understood. In this chapter, we introduce a mathematical framework linking epistasis to the redundancy of biological networks. The approach is based on network reliability, an engineering concept that allows for computing the probability of functional network operation under different network perturbations, such as the failure of specific components, which, in a genetic system, correspond to the knock-out or knock-down of specific genes. Using this framework, we provide a formal definition of epistasis in terms of network reliability and we show how this concept can be used to infer functional constraints in biological networks from observed genetic interactions. This formalism might help increase our understanding of the systemic properties of the cell that give rise to observed epistatic patterns.

4.1   Biological networks

A major goal of postgenomic biomedical research consists in understanding how the genetic components interact with each other to form living cells and organisms. The systems-wide approach requires both novel experimental techniques for mapping out such interactions and new mathematical models to describe and to analyze them. Interacting biological systems are often represented as networks (or graphs), where vertices correspond to components (e.g., genes, proteins, or metabolites) and edges correspond to pairwise interactions (e.g., activation, molecular binding, or chemical reaction). This abstract representation provides the conceptual basis for network biology, which aims at understanding the cell's functional organization and the complex behavior of living systems through biological network analysis (Strogatz 2001, Barabási & Oltvai 2004).

    Various experimental methods have been developed to measure physical interactions (molecular binding events) among proteins and several computational methods exist for predicting such interactions. These data give rise to protein–protein interaction (PPI) networks which are available from dedicated databases (Schwikowski et al. 2000, Xenarios et al. 2000, Jensen et al. 2009). Genetic interactions, or epistasis, refers to functional relationships between genes. Epistasis is a property of the underlying biochemical network. It describes the phenotypic effect of perturbing (e.g., knocking down

Systems Genetics: Linking Genotypes and Phenotypes, ed. F. Markowetz and M. Boutros. Published by Cambridge University Press. © Cambridge University Press 2015.

or knocking out) two genes separately versus jointly relative to the unperturbed system. Recent high-throughput screening (HTS) methods allow for measuring genetic interactions at a large scale, including epistatic miniarray profiling (Schuldiner et al. 2005), genetic interaction mapping (Decourty et al. 2008), and synthetic genetic analysis (Tong et al. 2001, Costanzo et al. 2010).

Current PPI networks are far from complete and it is hoped that the new HTS genetic interaction data will improve biological network reconstruction and our understanding of how physical interactions translate into phenotypes. Integration of physical and genetic interaction maps has been suggested in order to arrive at a more complete picture of the cell (Kelley & Ideker 2005, Beyer et al. 2007, Boone et al. 2007, Meur & Gentleman 2008, Dixon et al. 2009) and several computational methods have been proposed for predicting genetic interactions from PPI networks (Wong et al. 2004, Ulitsky & Shamir 2007, Jrvinen et al. 2008, Qi et al. 2008, Chipman & Singh 2009).

4.2   Epistasis

The notion of epistasis was originally developed in statistical genetics, where it refers to contributions to the phenotype that are not linear in the average effects of the single genes (Fisher 1918, Cordell 2002). Genetic interactions shape the structure and evolution of genetic systems (Phillips 2008). Fitness, the reproductive success of an individual, is the most relevant phenotype in evolutionary studies, because the epistatic interactions underlying fitness landscapes affect the course of evolution in various and often complicated ways (Wright 1931, Gavrilets 2004). Disease phenotypes are also of great interest and genetic interactions are likely to play an important role in mapping complex diseases in genome-wide association studies (Hoh & Ott 2003, Carlson et al. 2004).

Let us consider two biallelic genes and denote by a a mutation at the first gene locus and by b a mutation at the second locus. Then there exist four genotypes: the wild type $\emptyset$, the single mutants $\{a\}$ and $\{b\}$, and the double mutant $\{a, b\}$. We denote by w_g the phenotype of genotype g. Epistasis measures the deviation of the double-mutant phenotype from its expected value under the null hypothesis of no interaction between a and b. In a multiplicative null model, the expected value is $w_{\{a\}}w_{\{b\}}$ and epistasis can be defined as

$$\varepsilon_{ab}^{\times}(w) := w_{\emptyset}w_{\{a,b\}} - w_{\{a\}}w_{\{b\}}, \tag{4.1}$$

where the deviation has been measured as the difference between observed and expected phenotype. Other measures are conceivable, for example, the ratio $(w_{\emptyset}w_{\{a,b\}})/(w_{\{a\}}w_{\{b\}})$. Depending on the biological system and the measurement scale of the phenotype (Wagner et al. 1998), the null model can also be additive resulting in

$$\varepsilon_{ab}^{+}(w) := w_{\emptyset} + w_{\{a,b\}} - w_{\{a\}} - w_{\{b\}} \tag{4.2}$$

as an alternative definition of epistasis.

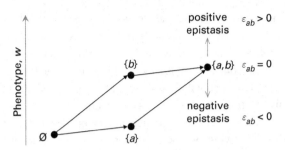

Figure 4.1   Schematic diagram illustrating the definition of epistasis. Positive (negative) epistasis ε_{ab} refers to a greater (smaller) effect of the double mutant $\{a, b\}$ on the phenotype than expected from the two single mutants $\{a\}$ and $\{b\}$ relative to the wild type Ø.

For both definitions, positive (or alleviating) epistasis ($\varepsilon_{ab} > 0$) indicates a greater phenotype of the double mutant as compared to the expectation based on the two single mutants relative to the wild type, whereas negative (or aggravating) epistasis ($\varepsilon_{ab} < 0$) indicates a smaller phenotype than expected under the null model (Figure 4.1). Multiplicative and additive epistasis as defined in Equations (4.1 and 4.2), respectively, are related by a logarithmic transformation of the phenotype, because

$$\varepsilon_{ab}^{\times}(w) = 0 \quad \Longleftrightarrow \quad \varepsilon_{ab}^{+}(\log w) = 0 \qquad (4.3)$$

and it is equivalent to measure multiplicative epistasis of w on a multiplicative scale and additive epistasis of $\log w$ on an additive scale:

$$\log w_{\emptyset} + \log w_{\{a,b\}} - \log w_{\{a\}} - \log w_{\{b\}} = \log \frac{w_{\emptyset} w_{\{a,b\}}}{w_{\{a\}} w_{\{b\}}}. \qquad (4.4)$$

The additive form (Eq. 4.2) allows for an intuitive geometric interpretation of epistasis, which can be generalized to higher-order gene interactions in a natural way (Beerenwinkel et al. 2007a, Beerenwinkel et al. 2007b, Poelwijk et al. 2007).

Negative epistasis can result from genetic redundancy, a constellation in which a certain biochemical function is redundantly encoded by two or more genes. Genetic redundancy can arise from gene duplication and, less likely, from convergent evolution. Mutations in individual redundant genes have little effect on fitness, but deleterious mutations in each of two redundant genes result in less fit individuals than expected (Hartman et al. 2001). This genetic buffering effectively removes selective pressure from redundant loci and accelerates genetic drift, which eventually leads to diverging functions. More precisely, redundancy is often referred to if the same function is performed by identical elements, whereas degeneracy denotes the ability of structurally different components to perform the same function under some conditions and different functions under other conditions (Tononi et al. 1999).

Degeneracy is prominent in many biological systems, including for example the genetic code, protein folds, and the immune system (Edelman & Gally 2001). It makes these systems robust to (environmental or genetic) perturbations, it correlates with the complexity of the system, and it has also been shown to increase evolvability (Whitacre

& Bender 2010*a*, Whitacre & Bender 2010*b*). Most biological networks display considerable degeneracy or redundancy (Ulitsky & Shamir 2007, Ma et al. 2008). Specifically, PPI networks often contain multiple pathways connecting two proteins or a set of proteins. If the connectivity is linked to a certain function, for example, the transduction of a specific cell signal, then these pathways are redundant components of the network for this function. For phenotypes depending on the connectivity, we can expect negative epistasis to be associated with degeneracy of the biological network.

In the following, we develop a mathematical framework to make this connection between epistasis and degeneracy precise. Pathway degeneracy, or redundancy, is formally described using the concept of network reliability, originally developed by network engineers. Within this framework epistatic gene interactions are defined as unexpected changes in reliability upon perturbing two genes.

4.3   Network reliability

Networks for transporting people, goods, energy, or information are indispensable infrastructure for all human activities. For example, electronic communication networks count among the most important networks, as their exponential growth over the past decades demonstrates. Since the components of such networks are subject to failure, engineers face the problem of designing, constructing, and operating networks that meet the required standards of reliability. Of particular interest is the estimation of how reliable a given network is in performing its function, provided some knowledge about the reliability of its components is available. In many cases, the functionality of the network can be expressed as the ability of its topology to support the network's operation. In other words, the network is functional if and only if certain connectivity properties are fulfilled. One of the simplest examples is two-terminal reliability, defined as the probability that there is at least one correctly functioning path in the network connecting a predefined source node to a predefined target node.

In order to address network reliability problems, engineers and mathematicians have developed a formal framework within which such problems can be formulated and solved. The network is modeled using the mathematical structure of a graph, in which the vertices (or nodes) represent the sources and sinks of whatever the network transports, and the edges represent transport channels or links between the sources or sinks. To account for the fact that sources and sinks as well as links might fail, the vertices and the edges in the graph are endowed with the probabilities that they are functioning at any given time. This approach leads to the concept of a probabilistic graph.

A graph $G = (V, E)$ consists of a set of vertices V and a set of edges $E \subseteq V \times V$. A probabilistic graph $G = (V, E, q, p)$ is a graph (V, E) endowed with vertex probabilities q_v ($v \in V$) and edge probabilities p_e ($e \in E$). We assume that these probabilities are independent for all $v \in V$ and all $e \in E$.

A probabilistic graph $G = (V, E, q, p)$ defines a statistical model over the set of all subgraphs of G as follows. A realization of G is a subgraph G' of (V, E), in symbols $G' \sqsubseteq G$, obtained from independent Bernoulli sampling of the vertices and edges of G

according to their probabilities q_v and p_e, respectively. Due to the assumed independence of vertex and edge probabilities, the probability of the subgraph $G' = (V', E')$ is

$$P(G') = \prod_{e \in E'} p_e \prod_{e \in E \backslash E'} (1 - p_e) \prod_{v \in V'} q_v \prod_{v \in V \backslash V'} (1 - q_v). \tag{4.5}$$

In this sense, we can regard G as a random variable with state space $\{G' \mid G' \sqsubseteq G\}$. The realization $G' = (V', E')$ represents the subnetwork that is still functioning after failure of all sources and sinks $v \notin V'$ and all links $e \notin E'$.

Using this formalism, we can introduce some of the most common network reliability problems. Since successful transmission of a message in a communication network depends on the connectivity of the realized subgraph, we first examine one of the simplest types of connectivity, namely two-terminal reliability. An instance of this problem consists of a probabilistic graph $G = (V, E, q, p)$, a source terminal $u \in V$, and a target terminal $v \in V \setminus \{u\}$. The solution is the probability

$$P(G \in C_{u,v}) = P\left(\{G' \sqsubseteq G \mid G' \in C_{u,v}\}\right), \tag{4.6}$$

where $C_{u,v} = C_{u,v}(G)$ is the set of all subgraphs of G that contain a path from u to v. This probability is called the two-terminal reliability. If the links in the network only allow transmission in one direction, then G is a directed graph and the paths in the definition of $C_{u,v}$ are directed paths. Otherwise, if the links transmit in either direction, G is undirected.

Using Eq. (4.5) we can calculate the two-terminal reliability as

$$P(G \in C_{u,v}) = \sum_{G' \in C_{u,v}(G)} P(G'). \tag{4.7}$$

This expression is a polynomial in the parameters q_v and p_e. If all probabilities are equal, $q_v = p_e = p$ for all $v \in V$ and $e \in E$, then it becomes an integer polynomial in one variable, called the reliability polynomial.

EXAMPLE 4.1   Let us consider the undirected probabilistic graph $G = (V, E, q, p)$ depicted in Figure 4.2a. We assume that the vertices are infallible, i.e., $q_v = 1$ for all $v \in V$, and that all edges have the same probability $p = p_e$ for all $e \in E$. Then the reliability polynomial $P(G \in C_{u,v})$ for the two-terminal reliability problem with source terminal u and target terminal v as shown in Figure 4.2a is

$$f := p^6 + 6p^5(1 - p) + 6p^4(1 - p)^2 + 2p^3(1 - p)^3 \tag{4.8}$$

and the reliability polynomial for the same two-terminal reliability problem in the network depicted in Figure 4.2b is

$$g := p^3(p^6 - 3p^3 + 3). \tag{4.9}$$

Figure 4.3 displays the network reliability in each of the two networks as a function of the edge probability p. We observe that $f < g$ for $p \in (0, 1)$, reflecting the higher reliability due to an additional connecting path of the network in Figure 4.2b as compared to the one in Figure 4.2a.   □

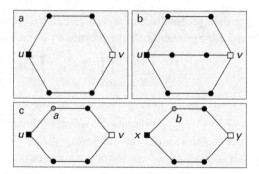

Figure 4.2   Network reliability problems. (a) Two-terminal network reliability problem requiring the connectivity of vertices u and v. (b) Same two-terminal network reliability problem, but the graph has been augmented by an additional u-to-v path increasing the reliability of the network. (c) Network reliability problem that requires the connectivity of u and v and of x and y. Because the pairs (u, v) and (x, y) lie in distinct connected components, the effects on connectivity (i.e., functionality) of removing vertex a and b from the graph are independent of each other. Thus, there is no epistatic interaction between the two events.

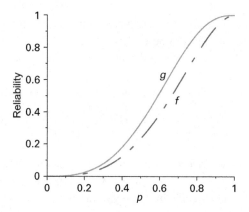

Figure 4.3   Two-terminal reliabilities f and g of the networks discussed in Example 4.1 and depicted in Figures 4.2a and b, respectively, as a function of the edge probability p. The incorporation of redundant paths increases the reliability of the network.

A number of additional network reliability problems have been studied (Colbourn 1987), including, for example, k-terminal reliability, which requires that k chosen terminals are mutually pairwise connected, and broadcasting, or all-terminal reliability, which requires that all terminals are pairwise connected. Many other reasonable network reliability problems can be defined or could arise from practical applications. Formally, once a model $G = (V, E, q, p)$ of the network has been chosen, a general mechanism to define a reliability problem is the following. A network operation is specified by defining a set $\mathrm{Op}(G) \subseteq \{G' \mid G' \sqsubseteq G\}$ of states considered to be functional. The set $\mathrm{Op}(G)$ is sometimes called a stochastic binary system and its elements are termed pathsets. The reliability problem consists of finding the probability $P(\mathrm{Op}(G))$ that a realization of G is operational.

A first naive algorithm to solve a network reliability problem is to enumerate all possible subgraphs, i.e., to determine the cardinality of the set Op(G), to decide whether a given state is a pathset or not (using, for example, efficient path finding or spanning tree algorithms), and to sum the probabilities of all pathsets. However, complete state enumeration requires the generation of all $2^{|V|+|E|}$ realizations of G implying that the running time of this algorithm would depend exponentially on the number of vertices and edges in the graph G.

Substantial effort has been put into finding more efficient algorithms for exact calculation of network reliability problems (Colbourn 1987), but efficient exact solutions seem unlikely, because the two-terminal reliability problem has been shown to be NP-hard (Ball 1980). Indeed, no algorithm of polynomial running time has been found that allows for the exact calculation of the probability $P(\text{Op}(G))$ of a given set of pathsets Op(G), unless very specific assumptions are made on the topology of the underlying probabilistic network (Colbourn 1987, Harms et al. 1995). Thus, in general, we cannot expect to be able to exactly compute network reliabilities.

Nevertheless, network reliability can be solved efficiently (exactly or approximately) for some classes of networks, including trees, full graphs, series-parallel graphs (Colbourn 1987), and channel graphs (Harms et al. 1995). Unfortunately, these classes of networks seem unrealistic as models of biological networks, which are often scale-free and have a small-world structure (Barabási & Oltvai 2004). Unlike communication networks, which are designed and engineered in order to obey certain topologies and reliability properties, biological networks have grown and formed during the course of evolution, often as the result of self-organization processes.

For approximate reliability computation, several Monte Carlo methods have been developed (Barlow & Proschan 1975, Fishman 1986, Cancela & El Khadiri 1995, Lomonosov & Shpungin 1999, Hui et al. 2003, Hui et al. 2005, Laumanns & Zenklusen 2009, Gertsbakh & Shpungin 2010). This approach has received increased attention in the last decade due to the power of modern computers and computing clusters. Monte Carlo simulation can provide an (unbiased) point estimate of reliability probabilities, which will converge to the actual value with increasing numbers of simulated samples.

4.4 Epistasis on networks

Probabilistic networks provide a formal framework that allows for a quantitative representation of network reliability capturing intuitive properties, for instance, that redundant paths increase the reliability of a network. The vertices and edges that constitute such redundant pathways must therefore play a complementary role such that simultaneous removal of pairs of nodes (or pairs of edges) either does not significantly affect the functionality of the network or seriously disrupts it. We now set out to describe such epistatic interactions formally in the context of probabilistic networks and network reliability.

We first extend the definition of what a network needs to do in order to carry out a particular function. We do this by introducing the concept of a pathway. For simplicity, we assume for the rest of this chapter that the vertices of the probabilistic graph G are infallible, i.e., $q_v \equiv 1$, and write $G = (V, E, p)$. Moreover, we will assume that G is undirected.

DEFINITION 4.2 (**Pathway**)   Let $G = (V, E)$ be an undirected graph and $\emptyset \neq A \subseteq B \subseteq V$ be two subsets of vertices. The pair (A, B) is called a *pathway* of G if A lies entirely in a connected component of the subgraph G_B induced in G by the vertex set B. $\square$

We consider sets of pairs $F := \{(A_i, B_i) \mid i = 1, \ldots, m\}$ and define the function of a network G' as the property that simultaneously each of the pairs (A_i, B_i) is a pathway of G'. For instance, in the case of the network depicted in Figure 4.2a, two-terminal connectivity is equivalent to the pair $(\{u, v\}, V)$ being a pathway of the graph. The two-terminal reliability problem is to compute the probability that $(\{u, v\}, V)$ is a pathway of a realization of G.

DEFINITION 4.3 (**Reliability**)   Let $G = (V, E, p)$ be an undirected probabilistic graph and F a set of pathways of G. The probability that all pairs $(A_i, B_i) \in F$ are simultaneously pathways of a realization of G is called the reliability of G with respect to F. We denote this probability as $\mathrm{Rel}_F(G)$. We set $\mathrm{Rel}_F(G) = 0$ if at least one pair $(A_i, B_i) \in F$ is not a pathway of G. $\square$

In light of this definition, we call the set of pathways F the function of the network and each pathway $(A_i, B_i) \in F$ a functional pathway of G if $\mathrm{Rel}_F(G) > 0$.

To define epistasis between vertices in a probabilistic network we consider the change in reliability upon removing the vertices separately versus jointly. For any vertex subset $S \subseteq V$ of a probabilistic graph $G = (V, E, p)$, the induced subgraph $G_{V \setminus S}$, denoted $G \setminus S$, is again a probabilistic graph inheriting the edge probabilities from G. For a given set F of functional pathways of G, the probability

$$w_S(F) := \mathrm{Rel}_F(G \setminus S) \tag{4.10}$$

is the network reliability after removing the vertices in S. When clear from the context, we will omit F and simply write w_S. The quantity (4.10) is the phenotype we are interested in.

Let a and b be two vertices of G. Epistasis between a and b measures the deviation of the double knock-out phenotype $w_{\{a,b\}}$ from the expected value based on the single knock-outs w_a and w_b, relative to the baseline phenotype $w_\emptyset = \mathrm{Rel}(G)$. For the reliability phenotype, the natural null model is multiplicative, because knock-outs in separate connected components of the network should not interact with each other. This is illustrated in the following example.

EXAMPLE 4.4   We consider the network model depicted in Figure 4.2c with all edge probabilities equal to p and the two pathways that require connectivity between u and v, and between x and y. Removal of vertex a and vertex b should not interact with respect to reliability. Indeed, we find

$$w_\emptyset w_{\{a,b\}} = p^6(2 - p^3)^2 \cdot p^6 = p^6(2 - p^3) \cdot p^6(2 - p^3) = w_{\{a\}} w_{\{b\}}. \tag{4.11}$$

In other words, $\varepsilon_{ab}^{\times} = 0$. On the other hand, $\varepsilon_{ab}^{+} = p^6(p-1)^2(p^2+p+1)^2 > 0$ for $p \in (0, 1)$. □

DEFINITION 4.5 (**Network epistasis**) Let $G = (V, E, p)$ be a probabilistic undirected graph, F a set of functional pathways of G, and $a, b \in V$. Then the (multiplicative) network epistasis between a and b with respect to F is

$$\varepsilon_{ab}^{\times}(w(F)) = w_{\emptyset}(F)w_{\{a,b\}}(F) - w_{\{a\}}(F)w_{\{b\}}(F). \tag{4.12}$$

When clear from the context, we will simply write ε_{ab}. □

For the network reliability problem of Figure 4.2a, it is easy to see that there is negative epistasis for vertices a and b ($\varepsilon_{ab} < 0$) if they are located on different parallel paths between u and v. Conversely, it can be shown that negative epistasis entails decomposability of the pathway, where a pathway (A, B) is called decomposable if there are different strict subsets $B_1, B_2 \subset B$ such that $B = B_1 \cup B_2$ and (A, B_i), $i = 1, 2$ are functional pathways of G. This result demonstrates that negative epistasis reveals redundant architecture of the pathway.

4.5 Inferring function from observed genetic interactions

For a given probabilistic graph G that describes a biological system of interest, the concept of epistasis on networks introduced above can be used not only to predict genetic interactions, but also for the inverse problem of inferring functional connectivity requirements from observed genetic interactions. Formally, for a fixed set of perturbations $\mathcal{P}$, each function F, defined as a set of pathways, induces an epistatic pattern,

$$F \mapsto \varepsilon(w(F)) = [\varepsilon_{ab}(w(F))]_{a,b \in \mathcal{P}}. \tag{4.13}$$

For experimentally observed network epistasis $e = [e_{ab}]_{a,b \in \mathcal{P}}$ based on a certain phenotype, we would like to find the pathway set F that gives rise to these measurements, i.e., the preimage of the data under the mapping (4.13). This function F is the set of pathways in G whose operation explains the measured phenotype under network perturbations.

In general, the inverse problem as stated above will not be well defined because the mapping (4.13) is not necessarily injective. In practice, the task is additionally complicated by experimental noise in the observed epistasis data. Often epistasis values are categorized as positive, negative, or no epistasis, a classification which is commonly the main goal of the experiment. This discretization results in a loss of information, but it can also reduce measurement noise.

For given epistasis data, a Monte-Carlo-based optimization approach can be used for solving the inverse problem described above. Let $f(e, \varepsilon(w(F)))$ be a cost function that quantifies how dissimilar the observed and the predicted epistasis patterns are. We seek the pathways with minimal cost by solving

$$\min_{F} f(e, \varepsilon(w(F))), \tag{4.14}$$

where F may run over all or possibly over a restricted set of pathway sets, for example those consisting only of a single pathway.

The cost function f can be defined in different ways reflecting the dynamic range of the measurements and the biological properties of interest. For discrete epistasis values, a simple choice of f is the fraction of knock-out pairs $(a, b) \in \mathcal{P} \times \mathcal{P}$ with different observed and predicted epistasis among all analyzed pairs.

For the predicted epistasis values $\varepsilon_{ab}(w(F))$, the network reliabilities $w_\emptyset$, w_a, w_b, and $w_{a,b}$ need to be computed. This can be done using Monte Carlo simulations of the intact and perturbed probabilistic networks. More specifically, we generate many random samples of the probabilistic graph $G \setminus S$, for all $S \in \{\emptyset, \{a\}, \{b\}, \{a, b\}\}$, and calculate the relative frequency of sampled networks satisfying the connectivity requirements imposed by F. In order to decide whether an epistatic interaction is positive, negative, or absent (zero), several simulated values need to be considered, which allow for detecting a significant difference between $w_\emptyset w_{a,b}$ and $w_a w_b$. For statistical testing, a non-parametric rank sum test can be applied. Moreover, using the central limit theorem, the reliability distributions can be approximated by normal distributions, such that their products obey normal product distributions and parametric tests can be applied.

In general, the non-linear discrete optimization problem (4.14) will be hard to solve exactly and one has to resort to heuristics for finding approximate solutions. However, for a small number of genes on the order of a few dozens, at least all single pathway functions can be enumerated exhaustively.

EXAMPLE 4.6 (**Epistasis among 26 yeast genes**)   St Onge et al. have quantitatively analyzed genetic interactions among 26 yeast genes conferring resistance to the DNA-damaging agent methyl methanesulfonate (St Onge et al. 2007). The phenotype of each deletion strain was its fitness, defined by its growth rate relative to that of wild type and measured as the reciprocal doubling time. Using the multiplicative model of epistasis (Eq. 4.12), the authors established for 323 pairs of genes out of the $\binom{26}{2} = 325$ possible pairs (two pairs were excluded due to technical difficulties) the sign of the epistatic interaction.

To define the probabilistic network G, we queried databases of protein–protein interactions such as PINA (Wu et al. 2008) and STRING (Szklarczyk et al. 2011) to locate the 26 genes within the network of physical protein–protein interactions. The probabilities were derived from STRING confidence values, which we assumed to be monotonically related to the actual probabilities of biophysical interaction. We focused on the immediate network neighbors of the 26 genes under consideration and trimmed this neighborhood by excluding genes connected with low-probability edges to obtain an ambient network consisting of the 26 genes plus 16 immediate neighbors (Figure 4.4).

With the topological and probabilistic information about the ambient network of protein–protein interactions, we set out to investigate whether a single pathway in this network could generate a pattern of epistatic relationships (calculated according to the formalism introduced above) similar to the experimentally observed pattern. We solved the cost minimization (4.14) with f the fraction of mismatching epistasis categories by exhaustive enumeration.

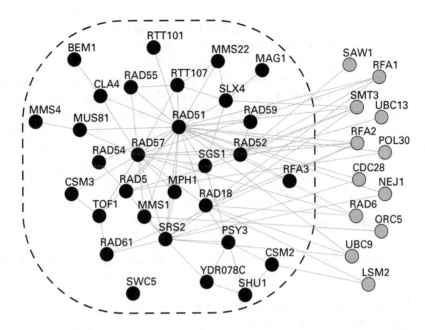

Figure 4.4 Ambient network surrounding the 26 gene products, which are displayed to the left of the vertical bar, studied in St Onge et al. (2007). Requiring that the black nodes are mutually connected generates a pattern of epistatic relationships very similar to the experimentally observed pattern. This similarity is highly unlikely to occur by chance. (Figure drawn with RedeR [Castro et al. 2012].)

The optimal pathway $F = \{(A, B)\}$ that we identified in this manner connects the 12 genes in the set $A := \{CDC28, LSM2, NEJ1, ORC5, POL30, RAD6, RFA1, RFA2, SAW1, SMT3, UBC13, UBC9\}$ (grey vertices in Figure 4.4) using the network spanned by the $30 + 12 = 42$ genes in $B := \{RAD54, RAD18, CSM3, SGS1, RFA3, SRS2, RAD55, BEM1, RAD52, RAD57, CLA4, YDR078C, RAD59, MMS22, MAG1, SHU1, PSY3, MMS1, MUS81, RTT107, RTT101, SLX4, RAD61, RAD5, CSM2, MMS4, RAD51, MPH1, SWC5, TOF1\} \cup A$. This pathway had a cost value of approximately 0.34. Locating this value in the empirical cost distribution derived from randomly generated epistasis values yields a p-value of less than $1.25 \cdot 10^{-4}$. □

4.6 Conclusions

In this chapter, we have shown that epistatic relationships, initially considered in statistical genetics, naturally arise in the framework of probabilistic graphs and network reliability. There is a close relationship between degeneracy or redundancy and epistasis, which can be made precise in this framework. As engineers have demonstrated in the design of reliable networks, degeneracies or redundancies are efficient mechanisms to introduce robustness and to increase reliability, which is reflected in negative epistatic relationships.

Given the multiple examples of epistatic relationships that have been discovered, measured, and quantified in various organisms, the question arises what the biochemical and molecular mechanisms are underlying the observed epistatic phenomena. As engineers have successfully used degeneracies or redundancies to increase the reliability and robustness of networks, one may speculate whether evolutionary processes have generated similar strategies at the level of biochemical networks inside the cell.

Here we have introduced a theoretical framework to address such questions and we have presented a statistical methodology for exploring potential explanations, at the level of biochemical networks, for the observed epistatic relationships.

References

Ball, M. O. (1980), 'Complexity of network reliability computations', *Networks* **10**, 153–165.

Barabási, A.-L. & Oltvai, Z. N. (2004), 'Network biology: understanding the cell's functional organization', *Nat Rev Genet* **5**(2), 101–113.

Barlow, R. E. & Proschan, F. (1975), *Statistical theory of reliability and life testing*, Holt, Rinehart and Winston, Inc., New York.

Beerenwinkel, N., Pachter, L. & Sturmfels, B. (2007*a*), 'Epistasis and shapes of fitness landscapes', *Stat Sinica* **17**, 1317–1342.

Beerenwinkel, N., Pachter, L., Sturmfels, B., Elena, S. & Lenski, R. (2007*b*), 'Analysis of epistatic interactions and fitness landscapes using a new geometric approach', *BMC Evol Biol* **7**(1), 60.

Beyer, A., Bandyopadhyay, S. & Ideker, T. (2007), 'Integrating physical and genetic maps: from genomes to interaction networks', *Nat Rev Genet* **8**(9), 699–710.

Boone, C., Bussey, H. & Andrews, B. J. (2007), 'Exploring genetic interactions and networks with yeast', *Nat Rev Genet* **8**(6), 437–449.

Cancela, H. & El Khadiri, M. (1995), 'A recursive variance-reduction algorithm for estimating communication-network reliability', *IEEE Trans Reliability* **44**(4), 595–602.

Carlson, C. S., Eberle, M. A., Kruglyak, L. & Nickerson, D. A. (2004), 'Mapping complex disease loci in whole-genome association studies', *Nature* **429**(6990), 446–452.

Castro, M., Wang, X., Fletcher, M., Meyer, K. & Markowetz, F. (2012), 'RedeR: R/Bioconductor package for representing modular structures, nested networks and multiple levels of hierarchical associations', *Genome Biol* **13**(4), R29.

Chipman, K. C. & Singh, A. K. (2009), 'Predicting genetic interactions with random walks on biological networks', *BMC Bioinform* **10**, 17.

Colbourn, C. J. (1987), *The combinatorics of network reliability*, Oxford University Press, New York.

Cordell, H. J. (2002), 'Epistasis: what it means, what it doesn't mean, and statistical methods to detect it in humans', *Hum Mol Genet* **11**(20), 2463–2468.

Costanzo, M., Baryshnikova, A., Bellay, J., Kim, Y., Spear, E. et al. (2010), 'The genetic landscape of a cell', *Science* **327**(5964), 425.

Decourty, L., Saveanu, C., Zemam, K., Hantraye, F., Frachon, E. et al. (2008), 'Linking functionally related genes by sensitive and quantitative characterization of genetic interaction profiles', *Proc Natl Acad Sci U S A* **105**(15), 5821–5826.

Dixon, S. J., Costanzo, M., Baryshnikova, A., Andrews, B. & Boone, C. (2009), 'Systematic mapping of genetic interaction networks', *Annu Rev Genet* **43**, 601–625.

Edelman, G. M. & Gally, J. A. (2001), 'Degeneracy and complexity in biological systems', *Proc Natl Acad Sci U S A* **98**(24), 13 763–13 768.

Fisher, R. A. (1918), 'The correlations between relatives on the supposition of Mendelian inheritance', *Trans R Soc Edinburgh* **52**, 399–433.

Fishman, G. S. (1986), 'A Monte Carlo sampling plan for estimating network reliability', *Oper Res* **34**(4), 581–594.

Gavrilets, S. (2004), *Fitness landscapes and the origin of species*, Princeton University Press, Princeton, NJ.

Gertsbakh, I. B. & Shpungin, Y. (2010), *Models of network reliability: analysis, combinatorics and Monte Carlo*, CRC Press, Boca Raton, FL.

Harms, D. D., Kraetzl, M., Colbourn, C. J. & Devitt, J. S. (1995), *Network reliability: experiments with a symbolic algebra environment*, CRC Press, Boca Raton, FL.

Hartman, J. L. t., Garvik, B. & Hartwell, L. (2001), 'Principles for the buffering of genetic variation', *Science* **291**(5506), 1001–1004.

Hoh, J. & Ott, J. (2003), 'Mathematical multi-locus approaches to localizing complex human trait genes', *Nat Rev Genet* **4**(9), 701–709.

Hui, K.-P., Bean, N., Kraetzl, M. & Kroese, D. (2003), 'The tree cut and merge algorithm for estimation of network reliability', *Probab Engrg Inform Sci* **17**(1), 23–45.

Hui, K.-P., Bean, N., Kraetzl, M. & Kroese, D. P. (2005), 'The cross-entropy method for network reliability estimation', *Ann Oper Res* **134**, 101–118.

Jensen, L. J., Kuhn, M., Stark, M., Chaffron, S., Creevey, C. et al. (2009), 'STRING 8: a global view on proteins and their functional interactions in 630 organisms', *Nucl Acids Res* **37**(Database issue), D412–D416.

Jrvinen, A. P., Hiissa, J., Elo, L. L. & Aittokallio, T. (2008), 'Predicting quantitative genetic interactions by means of sequential matrix approximation', *PLoS One* **3**(9), e3284.

Kelley, R. & Ideker, T. (2005), 'Systematic interpretation of genetic interactions using protein networks', *Nat Biotechnol* **23**(5), 561–566.

Laumanns, M. & Zenklusen, R. (2009), 'Computational complexity of impact size estimation for spreading processes on networks', *Eur Phys J B* **71**(4), 481–487.

Lomonosov, M. & Shpungin, Y. (1999), 'Combinatorics of reliability Monte Carlo', *Random Struct Algo* **14**(4), 329–343.

Ma, X., Tarone, A. M. & Li, W. (2008), 'Mapping genetically compensatory pathways from synthetic lethal interactions in yeast', *PLoS One* **3**(4), e1922.

Meur, N. L. & Gentleman, R. (2008), 'Modeling synthetic lethality', *Genome Biol* **9**(9), R135.

Phillips, P. C. (2008), 'Epistasis: the essential role of gene interactions in the structure and evolution of genetic systems', *Nat Rev Genet* **9**(11), 855–867.

Poelwijk, F. J., Kiviet, D. J., Weinreich, D. M. & Tans, S. J. (2007), 'Empirical fitness landscapes reveal accessible evolutionary paths', *Nature* **445**(7126), 383–386.

Qi, Y., Suhail, Y., Lin, Y., Boeke, J. D. & Bader, J. S. (2008), 'Finding friends and enemies in an enemies-only network: a graph diffusion kernel for predicting novel genetic interactions and co-complex membership from yeast genetic interactions', *Genome Res* **18**(12), 1991–2004.

Schuldiner, M., Collins, S. R., Thompson, N. J., Denic, V., Bhamidipati, A. et al. (2005), 'Exploration of the function and organization of the yeast early secretory pathway through an epistatic miniarray profile', *Cell* **123**(3), 507–519.

Schwikowski, B., Uetz, P. & Fields, S. (2000), 'A network of protein–protein interactions in yeast', *Nat Biotechnol* **18**(12), 1257–1261.

St Onge, R. P., Mani, R., Oh, J., Proctor, M., Fung, E. et al. (2007), 'Systematic pathway analysis using high-resolution fitness profiling of combinatorial gene deletions', *Nat Genet* **39**(2), 199–206.

Strogatz, S. H. (2001), 'Exploring complex networks', *Nature* **410**(6825), 268–276.

Szklarczyk, D., Franceschini, A., Kuhn, M., Simonovic, M., Roth, A. et al. (2011), 'The STRING database in 2011: functional interaction networks of proteins, globally integrated and scored', *Nucl Acids Res* **39**(Suppl 1), D561.

Tong, A. H., Evangelista, M., Parsons, A. B., Xu, H., Bader, G. D. et al. (2001), 'Systematic genetic analysis with ordered arrays of yeast deletion mutants', *Science* **294**(5550), 2364–2368.

Tononi, G., Sporns, O. & Edelman, G. M. (1999), 'Measures of degeneracy and redundancy in biological networks', *Proc Natl Acad Sci U S A* **96**(6), 3257–3262.

Ulitsky, I. & Shamir, R. (2007), 'Pathway redundancy and protein essentiality revealed in the *Saccharomyces cerevisiae* interaction networks', *Mol Syst Biol* **3**, 104.

Wagner, G. P., Laubichler, M. D. & Bagheri-Chaichian, H. (1998), 'Genetic measurement of theory of epistatic effects', *Genetica* **102–103**(1–6), 569–580.

Whitacre, J. & Bender, A. (2010*a*), 'Degeneracy: a design principle for achieving robustness and evolvability', *J Theor Biol* **263**(1), 143–153.

Whitacre, J. M. & Bender, A. (2010*b*), 'Networked buffering: a basic mechanism for distributed robustness in complex adaptive systems', *Theor Biol Med Model* **7**, 20.

Wong, S. L., Zhang, L. V., Tong, A. H. Y., Li, Z., Goldberg, D. S. et al. (2004), 'Combining biological networks to predict genetic interactions', *Proc Natl Acad Sci U S A* **101**(44), 15 682–15 687.

Wright, S. (1931), 'Evolution in Mendelian populations', *Genetics* **16**, 97–159.

Wu, J., Vallenius, T., Ovaska, K., Westermarck, J., Mäkelä, T. et al. (2008), 'Integrated network analysis platform for protein–protein interactions', *Nat Meth* **6**(1), 75–77.

Xenarios, I., Rice, D. W., Salwinski, L., Baron, M. K., Marcotte, E. M. et al. (2000), 'DIP: the database of interacting proteins', *Nucl Acids Res* **28**(1), 289–291.

5 Synthetic lethality and chemoresistance in cancer

Kimberly Maxfield and Angelique Whitehurst

Despite great strides in the development of anti-cancer strategies over the last 50 years, treatment regimens continue to cause significant toxicity and fail to fully eradicate disease. Enhancing the current state of therapy will require: (1) the expansion of available tumor selective and therapeutically tractable molecular targets, (2) the development of methods to provide a rational approach to identifying effective combinatorial drug cocktails, and (3) molecular markers that can accurately predict sensitive patient populations. To this end, efforts that reveal the molecular architecture supporting tumorigenic phenotypes are essential. RNA interference (RNAi)-mediated loss of function screens have emerged as a method for wholesale identification of tumor-specific dependencies that modulate chemoresponsiveness. Here, we provide a broad overview of how genome-scale RNAi screening is being implemented.

5.1 Cancer chemotherapy

5.1.1 Cytotoxic chemotherapy

Goodman and Gilman's 1946 discovery that lymphosarcomas respond to nitrogen mustard demonstrated that tumor cells may have an enhanced sensitivity to chemical poisons as compared to their normal counterparts. This finding revolutionized cancer treatment as it indicated that in addition to radiation and surgery, the only available modalities at the time, drugs could also be administered to reduce tumor burden (Goodman et al. 1946). Following on these initial observations, over the ensuing 50 years, an arsenal of cytotoxic agents were developed to treat a range of cancer types (Chabner & Roberts 2005, Strebhardt & Ullrich 2008).

The majority of these agents, as with the nitrogen mustard, share a common characteristic: they induce genomic damage. For example, agents such as cisplatin cause inter- and intrastrand DNA crosslinks. This DNA damage can lead to the inhibition of cell division by activating an arrest in the cell cycle to allow for DNA repair through the nucleotide excision repair (NER) pathway. This pathway is coupled to apoptotic programs that are activated if overwhelming damage is detected (Plunkett et al. 1995, Siddik 2003, Wang & Lippard 2005). Another class of chemotherapeutics are the anti-mitotics,

Systems Genetics: Linking Genotypes and Phenotypes, ed. F. Markowetz and M. Boutros. Published by Cambridge University Press. © Cambridge University Press 2015.

which include paclitaxel, docetaxel, and the vinca alkaloids. These agents prevent the ability of microtubules to bind and thus align chromosomes during mitosis. To prevent the mis-segregation of chromosomes to daughter cells, the spindle assembly checkpoint (SAC) is triggered to arrest mitosis. Again, if the damage is too great, apoptosis is activated (Horwitz et al. 1993, Jordan & Kamath 2007).

The tumorcidal activity of these agents demonstrates that tumor cells are unable to tolerate high levels of genomic damage and have the capacity to couple this damage to death. While normal cells would also be vulnerable, these agents do display some selectivity. Traditionally, this therapeutic window has been attributed to the frequent division of tumor cells, which is presumed to enhance their susceptibility to checkpoint activation as compared to normal counterparts. However, glances into the molecular etiology of cancer have revealed that tumor sensitivity to first-line chemotherapeutics is varied and complex (Aas et al. 1996, Gascoigne & Taylor 2008, Brito & Rieder 2009, Gascoigne & Taylor 2009). Discovering the basis for responsiveness would allow for the development of strategies to improve treatment.

In Goodman and Gilman's nitrogen mustard experiment, there was a transient, but dramatic, reduction in tumor burden. Within a few months, the tumors returned and were less sensitive to a subsequent nitrogen mustard regimen. These ephemeral responses were repeatedly observed in cancer patients receiving single-agent chemotherapy, suggesting that tumor cells were capable of rapidly acquiring traits that allow them to circumvent DNA damage (Goodman et al. 1946, Farber & Diamond 1948).

Efforts to prolong remission were initiated by Holland, Frei, and Freireich, who evaluated the effectiveness and toxicity of the simultaneous employment of multiple genome-damaging agents (Frei et al. 1958). Significantly, these combinatorial regimens enhanced the duration of remission and established that concurrent attacks on tumors may have a greater impact on increasing the time to relapse and resistance. These findings are the basis for the multifocal therapies that are the current first-line treatment for cancer. In some cases, such as acute lymphoblastic leukemia (ALL), these regimens have produced prolonged remission rates and even cures (Smith et al. 2010). In cancers such as Hodgkin's lymphoma and ovarian cancer, they have improved survival time from months to years (Chabner & Roberts 2005, Bast et al. 2007).

5.1.2   Challenges of cytotoxic chemotherapy

The widespread use of combinatorial chemotherapy is not without significant and well-known challenges. Despite the improvement of remission rates with poly-chemotherapeutic approaches, tumors often reoccur and are typically resistant to continued exposure to the initial therapeutic cocktail. The mechanisms by which tumors acquire resistance traits is not completely understood; however, one theory that has emerged is that genetic heterogeneity among a population of tumor cells may allow for the selection of variants that maintain traits to deflect cytotoxic insult (Dexter & Leith 1986). For example, recurrent tumors often express high levels of the p-glycoprotein (pgp) pump. pgp binds to specific chemical toxins, such as paclitaxel, and pumps them out of the cell, thereby reducing the intracellular concentrations of drugs and the intended genomic damage (Ambudkar et al. 2003). Cells within a population of tumors that maintain higher levels of the pump would have a selective advantage and therefore may deflect a chemotherapeutic attack. Another mechanism recently described is a dynamic epigenetic alteration that promotes a drug tolerant state through the activation of insulin-like growth factor 1R (IGF-1R). Direct inhibition of IGF-1R or disruption of epigenetic alterations through histone deacetylase (HDAC) inhibitor exposure reverses resistance in this population of cells (Sharma et al. 2010a). Given the ability of a small population of tumor cells to resist therapeutic insult, there is a need to enhance the initial killing capacity of a chemotherapeutic agent and also to identify methods to rapidly address recurrent tumors.

Another notorious obstacle to effective treatment is the discordant responses among cancer patients to identical therapeutic regimens. For example, testicular cancer is nearly curable in response to cisplatin-based therapy (Einhorn 2002). However, ovarian and lung cancer, which are treated with similar agents, have poor 5-year survival rates (Schiller et al. 2002, Bast et al. 2007). Similarly, the responses within a population of patients with the same disease may be divergent. In triple negative breast cancer, a cohort of patients responds well to first-line chemotherapeutics, while only minimal changes in other patients are observed (Isakoff 2010). This heterogeneity of response suggests that the characteristic of rapid cell division is not the only susceptibility trait for chemotherapy-induced death. Perhaps, intrinsic tumor cell properties can modulate both the acquisition and coupling of genomic disruption to death. One implicated specifier of response is p53 status. Tumors with a wild-type p53 often display a heightened sensitivity to chemotherapeutics, likely because p53 senses DNA damage and can modulate an apoptotic response (Vogelstein et al. 2000, Sullivan et al. 2004). The identification of other biomarker(s) for response may allow for both the prediction and tailoring of specific therapeutic regimens to highly sensitive populations. Increasing our knowledge of the drug response network would significantly enhance the ability to identify these biomarkers.

A major problematic issue in the use of current cytotoxic-based therapy is the numerous and potentially life-threatening adverse effects on normal tissues. By targeting cell division, these agents kill fast-growing cells in the intestine and hair follicles, leading to the nausea, vomiting, and hair loss, which are intimately associated with our image of the treated patient. More problematic, however, are side effects such as myelosuppression, which can lead to hemorrhage or infection. Typically, these events subside once a chemotherapeutic regimen is discontinued. However, peripheral neuropathy, which is a common side effect of taxane-based therapy, can be irreversible (Rowinsky & Donehower 1995). Thus, the collateral damage of first-line chemotherapeutics on normal tissue significantly reduces the dose ceiling of agents that may otherwise be quite effective in destroying tumor cells.

5.1.3   The new generation of targeted agents

While chemotherapies that damage DNA have gained ground in the treatment of cancer, focus has recently shifted to the identification and inhibition of the growth-promoting proteins that drive uncontrolled proliferation in tumor cells. These so-called oncoproteins may be mutated in tumors and thereby present a unique, tumor-specific target. Indeed, inhibition of the *BCR-Abl* oncogene in chronic myelogenous leukemia (CML) with imatinib (Gleevec) has demonstrated that disrupting oncoprotein function can produce prolonged remission rates (Druker 2008). Similarly, in breast cancer, long-term treatment with the estrogen receptor (ER) agonist tamoxifen extends survival time in patients with ER positive tumors (Early Breast Cancer Trialists' Collaborative Group [EBCTCG] 2005). In addition, the HER2/neu inhibitory antibody trastuzamab significantly improves outcomes of patients harboring tumors driven by the HER2 oncogene (Moasser 2007).

The use of these agents is particularly attractive because they have a tumor-specific target that is detectable. Thus, sensitive patient populations can be identified and impacts on normal tissues are minimized. While targeted therapies hold great promise, there are challenges that still need to be overcome. First, targeted agents often inhibit growth, but may not kill cancer cells. Thus, they may need to be combined with the genome-damaging chemotherapeutics that can activate apoptosis. Identifying the correct regimens that will enhance tumor-killing activity and reduce side effects is a trial-and-error process as exemplified in the identification of appropriate trastuzumab regimens (Slamon et al. 2001). Second, resistance, whether by the acquisition of mutations that prevent drug binding or the mutation of growth-promoting properties in parallel pathways, is frequent, but not always predictable ab initio. Finally, many known targets, such as RAS, have proven intractable to direct pharmacological inhibition to date (Cox & Der 2010).

To address the current challenges facing cancer treatment a number of steps must be taken. First, expansion of the pool of tumor-specific and therapeutically addressable molecular targets would greatly enhance the opportunity to develop inhibitors that exhibit selective tumorcidal activity. Additionally, knowledge of the molecular components required to modulate chemotherapeutic response would allow for the prediction of sensitive populations, the selection of rational drug combinations and the anticipation of resistance mechanisms. Recent advances in genomics have been leveraged to isolate disease initiating and/or supporting lesions in cancer. These techniques have revealed a constellation of genetic, epigenetic, and signaling alterations that are correlated with the cancer phenotype. However, knowledge of the causal associations between the majority of genes in the genome and tumorigenic phenotypes remains paltry (Fedorov et al. 2010).

5.2   Employing small interfering RNA (siRNA) to identify modifiers of chemotherapeutic responsiveness

5.2.1   RNAi: from biological mechanism to screening tool

Until 2001, studying gene function in the context of a disease state was a long, protracted process characterized by stepwise hypothesis generation and testing. The means to study the functional role of a gene in mammalian cell culture systems was significantly accelerated by the discovery of gene silencing biology in lower metazoans and the adaptation of this process to mammalian systems. This technique, known as siRNA, allows for the rapid depletion of the expression a specific protein (Fire et al. 1998, Elbashir et al. 2001). siRNA-mediated gene silencing in mammalian tissue culture systems involves the lipid- or viral-based delivery of short 19-nucleotide RNA molecule, or siRNA to cells. These siRNAs are designed to bind to the mRNA encoded by a single gene of interest inducing a stall in translation and ensuing degradation of the mRNA transcript. Subsequently, a reduction in the level of the encoded protein can be observed. Importantly, rarely does siRNA produce a complete obliteration of mRNA and thus protein expression is typically reduced but not eliminated.

The annotation of the complete coding sequence of the human genome promoted the construction of siRNA collections, either synthetic or genetically encoded, that target each gene in the genome. Accordingly, there has been great interest in leveraging these pangenomic siRNA libraries to uncover the essential gene products contributing to a range of oncogenic processes (Sachse & Echeverri 2004, Westbrook et al. 2005, Bartz et al. 2006, Mullenders & Bernards 2009). While RNAi has proven to be a robust method to uncover potential functional elements supporting biological systems, the discovery power of large-scale RNAi screens is restricted to phenotypes that can be measured by single-gene depletion at the cell autonomous level. Resolution of key tumor cell vulnerabilities can be severely limited by gene products that continue to function at the reduced levels RNAi induces and by intrinsic redundancy among gene products supporting a given biological process. One solution for reducing this false negative space, commonly applied in genetic model organisms, is to employ a collateral chemical or genetic perturbation. This ancillary perturbation generates a sensitized background for more effective phenotypic penetrance of single-gene depletions (Novick et al. 1989, Bender & Pringle 1991, Guarente 1993, Hartwell et al. 1997, Lum et al. 2004, Lehner et al. 2006).

With respect to cancer, the implementation of chemical synthetic lethal screens in the context of first-line therapeutics is an obvious first step to revealing the molecular architecture mediating drug responsiveness. However, modeling the components that support human diseases such as cancer requires screens to be performed in a biological context that represents the disease state. The staggering collection of tumor-derived cell lines developed over the last 60 years presents a system in which to deduce the minimal tumorigenic platform modulating drug response. These cells are the fully evolved (and most lethal) disease state and harbor the corrupted signaling system that drives uncontrolled growth. Thus, tumor-derived cell lines represent an ideal system to collect the causal associations between gene products (Sharma et al. 2010*b*). To this end, this strategy has been leveraged to identify proteins that modulate responsiveness to first-line chemotherapeutics, targeted therapeutics, and experimental compounds. Alternatively, siRNA-based strategies in the presence of high doses of the chemotherapeutics have been used to identify resistance mechanisms. Here we will discuss the strategies used for these different analysis, as well as key findings from these efforts.

5.2.2   Chemical–genetic synthetic lethal screening platforms

Synthetic lethality is defined as a relationship between two genes, in which their simultaneous, but not independent, disruption induces a significant consequence on viability.

In model organisms, such as yeast, a synthetic lethal screen begins with a viable, single-gene deletion strain, which is used to systematically identify independent gene deletions that confer lethality. The widespread use of this technique has revealed the existence of numerous synthetic lethal interactions, typically indicating a functional connection between the encoded gene products. Importantly, the existence of synthetic lethality suggests that organisms can compensate for deficiencies, such as the loss of a single-gene product, to maintain viability (Hartman et al. 2001, Boone et al. 2007). The identification of single-gene heterozygous yeast deletion mutants, which display enhanced sensitivity to synthetic compounds, similarly demonstrated that organisms have the capacity to buffer response to chemical insults (Giaever et al. 1999, Lum et al. 2004).

The development of siRNA as a tool to manipulate gene expression in mammalian cells thus provided a mechanism to perform similar high-throughput screens in higher organisms. With respect to cancer, the chemical–genetic approach brings the opportunity to reveal those gene products that modulate responsiveness to first-line anti-neoplastic agents. Importantly, the pioneering finding that heterozygous yeast deletion mutants could confer sensitivity to a therapeutic agent empirically demonstrated the feasibility of employing hypomorphic approaches, such as siRNA, to uncover chemosensitizer loci (Lum et al. 2004). If targetable, isolated proteins may represent mechanisms that would enhance the narrow therapeutic window of current chemotherapeutic regimens.

The platform most extensively employed for genome-scale chemosensitizer screens in mammalian cells uses a library of synthetically produced siRNAs that are arrayed on multi-well plates. These libraries are available from commercial sources and may target the entire genome or can be customized for a specific subclass of genes, such as kinases, G-protein-coupled receptors, or E3-ubiquitin ligases. Multiple, independent siRNAs that target the same gene are pooled together in a single well creating a one-gene/one-well format. This framework necessitates the use of a high-throughput robotic infrastructure, similar to that used for chemical compound screening, to allow for efficient plate handling, reagent dispensing, and assay measurements.

High-efficiency delivery of siRNAs to cells is achieved through a reverse transfection method in which siRNAs are complexed with a delivery agent. A trypsinized cell suspension is then exposed to the siRNA–lipid mixture. This 'reverse' delivery method can achieve up to 99% transfection efficiency in many diverse settings, including primary, immortalized, and tumor-derived cell lines from both mouse and humans (Whitehurst & White 2006, Whitehurst et al. 2007, Hu et al. 2009, Whitehurst et al. 2010). While transfection efficiency is quite high using this protocol, the depletion of protein levels, or knockdown efficiency, is dependent upon the half-life of each protein. Empirical monitoring of protein levels reveals that reduced expression is typically achieved between 48 and 72 hours after transfection. However, there are notable exceptions, such as caveolin, which require prolonged siRNA exposure due to a long protein half-life.

The chemical–genetic synthetic lethality screen seeks to identify gene products that may buffer response to a specific therapeutic compound. Thus, the strategy is to apply an otherwise innocuous dose of an agent of interest to siRNA-transfected cells and evaluate the consequences of this combination on cell viability. To minimize single-agent activity, a dose of drug that does not produce a detectable impact on viability, but is within 1 log of the EC50 can be used. In general, drug exposure is typically initiated between 24 and 72 hours post transfection to coincide with the estimated time at which protein knockdown is most likely to occur. Because synthetic lethal approaches are specifically attempting to identify gene products whose depletion is only relevant in the presence of the peturbagen, parallel analysis under both treated and untreated conditions is carried out.

A number of different endpoint assays have been used to measure activity of siRNAs in chemosensitizer screens. Selection criteria for assays are based on both the phenotype of interest to the investigator and adaptability of an assay to a high-throughput format. To this end, the most extensively used endpoint assays are those that measure cell viability. A number of luminescence- or fluorescence-based reagents, such as CellTiter-Glo, are available. These reagents are ideal for high-throughput screening because they can be delivered directly to cells and require only minimal incubation times prior to data collection.

Because viability assessment does not provide information on whether cells have altered proliferation or survival rates, reagents that measure apoptotic cell death, such as Apo-ONE, are useful. These assays can be significantly more expensive and require prolonged incubation times and manipulations that may make their application in genome-scale investigations challenging. Thus, their use is most feasible in screens that consider only small collections of siRNAs or as a secondary screening strategy to large genome-scale initiatives.

Single-cell behaviors, including those only assessable in real time, can be measured using high-content microscopes capable of multi-spectral imaging. For example, screens to identify modifiers of mitotic progression have been performed by measuring DNA, centrosome, or tubulin status (Draviam et al. 2007, Neumann et al. 2010). DNA synthesis rates can also be determined by automated analysis of the incorporation of fluorescently conjugated analogues of thymidine, such as bromodeoxyuridine or 5-ethynyl-2-deoxyuridine to measure DNA synthesis rates. Ultimately, the choice of endpoint assay is dependent upon the scale of the investigation, the question being asked, and the availability of high-throughput plate readers and/or microscopes.

5.2.3   Data analysis of arrayed screening formats

The overarching theme for data analysis strategies is to generate a set of lead siRNAs for subsequent follow-up experimentation. While significant attention has been paid to best practices for analyzing screen data, there is no widely accepted method (Birmingham et al. 2009). This lack of convention is due to differences in the scale of investigations, the sources of error for each screen, and ultimately the resources available to retest candidates. Thus, the method implemented is best determined on a screen-by-screen basis with the goal of minimizing false positive and negative rates. For example, the multi-well format introduces location effects that may influence the magnitude of a phenotype. Outer wells on plates can have reduced cell viability due to evaporation of media. Given that this location bias can be idiosyncratic to a particular screen, the presence of position effects should be evaluated for each screen. If identified, the issue can be minimized by data smoothing algorithms that normalize data based on the median values for a specific position. Following the data normalization step, an algorithm to identify outliers is applied. The algorithm(s) chosen typically take into account both the magnitude of the phenotype as well as the reproducibility of the biological replicates and the type of the secondary screening effort to be undertaken (Whitehurst et al. 2007).

Large secondary screens allow the user to sample a broad spectrum of strong to weak phenotypes presenting the best opportunity to identify targets of interest, but can become prohibitively expensive. Smaller-scale retesting strategies, while less expensive, may restrict the investigator to only those siRNAs that produced the most significant phenotypes, thereby increasing false negative space. Reduction of this false negative space may be achieved by sampling different phenotypic magnitudes for highly reproducible siRNAs, thereby establishing an empirical cutoff to enrich for true positives (Figure 5.1).

5.2.4   shRNA screening libraries

In addition to synthetically produced siRNA libraries, genetically encoded short hairpin RNA (shRNA) collections have been developed (Brummelkamp et al. 2002, Chang et al. 2006, Yang et al. 2011). A number of libraries are available that contain shRNAs integrated in a variety of different vector backbones. These shRNAs are packaged to generate virus for transduction into dividing cells. The virus can be arrayed on multi-well plates for an arrayed screening format, similar to the synthetic-based libraries mentioned above. Alternatively, a population of viruses, harboring shRNAs targeting different genes, can be pooled and transduced into target cells. Because the shRNAs are integrated into the genome, this system allows for the prolonged depletion of a specific gene product. In the chemical–genetic context, this screening strategy has been used to identify genes whose depletion either enhances resistance or increases sensitivity to first-line therapeutics. In the former scenario, a toxic dose of a therapeutic is typically applied for a prolonged period of time. Cells that are resistant are positively selected and the shRNAs they harbor are decoded by either microarray or sequencing. Alternatively, dropout screens have been performed. In this context, negative selection with a chemical agent will kill cells that are unable to withstand prolonged drug exposure. The abundance and identity of remaining shRNAs are collected and matched to those in the original library to identify shRNAs lost during selection (Luo et al. 2008, Silva et al. 2008, Sims et al. 2011). One drawback of the pooled format is that the ability to measure diverse phenotypes is somewhat limited. Thus, the choice of platform, either shRNA or siRNA, is dependent upon both the initiating question and the available resources to the investigator.

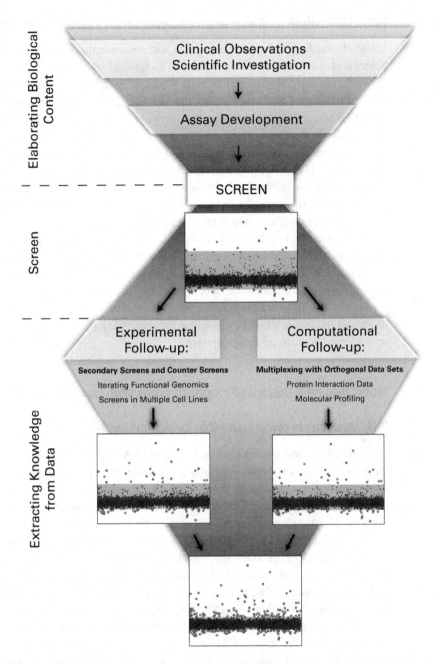

Figure 5.1   Overview of functional screen development and knowledge generation. Top: Initial observations and investigations are used to develop a platform for siRNA screening. Middle: Screens return annotation for phenotypes for each gene. Outlier siRNAs are identified for further analysis (large circles in white shading). Bottom: Follow-up analysis using counter screens and orthogonal data sets allows for the isolation of biologically relevant, but phenotypically milder, siRNAs. (Illustration by Angela Diehl, UT-Southwestern Medical Center.) A black and white version of this figure will appear in some formats. For the color version, please refer to the plate section.

5.2.5   Large-scale siRNA chemical–genetic screens return landmark components

While chemical–genetic screens in yeast had demonstrated that a heterozygous yeast deletion mutant could confer sensitivity to a chemical toxin, whether a siRNA-induced hypomorphic state in mammalian cells would be sufficient for therapeutic sensitivity was not a foregone conclusion. Thus, the identification of landmark components of the biological targets of chemotherapeutic agents in initial RNAi screens was a welcome observation. In the case of DNA damage, genome-wide synthetic lethal screens with the DNA crosslinking agent cisplatin returned proteins associated with the DNA damage checkpoint and DNA repair (Bartz et al. 2006). Similarly, genome-wide screens with low dose paclitaxel, an inhibitor of microtubule dynamics, isolated multiple members of the gamma-tubulin ring complex, which nucleates mitotic microtubules (Whitehurst et al. 2007). Another familiar component is the pgp drug efflux pump, whose depletion had been demonstrated to enhance intracellular accumulation of therapeutic agents. Accordingly, an shRNA-based array screen for taxol collaborators identified pgp (Ji et al. 2007). These initial findings empirically indicated that similar to yeast systems, a mammalian hypomorph was sufficient to enhance sensitivity to a therapeutic agent. Furthermore, the capacity of an siRNA screen to return the biological targets of the therapeutic toxin used illustrated the power of siRNA to reveal the underlying molecular architecture, known or unknown, modulating drug response.

5.3   Mobilizing new therapeutic opportunities with large-scale RNAi screens

5.3.1   RNAi screening to reveal actionable therapeutic targets

One immediate and translational aspect of siRNA-based chemical genetic screens is the identification of pharmacologically addressable targets that will synergize with first-line chemotherapeutic agents. To this end, one strategy has been to focus specifically on targets returned from chemosensitizer screens that are therapeutically addressable. For example, Bauer et al. employed a collection of shRNAs that target proteins for which chemical inhibitors already exist. The authors identified an association between the disruption of either TGF-beta signaling, SP1-mediated transcription, or the phosphatase PPM1D, and paclitaxel. Accordingly, chemical inhibition of these proteins or pathways synergizes with paclitaxel in 2D and 3D culture (Bauer et al. 2010). Similarly, by employing a collection of siRNAs targeting kinases, Swanton and colleagues implicated the collagen kinase COL4A3BP, which modulates ceramide transport between the endoplasmic reticulum (ER) and Golgi apparatus, as a modulator of paclitaxel responsiveness. The capture of a protein involved in ER maintenance suggested that chemical perturbation of the same process may increase sensitivity to paclitaxel. Indeed the authors found that tunicamycin, which induces ER stress, enhances paclitaxel sensitivity in breast cancer cells (Swanton et al. 2007). In the same vein, a genome-wide paclitaxel synthetic lethal screen returned a subunit of the vacuolar ATPase, which is critical for lysosomal function. Accordingly, an inhibitor of this ATPase enhanced sensitivity to paclitaxel. In this same screen, individual depletion of multiple components of the proteasome enhanced paclitaxel sensitivity (Whitehurst et al. 2007). Given the clinical availability of the proteasome inhibitor bortezomib, this combination is currently under clinical evaluation (Ma et al. 2007, Jatoi et al. 2008, Ramaswamy et al. 2010, Mehnert et al. 2011).

Additionally, siRNA screening has revealed combinatorial regimens that induce sensitivity in tumors that are otherwise resistant to targeted agents. BRAF is a growth-promoting oncogene, which when mutated (typically a V600E) can promote uncontrolled proliferation through constitutive activation of a MAP kinase pathway (Davies et al. 2002, Wan et al. 2004). An inhibitor of this mutated protein, vemurafenib (PLX3397), has found clinical success in melanoma, where this mutation is quite frequent (Chapman et al. 2011). However, colon cancers harboringV600E mutations are not sensitive to vemurafenib (Yang et al. 2012). Prahallad and colleagues addressed the molecular basis for this resistance by employing a dropout shRNA screen in the presence of vemurafenib. This screen found that upon inhibition of BRAF, the epidermal growth factor receptor (EGFR) becomes activated to continue the transmission of growth signals through the MAP kinase pathway. Combining EGFR inhibition, using either erlotinib or cetuximab, with vemurafenib in mouse models attenuated tumor growth. Importantly, these findings suggest the rational combination of two clinically approved agents to improve colon cancer treatment in patients with a V600E mutation (Prahallad et al. 2012). The examples highlighted here demonstrate that siRNA screens can return a single member of a protein complex or pathway and thereby implicate a pharmacological entry point to enhance chemosensitivity. When inhibitory agents are available, particularly if they are already in clinical use, RNAi discovery efforts can streamline the identification of effective combinatorial regimens.

5.3.2   Chemical synthetic lethal RNAi screens reveal tumor-cell-selective vulnerabilities

One important strategy for improving anti-cancer therapy is the identification of new molecular targets that can enhance the response of tumors to first-line agents. Ideal targets are essential for tumor cell growth and survival, but innocuous when inhibited in normal cells. Because genome-scale, and in particular genome-wide, siRNA screens sample a large fraction of the genome, they are an ideal discovery method to identify these tumor-selective dependencies in tumor cells. For example, in a screen for paclitaxel sensitizers in non-small-cell lung cancer (NSCLC) cells, a cohort of genes known as cancer-testes antigens (CT antigens) were shown to be required for cell growth specifically in the presence of taxol but not in tumor cells growing in the presence of diluent. CT antigens are expressed anomalously in tumors and because of the blood–testes barrier, they evoke an immune response in cancer patients. While studies have focused on their exploitation for cancer immunotherapy, any potential functional role in tumors has not been investigated. Their identification in an unbiased genome-wide screen for paclitaxel sensitizers revealed that these proteins can be functional in supporting tumorigenic phenotypes and, in particular, mediating response to anti-mitotic insult. Indeed, follow-up studies revealed that one member of this cohort, acrosin binding protein (ACRBP), associates with the mitotic protein nuclear mitotic apparatus protein 1, NUMA1, whose elevated expression levels can lead to chromosome alignment errors in mitosis. ACRBP negatively regulates NUMA1 levels, thereby preventing an error-prone mitosis. Loss of ACRBP exacerbates the impact of paclitaxel because the level of chromosome mis-segregation is elevated. ACRBP expression is not found at high frequency in NSCLC tumors, though nearly 60% of ovarian tumors have detectable protein expression (Tammela et al. 2006). Indeed, high ACRBP-expressing tumors portend a poorer outcome than low-expressing tumors (Whitehurst et al. 2010). This example demonstrates not only the discovery power of siRNA screens to pinpoint tumor-selective dependencies that would otherwise not be investigated, but also how findings from a single screen may be extended to diverse tumorigenic settings.

Similarly, large-scale screening may also reveal tumor-selective dependencies that mediate response to targeted therapies. Poly ADP-ribose polymerase (PARP) is an enzyme that enhances the repair of DNA single-strand breaks. PARP inhibitors have begun to gain clinical use as they present an additional mechanism to elevate DNA damage in tumor cells, particularly those treated with an ancillary genome-disrupting agent. To identify molecular mechanisms that enhance PARP activity, Turner et al. screened a kinome library to identify components that enhance sensitivity to PARP inhibitors. This screen revealed that a neuronal kinase, CDK5, mediates PARP responsiveness. Furthermore, CDK5 was demonstrated to be essential for remediating single-strand breaks (Turner et al. 2008). Both of these examples practically demonstrated the ability of genome-scale screens to identify novel modulators of chemoresponsiveness that might otherwise not even be considered. Furthermore, the observation that, in both of these cases, the protein is likely anomalously expressed in tumors reinforces the emerging paradigm that tumor cells may become dependent on gene products that are the results of the activation of aberrant gene expression programs activated during tumorigenesis.

5.3.3   Biomarker identification

As mentioned above, a particularly confounding problem in applying chemotherapy is the heterogeneity and unpredictability of response within a patient population that appear to have similar tumor types. This complication has driven forth efforts to identify molecular markers that may indicate sensitivity of tumors, thereby allowing for the therapy to be tailored to a specific tumor. siRNA screening may enhance the discovery of these biomarkers, because it may reveal specific components that are upregulated in tumors and deflect response. For example Bartz et al. revealed that loss of *BRCA1* could enhance sensitivity to cisplatin, potentially due to an already compromised DNA repair system. Extending these studies demonstrated that cisplatin was more sensitive in tumor cells with a *BRCA1* mutation (Bartz et al. 2006). While this case is somewhat of an obvious connection, it clearly demonstrates the significance of performing screens, particularly in multiple diverse oncogenic backgrounds, to determine if sensitivity segregates with specific molecular alterations.

5.3.4   Use of shRNA barcode screens to identify resistance mechanisms

The ability of tumor cells to adapt to harsh environments is notorious and presents an enormous challenge to preventing relapse of disease. As described above, a number of molecular mechanisms that promote the evasion of drug toxicity have been discovered including the upregulation of drug pumps, mutations that inhibit drug binding to targets, and the activation of pathways that circumvent the dependence on a targeted protein (Ambudkar et al. 2003, Turner et al. 2008). However, predicting how a tumor will gain resistance to an agent is typically performed in retrospect by examining the characteristics of a resistant tumor. In vitro methods to identify routes for achieving resistant would promote the development of strategies to thwart recurrent tumors more rapidly. Functional genomics strategies for resistance traits require the stable, long-term depletion of a protein in the presence of a prolonged drug exposure to identify variants that can deflect the cytotoxic insult. Thus, pooled shRNA screens that use positive selection have been employed for this avenue of research. For example, Berns and colleagues applied Herceptin in this setting and identified that the loss of the tumor-suppressor phosphatase PTEN leads to resistance. Based on this finding, the authors analyzed microarray data in breast-cancer patients and identified a correlation between loss of this gene and Herceptin resistance and were able to predict populations at risk for developing resistance. Subsequently, an inhibitor of PI3K signaling could resensitize tumors to Herceptin (Berns et al. 2007). Similarly a positive selection screen demonstrated that the tumor cells could evolve resistance to the HDAC inhibitor SAHA, by increasing proteasome-mediated protein degradation. The authors pinpointed HR23B as a negative regulator of proteasome activity whose depletion drives HDAC inhibition. Indeed inhibition of the proteasome with bortezomib rescued SAHA resistance. Examination of expression data from cutaneous T cell lymphoma (CTCL), which is HDAC inhibitor sensitive, revealed high levels of HR23B (Fotheringham et al. 2009). The clinical validation of these empirical findings demonstrates that cell culture assays can identify resistance mechanisms that are acquired in vivo. Ultimately, the identification of these mechanisms may allow for the prediction of patients at risk of developing resistance and also additional therapies to thwart recurrent disease.

5.4   Conclusions

While current chemotherapeutic regimens can improve outcomes and even cure patients, there are significant obstacles to increasing their effectiveness and decreasing adverse events on patients. To overcome these challenges, our knowledge of the functional landscape that modulates drug resistance will need to continuously be broadened. Over the last 10 years, siRNA screening has provided a glimpse into the tumor cell vulnerabilities that may be exploited for improved therapy. In some cases, these screens have revealed cell biological processes such as the ER or proteasome that may be directly targetable in a wide range of tumors. On the other hand, siRNA approaches have demonstrated how a specific genetic alteration, whether a mutation in *BRCA1* or the expression of ACRBP, can influence drug responsiveness in a context-dependent manner. Furthermore, mechanisms to achieve resistance are beginning to emerge from shRNA screens. In the future, screening platforms will become increasingly commonplace, allowing researchers to begin to gain a more comprehensive picture of how tumor cells respond and deflect stress.

References

Aas, T., Borresen, A. L., Geisler, S., Smith-Sorensen, B., Johnsen, H. et al. (1996), 'Specific P53 mutations are associated with de novo resistance to doxorubicin in breast cancer patients', *Nature Medicine* **2**(7), 811–14.

Ambudkar, S. V., Kimchi-Sarfaty, C., Sauna, Z. E. & Gottesman, M. M. (2003), 'P-glycoprotein: from genomics to mechanism', *Oncogene* **22**(47), 7468–85.

Bartz, S. R., Zhang, Z., Burchard, J., Imakura, M., Martin, M. et al. (2006), 'Small interfering RNA screens reveal enhanced cisplatin cytotoxicity in tumor cells having both BRCA network and TP53 disruptions', *Molecular and Cellular Biology* **26**(24), 9377–86.

Bast, R. C., J., Brewer, M., Zou, C., Hernandez, M. A., Daley, M. et al. (2007), 'Prevention and early detection of ovarian cancer: mission impossible?', *Recent Results in Cancer Research* **174**, 91–100.

Bauer, J. A., Ye, F., Marshall, C. B., Lehmann, B. D., Pendleton, C. S. et al. (2010), 'RNA interference (RNAi) screening approach identifies agents that enhance paclitaxel activity in breast cancer cells', *Breast Cancer Research* **12**(3), R41.

Bender, A. & Pringle, J. R. (1991), 'Use of a screen for synthetic lethal and multicopy suppressee mutants to identify two new genes involved in morphogenesis in *Saccharomyces cerevisiae*', *Molecular and Cellular Biology* **11**(3), 1295–305.

Berns, K., Horlings, H. M., Hennessy, B. T., Madiredjo, M., Hijmans, E. M. et al. (2007), 'A functional genetic approach identifies the PI3K pathway as a major determinant of trastuzumab resistance in breast cancer', *Cancer Cell* **12**(4), 395–402.

Birmingham, A., Selfors, L., Forster, T., Wrobel, D., Kennedy, C. et al. (2009), 'Statistical methods for analysis of high-throughput RNA interference screens', *Nature Methods* **6**(8), 569–75.

Boone, C., Bussey, H. & Andrews, B. J. (2007), 'Exploring genetic interactions and networks with yeast', *Nature Reviews Genetics* **8**(6), 437–49.

Brito, D. A. & Rieder, C. L. (2009), 'The ability to survive mitosis in the presence of microtubule poisons differs significantly between human nontransformed (RPE-1) and cancer (U2OS, HeLa) cells', *Cell Motility and the Cytoskeleton* **66**(8), 437–47.

Brummelkamp, T. R., Bernards, R. & Agami, R. (2002), 'A system for stable expression of short interfering RNAs in mammalian cells', *Science* **296**(5567), 550–3.

Chabner, B. A. & Roberts, T. G. J. (2005), 'Timeline: chemotherapy and the war on cancer', *Nature Reviews Cancer* **5**(1), 65–72.

Chang, K., Elledge, S. J. & Hannon, G. J. (2006), 'Lessons from Nature: microRNA-based shRNA libraries', *Nature Methods* **3**(9), 707–14.

Chapman, P. B., Hauschild, A., Robert, C., Haanen, J. B., Ascierto, P. et al. (2011), 'Improved survival with vemurafenib in melanoma with BRAF V600E mutation', *New England Journal of Medicine* **364**(26), 2507–16.

Cox, A. D. & Der, C. J. (2010), 'Ras history: the saga continues', *Small GTPases* **1**(1), 2–27.

Davies, H., Bignell, G. R., Cox, C., Stephens, P., Edkins, S. et al. (2002), 'Mutations of the BRAF gene in human cancer', *Nature* **417**(6892), 949–54.

Dexter, D. L. & Leith, J. T. (1986), 'Tumor heterogeneity and drug resistance', *Journal of Clinical Oncology* **4**(2), 244–57.

Draviam, V. M., Stegmeier, F., Nalepa, G., Sowa, M. E., Chen, J. et al. (2007), 'A functional genomic screen identifies a role for TAO1 kinase in spindle-checkpoint signalling', *Nature Cell Biology* **9**(5), 556–64.

Druker, B. J. (2008), 'Translation of the philadelphia chromosome into therapy for CML', *Blood* **112**(13), 4808–17.

Early Breast Cancer Trialists' Collaborative Group (EBCTCG) (2005), 'Effects of chemotherapy and hormonal therapy for early breast cancer on recurrence and 15-year survival: an overview of the randomised trials', *Lancet* **365**(9472), 1687–717.

Einhorn, L. H. (2002), 'Curing metastatic testicular cancer', *Proceedings of the National Academy of Sciences of the United States of America* **99**(7), 4592–5.

Elbashir, S. M., Harborth, J., Lendeckel, W., Yalcin, A., Weber, K. et al. (2001), 'Duplexes of 21-nucleotide RNAs mediate RNA interference in cultured mammalian cells', *Nature* **411**(6836), 494–8.

Farber, S. & Diamond, L. K. (1948), 'Temporary remissions in acute leukemia in children produced by folic acid antagonist, 4-aminopteroyl-glutamic acid', *New England Journal of Medicine* **238**(23), 787–93.

Fedorov, O., Muller, S. & Knapp, S. (2010), 'The (un)targeted cancer kinome', *Nature Chemical Biology* **6**(3), 166–169.

Fire, A., Xu, S., Montgomery, M. K., Kostas, S. A., Driver, S. E. et al. (1998), 'Potent and specific genetic interference by double-stranded RNA in *Caenorhabditis elegans*', *Nature* **391**(6669), 806–11.

Fotheringham, S., Epping, M. T., Stimson, L., Khan, O., Wood, V. et al. (2009), 'Genome-wide loss-of-function screen reveals an important role for the proteasome in HDAC inhibitor-induced apoptosis', *Cancer Cell* **15**(1), 57–66.

Frei, E., Holland, J., Schneiderman, M., Pinkel, D., Selkirk, G. et al. (1958), 'A comparative study of two regimens of combination chemotherapy in acute leukemia', *Blood* **13**, 1126–48.

Gascoigne, K. E. & Taylor, S. S. (2008), 'Cancer cells display profound intra- and interline variation following prolonged exposure to antimitotic drugs', *Cancer Cell* **14**(2), 111–22.

Gascoigne, K. E. & Taylor, S. S. (2009), 'How do anti-mitotic drugs kill cancer cells?', *Journal of Cell Science* **122**(15), 2579–85.

Giaever, G., Shoemaker, D. D., Jones, T. W., Liang, H., Winzeler, E. A. et al. (1999), 'Genomic profiling of drug sensitivities via induced haploinsufficiency', *Nature Genetics* **21**(3), 278–83.

Goodman, L., Wintrobe, M. M., Dameshek, W., Goodman, M. & Gilman, A. (1946), 'Nitrogen mustard therapy', *Journal of the American Medical Association* **132**(3), 126–32.

Guarente, L. (1993), 'Synthetic enhancement in gene interaction: a genetic tool come of age', *Trends in Genetics* **9**(10), 362–6.

Hartman, J. L. t., Garvik, B. & Hartwell, L. (2001), 'Principles for the buffering of genetic variation', *Science* **291**(5506), 1001–4.

Hartwell, L. H., Szankasi, P., Roberts, C. J., Murray, A. W. & Friend, S. H. (1997), 'Integrating genetic approaches into the discovery of anticancer drugs', *Science* **278**(5340), 1064–8.

Horwitz, S. B., Cohen, D., Rao, S., Ringel, I., Shen, H. J. et al. (1993), 'Taxol: mechanisms of action and resistance', *Journal of the National Cancer Institute Monographs* **15**(15), 55–61.

Hu, G., Kim, J., Xu, Q., Leng, Y., Orkin, S. H. et al. (2009), 'A genome-wide RNAi screen identifies a new transcriptional module required for self-renewal', *Genes & Development* **23**(7), 837–48.

Isakoff, S. J. (2010), 'Triple-negative breast cancer: role of specific chemotherapy agents', *Cancer Journal* **16**(1), 53–61.

Jatoi, A., Dakhil, S. R., Foster, N. R., Ma, C., Rowland, K. M., et al. (2008), 'Bortezomib, paclitaxel, and carboplatin as a first-line regimen for patients with metastatic esophageal, gastric, and gastroesophageal cancer: phase ii results from the North Central Cancer Treatment Group (N044B)', *Journal of Thoracic Oncology* **3**(5), 516–20.

Ji, D., Deeds, S. L. & Weinstein, E. J. (2007), 'A screen of shRNAs targeting tumor suppressor genes to identify factors involved in A549 paclitaxel sensitivity', *Oncology Reports* **18**(6), 1499–505.

Jordan, M. A. & Kamath, K. (2007), 'How do microtubule-targeted drugs work? An overview', *Current Cancer Drug Targets* **7**(8), 730–42.

Lehner, B., Crombie, C., Tischler, J., Fortunato, A. & Fraser, A. G. (2006), 'Systematic mapping of genetic interactions in *Caenorhabditis elegans* identifies common modifiers of diverse signaling pathways', *Nature Genetics* **38**(8), 896–903.

Lum, P. Y., Armour, C. D., Stepaniants, S. B., Cavet, G., Wolf, M. K. et al. (2004), 'Discovering modes of action for therapeutic compounds using a genome-wide screen of yeast heterozygotes', *Cell* **116**(1), 121–37.

Luo, B., Cheung, H. W., Subramanian, A., Sharifnia, T., Okamoto, M. et al. (2008), 'Highly parallel identification of essential genes in cancer cells', *Proceedings of the National Academy of Sciences of the United States of America* **105**(51), 20 380–5.

Ma, C., Mandrekar, S. J., Alberts, S. R., Croghan, G. A., Jatoi, A. et al. (2007), 'A phase I and pharmacologic study of sequences of the proteasome inhibitor, bortezomib (PS-341, Velcade), in combination with paclitaxel and carboplatin in patients with advanced malignancies', *Cancer Chemotherapy and Pharmacology* **59**(2), 207–15.

Mehnert, J. M., Tan, A. R., Moss, R., Poplin, E., Stein, M. N. et al. (2011), 'Rationally designed treatment for solid tumors with MAPK pathway activation: a phase I study of paclitaxel and bortezomib using an adaptive dose-finding approach', *Molecular Cancer Therapeutics* **10**(8), 1509–19.

Moasser, M. M. (2007), 'Targeting the function of the HER2 oncogene in human cancer therapeutics', *Oncogene* **26**(46), 6577–92.

Mullenders, J. & Bernards, R. (2009), 'Loss-of-function genetic screens as a tool to improve the diagnosis and treatment of cancer', *Oncogene* **28**(50), 4409–20.

Neumann, B., Walter, T., Heriche, J. K., Bulkescher, J., Erfle, H. et al. (2010), 'Phenotypic profiling of the human genome by time-lapse microscopy reveals cell division genes', *Nature* **464**(7289), 721–7.

Novick, P., Osmond, B. C. & Botstein, D. (1989), 'Suppressors of yeast actin mutations', *Genetics* **121**(4), 659–74.

Plunkett, W., Huang, P., Xu, Y. Z., Heinemann, V., Grunewald, R. et al. (1995), 'Gemcitabine: metabolism, mechanisms of action, and self-potentiation', *Seminars in Oncology* **22**(4 Suppl 11), 3–10.

Prahallad, A., Sun, C., Huang, S., Di Nicolantonio, F., Salazar, R. et al. (2012), 'Unresponsiveness of colon cancer to BRAF(V600E) inhibition through feedback activation of EGFR', *Nature* **483**(7387), 100–3.

Ramaswamy, B., Bekaii-Saab, T., Schaaf, L. J., Lesinski, G. B., Lucas, D. M. et al. (2010), 'A dose-finding and pharmacodynamic study of bortezomib in combination with weekly paclitaxel in patients with advanced solid tumors', *Cancer Chemotherapy and Pharmacology* **66**(1), 151–8.

Rowinsky, E. K. & Donehower, R. C. (1995), 'Paclitaxel (taxol)', *New England Journal of Medicine* **332**(15), 1004–14.

Sachse, C. & Echeverri, C. J. (2004), 'Oncology studies using siRNA libraries: the dawn of RNAi-based genomics', *Oncogene* **23**(51), 8384–91.

Schiller, J. H., Harrington, D., Belani, C. P., Langer, C., Sandler, A. et al. (2002), 'Comparison of four chemotherapy regimens for advanced non-small-cell lung cancer', *New England Journal of Medicine* **346**(2), 92–8.

Sharma, S. V., Lee, D. Y., Li, B., Quinlan, M. P., Takahashi, F. et al. (2010*a*), 'A chromatin-mediated reversible drug-tolerant state in cancer cell subpopulations', *Cell* **141**(1), 69–80.

Sharma, S. V., Haber, D. A. & Settleman, J. (2010*b*), 'Cell line-based platforms to evaluate the therapeutic efficacy of candidate anticancer agents', *Nature Reviews Cancer* **10**(4), 241–53.

Siddik, Z. H. (2003), 'Cisplatin: mode of cytotoxic action and molecular basis of resistance', *Oncogene* **22**(47), 7265–79.

Silva, J. M., Marran, K., Parker, J. S., Silva, J., Golding, M. et al. (2008), 'Profiling essential genes in human mammary cells by multiplex RNAi screening', *Science* **319**(5863), 617–20.

Sims, D., Mendes-Pereira, A. M., Frankum, J., Burgess, D., Cerone, M. A. et al. (2011), 'High-throughput RNA interference screening using pooled shRNA libraries and next generation sequencing', *Genome Biology* **12**(10), R104.

Slamon, D. J., Leyland-Jones, B., Shak, S., Fuchs, H., Paton, V. et al. (2001), 'Use of chemotherapy plus a monoclonal antibody against HER2 for metastatic breast cancer that overexpresses HER2', *New England Journal of Medicine* **344**(11), 783–92.

Smith, M. A., Seibel, N. L., Altekruse, S. F., Ries, L. A. G., Melbert, D. L. et al. (2010), 'Outcomes for children and adolescents with cancer: challenges for the twenty-first century', *Journal of Clinical Oncology* **28**(15), 2625–34.

Strebhardt, K. & Ullrich, A. (2008), 'Paul Ehrlich's magic bullet concept: 100 years of progress', *Nature Reviews Cancer* **8**(6), 473–80.

Sullivan, A., Syed, N., Gasco, M., Bergamaschi, D., Trigiante, G. et al. (2004), 'Polymorphism in wild-type p53 modulates response to chemotherapy in vitro and in vivo', *Oncogene* **23**(19), 3328–37.

Swanton, C., Marani, M., Pardo, O., Warne, P. H., Kelly, G. et al. (2007), 'Regulators of mitotic arrest and ceramide metabolism are determinants of sensitivity to paclitaxel and other chemotherapeutic drugs', *Cancer Cell* **11**(6), 498–512.

Tammela, J., Uenaka, A., Ono, T., Noguchi, Y., Jungbluth, A. A. et al. (2006), 'OY-TES-1 expression and serum immunoreactivity in epithelial ovarian cancer', *International Journal of Oncology* **29**(4), 903–10.

Turner, N. C., Lord, C. J., Iorns, E., Brough, R., Swift, S. et al. (2008), 'A synthetic lethal siRNA screen identifying genes mediating sensitivity to a PARP inhibitor', *EMBO Journal* **27**(9), 1368–77.

Vogelstein, B., Lane, D. & Levine, A. J. (2000), 'Surfing the p53 network', *Nature* **408**(6810), 307–10.

Wan, P. T., Garnett, M. J., Roe, S. M., Lee, S., Niculescu-Duvaz, D. et al. (2004), 'Mechanism of activation of the RAF–ERK signaling pathway by oncogenic mutations of B-RAF', *Cell* **116**(6), 855–67.

Wang, D. & Lippard, S. J. (2005), 'Cellular processing of platinum anticancer drugs', *Nature Reviews Drug Discovery* **4**(4), 307–20.

Westbrook, T. F., Stegmeier, F. & Elledge, S. J. (2005), 'Dissecting cancer pathways and vulnerabilities with RNAi', *Cold Spring Harbor Symposia on Quantitative Biology* **70**, 435–44.

Whitehurst, A. W. & White, M. A. (2006), 'Harnessing RNAi for analyses of Ras signaling and transformation', *Methods in Enzymology* **407**, 259–68.

Whitehurst, A. W., Bodemann, B. O., Cardenas, J., Ferguson, D., Girard, L. et al. (2007), 'Synthetic lethal screen identification of chemosensitizer loci in cancer cells', *Nature* **446**(7137), 815–19.

Whitehurst, A. W., Xie, Y., Purinton, S. C., Cappell, K. M., Swanik, J. T. et al. (2010), 'Tumor antigen acrosin binding protein normalizes mitotic spindle function to promote cancer cell proliferation', *Cancer Research* **70**(19), 7652–61.

Yang, H., Higgins, B., Kolinsky, K., Packman, K., Bradley, W. D. et al. (2012), 'Antitumor activity of BRAF inhibitor vemurafenib in preclinical models of BRAF-mutant colorectal cancer', *Cancer Research* **72**(3), 779–89.

Yang, X., Boehm, J. S., Salehi-Ashtiani, K., Hao, T., Shen, Y. et al. (2011), 'A public genome-scale lentiviral expression library of human ORFs', *Nature Methods* **8**(8), 659–61.

6 Joining the dots: network analysis of gene perturbation data

Xin Wang, Ke Yuan, and Florian Markowetz

How to link genotypes and phenotypes is a long-standing question in modern biology. Modern high-throughput approaches are key technologies at the forefront of genetic research. They enable the analysis of a biological response to thousands of experimental perturbations and require a tight collaboration between experimental and computational scientists. Perturbation studies and computational approaches have revolutionized research in functional genomics and genetics and promise to lay the foundation for personalized medicine. For modern high-throughput technologies, computation is as important as experimentation. Genome-wide image-based RNA interference (RNAi) screens, for example, are only feasible because of computational techniques. Computational skills to analyse the data have become as important as experimental skills to generate the data.

Design and analysis of phenotyping screens depend on the number of genes perturbed and the richness of the phenotype observed (Figure 6.1). At one extreme are high-throughput screens with single reporters, e.g. a genome-wide screen for new components of a pathway. At the other extreme are perturbations of individual genes with very rich phenotypes, e.g. assessing the effects of a single gene perturbation on several molecular levels over time. Between these two extremes lie a variety of possible screen designs. Two widely used scenarios are small-scale perturbations (<20 genes) of a single target pathway with rich readouts, e.g. a global transcriptional profile, and medium-scale perturbations (hundreds of genes) with multi-parametric readouts, e.g. cell morphology or growth in different media. In the following we will discuss statistical and computational methodologies for functional analysis in all four scenarios.

6.1 Scenario 1: Genome-wide screens with single reporters

RNAi screens have been frequently and successfully applied for functional profiling of genes on a large scale (Boutros & Ahringer 2008). The vast majority of these applications use a single phenotype (e.g. cell viability, growth rate, activity of reporter constructs) to characterize the function of genes in specific biological pathways. This approach has been extensively used to identify essential genes for particular signalling

Systems Genetics: Linking Genotypes and Phenotypes, ed. F. Markowetz and M. Boutros. Published by Cambridge University Press. © Cambridge University Press 2015.

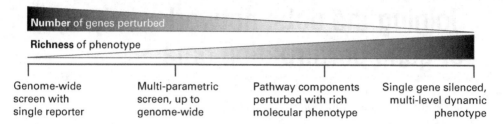

Figure 6.1   Design and analysis of phenotyping screens depends on the number of genes perturbed and the richness of the phenotype observed.

pathways (Boutros et al. 2004, Müller et al. 2005, Orvedahl et al. 2011, Kessler et al. 2012). Recently, genome-scale pooled short hairpin RNA (shRNA) screening has been applied to study genetic vulnerabilities across various cancer cell lines (Cheung et al. 2011).

The preprocessing of screening data, including quality control, normalization and phenotype scoring (Zhang et al. 1999, Li & Wong 2001, Brideau et al. 2003, Hahne et al. 2006, Malo et al. 2006, Zhang et al. 2006, Birmingham et al. 2009), has been widely addressed and methods are implemented in widely used software packages like cellHTS2 (Boutros et al. 2006). Understanding the underlying biological mechanisms, however, requires higher-level bioinformatic analyses (Markowetz 2010). Here we introduce our analysis strategies to link observed phenotypes to cellular pathways and networks.

Our goal was to speed up the phase of unbiased and hypothesis-free data exploration and quickly reach a phase of hypothesis-driven follow-up. This is why we developed HTSanalyzeR (Wang et al. 2011), a flexible software to build integrated analysis pipelines for high-throughput screening (HTS) data. HTSanalyzeR directly connects to commonly used preprocessing packages for HTS data (like cellHTS2, Pelz et al. 2010), as well as more general packages (like limma, Smyth 2005), and presents its results as HTML pages and network plots (Figure 6.2). HTSanalyzeR visualizes graphs using RedeR (Castro et al. 2012), an R package combined with a Java core engine to bridge the gap between network visualization and analysis. The package contains overrepresentation analysis, gene set enrichment analysis, comparative gene set analysis and rich subnetwork identification.

Gene set overrepresentation (Beißbarth & Speed 2004) and enrichment analysis (GSEA, Subramanian et al. 2005) are simple ways to map phenotypic data to biological functions and pathways. Both approaches quantify the association between genes with strong phenotypes and a collection of predefined gene sets (e.g. Gene Ontology (Ashburner et al. 2000), KEGG (Ogata et al. 1999) or MSigDB (Liberzon et al. 2011)). Overrepresentation analyses rely on Fisher's exact test or a hypergeometric test (Falcon & Gentleman 2007) to rank gene sets by their overlap with the set of genes showing the strongest phenotype (the so-called *hits*). Without making a hard cutoff to select hits, GSEA (Subramanian et al. 2005) identifies functional gene sets that show a significant shift to extreme phenotypes.

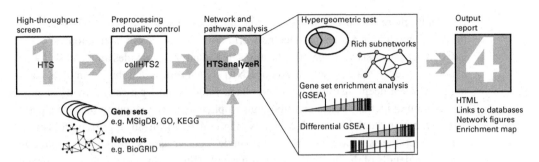

Figure 6.2   HTSanalyzeR takes as input HTS data that has already been preprocessed, normalized and quality checked, e.g. by cellHTS2. HTSanalyzeR then combines the HTS data with gene sets and networks from freely available sources and performs three types of analysis: (i) hypergeometric tests for overlap between hits and gene sets; (ii) gene set enrichment analysis (GSEA) for concordant trends of a gene set in one phenotype; (iii) differential GSEA to identify gene sets with opposite trends in two phenotypes; and (iv) identification of subnetworks enriched for hits. The results are provided to the user as figures and HTML tables linked to external databases for annotation. (Reproduced with permission from Wang et al. 2011.)

Gene set analysis reduces complexity and improves interpretability of results by moving from single genes to functionally related gene sets. This type of analysis is ideal for a comprehensive first overview. However, enrichment analysis loses its value for complexity reduction if the number of gene sets becomes too big. It is quite tedious to inspect dozens or even hundreds of significant gene sets and speculate about the underlying biology. Additionally, interpretation of gene set results is hampered by overlap and redundancies between gene sets. Statistical methods have been adapted to account for dependencies within the Gene Ontology (GO) graph (Alexa et al. 2006, Bauer et al. 2008) but not yet for more general collections of gene lists like MSigDB (Liberzon et al. 2011). HTSanalyzeR addresses this issue by representing gene set results as enrichment maps (Merico et al. 2010), a graph structure where nodes represent gene sets coloured according to their significance and edges between nodes indicate the overlap between gene sets. This visualization directly shows whether the significance of gene sets is driven by a set of shared genes, or whether they represent independent biological themes.

A conceptual limitation of overrepresentation and enrichment methods is that they rely on known gene sets and cannot uncover new pathways or components. Enrichment methods treat pathways as bags of unconnected genes without considering connections within and between pathways. Thus, enrichment methods can only deliver a very crude picture of the cell. To gain a more direct view of the cooperativity between genes, functional modules can be identified from networks that encode pairwise associations such as functional, physical or genetic interactions. These networks can be obtained from databases like STRING (Jensen et al. 2009) and BioGRID (Breitkreutz et al. 2008), which are generated through literature curation and high-throughput data integration. Several computational methods are available to map phenotyping data to a network and search for subnetworks enriched for strong phenotypes (reviewed in Markowetz 2010).

HTSanalyzeR relies on BioNet (Beisser et al. 2010) to search for the highest scoring functional module in a user-provided network, typically of protein interactions.

Good data analysis asks specific questions. HTSanalyzeR provides all the tools for an hypothesis-free and unbiased analysis of RNAi screen results and thus is an ideal starting point for a deeper exploration of the data. Single reporter assays provide enough information to identify genes contributing to a phenotype and mapping them to a pathway or functional unit in the cell. Methods to infer networks from single reporter assays exist (Kaderali et al. 2009) but in general single reporter assays are not rich enough to dissect the details of the internal wiring of the cell. This is why a major focus of current research is on developing and quantifying rich phenotypes of gene perturbations.

6.2   Scenario 2: Single gene silenced, multi-level dynamic phenotype

The other extreme of screen designs are perturbations of single genes with readouts on different molecular levels over time. An example is a dynamic multi-level study on Nanog, a key stem cell regulator (Lu et al. 2009). For five days after silencing changes in histone acetylation, PolII binding, mRNA expression and nuclear protein expression were monitored. These data demonstrate how a single genetic perturbation leads to progressive widespread changes in several molecular regulatory layers, and provide a dynamic view of information flow in the epigenome, transcriptome and proteome.

A follow-up study focused on the temporal and spatial dynamics of histone acetylation data and its correlation with gene expression (Markowetz et al. 2010). Compared to the rest of the genome, embryonic stem cell (ESC)-specific genes show significantly more acetylation signal and a much stronger decrease in acetylation over time, which is often not reflected in a concordant expression change. These results shed light on the complexity of the relationship between histone acetylation and gene expression and are a step forward to dissect the multilayer regulatory mechanisms that determine stem cell fate.

These studies are examples of very rich molecular phenotypes for individual genes. In the near future there will be many more studies like this on more and more genes. These data will provide novel challenges for data analysis and network reconstruction on how to integrate the effects in different molecular levels into a coherent picture of the cell.

6.3   Scenario 3a: Pathway components perturbed with global transcriptional phenotypes

Between the two extremes we have discussed in the last two sections lies a wide variety of possible screen designs. An important example are perturbation experiments on a small number of components (or candidates) involved in a specific signalling pathway. The goal is to infer the position of the components in the pathway and find their interaction partners.

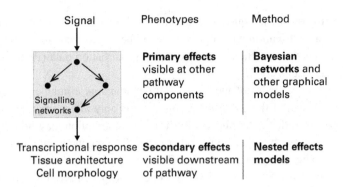

Signal	Phenotypes	Method
	Primary effects visible at other pathway components	**Bayesian networks** and other graphical models
Transcriptional response Tissue architecture Cell morphology	**Secondary effects** visible downstream of pathway	**Nested effects models**

Figure 6.3   Reconstructing signalling pathways from perturbation data. Bayesian networks infer causal relations from primary effects of interventions. Nested effects models (NEMs), however, reconstruct signalling relationships from various indirect downstream effects such as expression of reporter genes, phenotypic changes in cell morpology or tissue architecture.

Analysis of pathway-specific studies depends on what information the observed phenotypes provide. We distinguish between *primary effects* visible at other pathway components, and *secondary effects* visible at downstream reporters (Figure 6.3). In the first case, where perturbation effects on the abundance and activity states of pathway components are observed (as in Sachs et al. 2005) statistical network reconstruction methods can be applied (Markowetz & Spang 2007) and we will discuss *Bayesian networks* in the next section.

Often, however, phenotypes measure transcriptional response or cell morphologies and not protein activity states. These phenotypes are downstream of the perturbed pathway and only allow an indirect view of the activity of the pathway components. The situation has been likened to making inference from the shadows on the wall of Plato's cave (Shimoni et al. 2010). In this case, classical network reconstruction methods often fail, and we discuss an alternative inference method called *Nested effects model* that is particularly tailored to inference from secondary effects.

6.3.1   Inference from primary effects: Bayesian networks

Bayesian networks are probabilistic graphical models that represent random variables as nodes and encode conditional dependencies by edges in a directed acyclic graph (DAG) (Pearl 1988). Bayesian networks are used extensively for causal modelling (Pearl 2000) and to predict the effects of perturbations (Maathuis et al. 2010). In computational biology, Bayesian networks have been introduced to infer transcriptional regulatory networks from the expression profiles (Friedman et al. 2000, Pe'er et al. 2001, Friedman 2004). Transcriptional networks are still the main application and only few examples modelling signalling pathways exist (Sachs et al. 2005). The chapter by Sun et al. in this volume (Chapter 8) contains a more detailed description of Bayesian networks.

In Bayesian networks, perturbations to individual genes can be modelled as 'ideal interventions' (Pearl 2000). In this model, a perturbation is thought to completely control the target gene and deterministically force it to a target state. In a Bayesian network this is equivalent to cutting all edges into the target node and replacing the local distribution with a point mass at the target state. This model is used in almost all applications of Bayesian networks to perturbation data (Pe'er et al. 2001, Pe'er 2005, Sachs et al. 2005).

Pearl's model of ideal interventions contains a number of idealizations. The most important of these are that manipulations only affect single genes and that results can be controlled deterministically. The first assumption may not be true if there are compensatory effects involving other genes. The second assumption is also very limiting in realistic biological scenarios. Often the experimentalist lacks knowledge about the exact size of perturbation effects. To cope with this uncertainty, Markowetz et al. (2005b) introduced soft interventions as a generalization of ideal interventions. Variables are 'pushed' in the direction of target states without fixing them. This idea is formalized in a Bayesian framework based on conditional Gaussian networks that can contain discrete and continuous nodes.

Since Bayesian networks model the dependencies between network components, they can only be applied when *primary effects* of perturbations on other pathway components are visible in the data. However, for signalling pathways these primary effects (e.g. activation states of proteins when silencing a kinase) are very often difficult to obtain on a large scale. Instead, *secondary effects* such as changes in downstream genes, cell morphology and tissue architecture that are indirect evidence of pathway activity are usually easier to measure (Figure 6.3). In the following section, we will introduce nested effects models to infer signalling networks from secondary effects of gene perturbation.

6.3.2   Inference from secondary effects: nested effects models

Nested effects models (NEMs) are a statistical approach that is specifically tailored to reconstruct features of pathways from perturbation effects in downstream reporters (Markowetz et al. 2005a, Markowetz et al. 2007). In contrast to other graphical models, which are all based on measures of pairwise association (e.g. coexpression networks) and encode conditional independence relations (e.g. Bayesian networks), NEMs describe *subset relationships* between observed downstream effects of perturbations.

Figure 6.4 shows a toy signalling pathway consisting of a kinase (A) and three transcription factors (B, C and D), which directly regulate reporter genes (1 to 10). Phenotypic data generally do not include the states of proteins A to D after perturbations (Figure 6.4B). This is because phenotypes like gene expression or cell morphologies are downstream of the pathway of interest and often do not contain much information on the activity states of the proteins in the pathway.

NEMs assume that perturbing upstream genes may impact a global process, while silencing downstream genes only affect local subprocesses. This results in a subset pattern in the observed data. For example, for the pathway in Figure 6.4A the perturbation effects (expression changes in reporter genes) of genes B, C and D are subsets of the effect of gene A: perturbing A has an effect on all reporters (1–10), while perturbing B only affects reporters 1–4, perturbing C affects reporters 5–10 and perturbing D affects reporters 5–8 (Figure 6.4C).

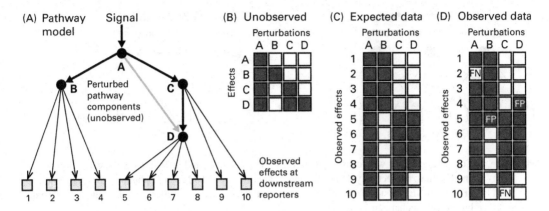

Figure 6.4   (A) A schematic figure of nested effects models. Signalling genes (S-genes) perturbed and effect reporter genes (E-genes) are shown in black round circles and grey rectangles, respectively. The thick black arrows between S-genes denote their signalling relationships, while the thin black arrows between S-genes and E-genes represent transcriptional regulations. The thick grey arrow between S-genes A and D is a transitive edge indicating an indirect effect of A on D (via C). (B) The state of S-genes after perturbation of other pathway components is generally unobservable, and thus NEMs assume that visible changes are only observable in the downstream reporters. (C) Expected states of E-genes (black, effect; white, no effect) after perturbing S-genes. (D) In real biological applications, the noisy measurement of E-gene expression may include false negatives (FNs) and false positives (FPs).

Reading these subset patterns off a given graph is easy; NEMs however start from observed data (Figure 6.4D), which is a noisy version of the data expected for a pathway (Figure 6.4C) and infer the most likely pathway structure that can explain the subset patterns. To distinguish between direct and indirect signalling relationships, NEMs perform *transitive reduction* (Wagner 2001) to remove direct shortcut edges (e.g. the edge from A to D) between genes that are also connected by a path (Markowetz et al. 2007).

NEMs versus cause–effect graphs

It is illustrative to compare NEMs to 'cause–effect graphs' that simply link perturbed genes to differentially expressed downstream genes (Rung et al. 2002). While cause–effect graphs do not infer the internal structure between signalling genes, NEMs reconstruct both signalling relations between signalling genes and regulatory relations between signalling genes and reporter genes (Figure 6.5). As the redundant edges in cause–effect graphs can be resolved by distinguishing between direct and indirect effects, NEMs additionally return much sparser graphs.

The likelihood of a NEM

In the original formulation of NEMs (Markowetz et al. 2005*a*), genes coding for proteins that function in a signalling network are called 'signalling genes' (*S-genes*), while genes used to monitor perturbation effects of S-genes are named 'effect reporter genes' (*E-genes*). A signalling network is modelled by $G = (\mathcal{V}, \mathcal{E})$, in which $\mathcal{V}$ is the set of S-genes and $\mathcal{E}$ includes all interactions between them. Let $D = [d_{ik}]_{m \times l}$ be observed

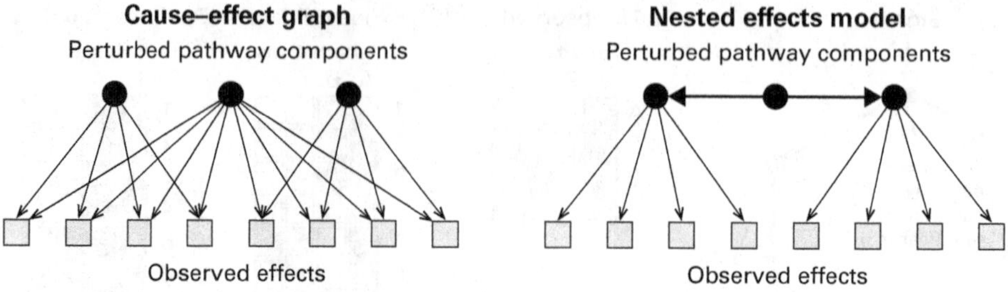

Cause–effect graph

Perturbed pathway components

Nested effects model

Perturbed pathway components

Observed effects

Observed effects

Figure 6.5   Comparing NEMs with cause–effect graphs. Black circles and grey squares represent S-genes and E-genes, respectively. Perturbed genes are directly linked to their phenotypes in cause–effect graphs. In NEMs, signalling relations and regulatory relations are decoupled to reveal underlying biological mechanism.

perturbation effects (Figure 6.4D), where d_{ik} is the effect of perturbation k on effect reporter i.

Let $\Theta = \{\theta_i\}_1^m$, $\theta_i \in \{1,\ldots,n\}$ be a set of parameters indicating the positions of E-genes (i.e. the thin arrows between S-genes and E-genes in Figure 6.4A). E-gene i is directly regulated by S-gene j if $\theta_i = j$. For a single E-gene i under perturbation k, the probability to observe d_{ik} given G and its position $\theta_i = j$ can be computed by:

$$P(d_{ik}|D, \theta_i = j) = \begin{cases} \begin{matrix} d_{ik} = 1 & d_{ik} = 0 \\ \alpha & 1 - \alpha \\ 1 - \beta & \beta \end{matrix} & \begin{matrix} \\ S_{jk} = 0 \\ S_{jk} = 1 \end{matrix} \end{cases}, \tag{6.1}$$

where α and β are global false positive and false negative rate of D that can often be estimated from control experiments; S_{jk} is the state of S-gene j upon perturbation k (Figure 6.4B).

Since E-gene positions are generally unknown, the likelihood of the signalling network G given the observation D is computed by marginalization over Θ:

$$P(D|G) = \int P(D|G, \Theta)P(\Theta|G)d\Theta$$

$$= \frac{1}{n^m} \prod_{i=1}^{m} \sum_{j=1}^{n} \prod_{k=1}^{l} P(d_{ik}|G, \theta_i = j), \tag{6.2}$$

where the first product is over all effect reporters under the assumption that reporters are independent of each other, while the second product is over all replicates under the assumption that replicates are independent of each other. During marginalization, each effect reporter is 'attached' to all pathway components; we thus implicitly take multiple regulators into account (but not complex interactions between them).

Experimental design for NEMs

NEMs are particularly suited for the reconstruction of cytoplasmic signalling pathways, where the effects of perturbing one protein on all other proteins in the pathway is often not contained in large-scale data, which mostly measure transcriptional and not proteomic effects. NEMs draw their strength from very controlled and focused experimental designs, where a target pathway can be turned on and off, and perturbations of components in the activated pathway have an impact on the transcription of downstream genes or cell morphology.

To achieve the best performance of NEMs, the experimental design requires the following. First, perturbed genes are known to be involved in the target pathway. Second, the signalling pathway should be perturbed when stimulated, such that perturbing any one of the components blocks (parts of) the signal and leads to observable phenotypes. Third, positive controls (expression profiles while pathway is active/stimulated) and negative controls (expression profiles while pathway is inactive/unstimulated) are necessary to optimally select E-genes and define effects.

The set of E-genes is then defined as those genes differentially expressed between the positive and negative controls. An E-gene is considered to show an effect when its expression is more similar to the one it shows in the negative control than the postive control (Markowetz et al. 2005a).

Extensions of NEMs

The extensions of NEMs and their contributions to the community are summarized in Table 6.1. The original NEM was established by Markowetz et al. (in Markowetz et al. 2005a), where discretized expression profiles of reporter genes were taken as input and the best scoring pathway was found by exhaustive enumeration. In the following years, a variety of heuristic search algorithms were proposed such as pairwise and triples inference (Markowetz et al. 2007), module networks (Fröhlich et al. 2007, Fröhlich et al. 2008b) and alternating maximum *a posteriori* (MAP) optimization (Tresch & Markowetz 2008). The definition of effects has also been extended, for example by modelling differential expression as mixture model of *p*-values (Fröhlich et al. 2007) or log-ratios (Tresch & Markowetz 2008).

Maximum likelihood estimate was adopted in the original NEM for inference, while the model itself can be naturally extended by incorporating priors of the signalling network and/or the regulatory structure between signalling genes and reporter genes. Fröhlich et al. extended NEMs to take into account prior knowledge about network structure, and applied Bayesian regularization using Akaike information criterion (AIC) (Fröhlich et al. 2007). Maximum *a posteriori* estimation was also proposed to incorporate prior network structure in Tresch & Markowetz (2008) and Fröhlich et al. (2008b). Markov chain Monte Carlo (MCMC) sampling, combined with expectation maximization (EM), was recently proposed and demonstrated an improved network inference performance by Niederberger et al. (2012).

The models introduced above do not distinguish between different types of interaction (e.g. activation or inhibition). This limitation was addressed in Vaske et al. (2009), where a generalized NEM was proposed based on a factor graph (FG-NEM) that encodes the sign of interactions on edges. They demonstrated that with such an extension NEMs are even more powerful for inferring regulatory networks (Vaske et al. 2009, House et al. 2010).

Table 6.1 Nested effects models and extensions

Name	Reference	Contributions
NEMs	Markowetz et al. (2005*a*)	Original NEM, application to *Drosophila* immune response
	Markowetz et al. (2007)	Heuristic inference: pairwise and triples
	Tresch & Markowetz (2008)	Generalized model space and likelihood, prior incorporation, feature selection
	Fröhlich et al. (2007)	Generalized likelihood, prior incorporation, application to breast cancer
	Fröhlich et al. (2008*a*)	Heuristic inference: module networks
	Fröhlich et al. (2008*b*)	Bioconductor package
	Niederberger et al. (2012)	MCMC inference, application to the mediator signalling network
FG-NEMs	Vaske et al. (2009)	Inference by factor graphs, signed edges, application to colon cancer invasion
	House et al. (2010)	Application to colon cancer invasion
DNEMs	Anchang et al. (2009)	Dynamic network inference, application to embryonic stem cell development
dynoNEMs	Fröhlich et al. (2011)	Dynamic network inference, feedback loops resolved
	Failmezger et al. (2013)	Application of dynoNEMs to time-lapse microscopy-based screens of cell morphologies
HMNEMs	Wang et al. (2014)	Evolving network inference, MCMC algorithm, application to neutrophil polarization and stem cell differentiation

The models described above deal with static snapshots of perturbation phenotypes and predict interactions between genes at a certain time point. However, processes in the cell are dynamic in nature and contain feedback loops that cannot be disentangled from static data.

Recent extensions adapted NEMs to time-series observations of perturbation effects (Table 6.1). Anchang et al. (2009) extended NEMs to model dynamics within signalling pathways under the assumption that the observed dynamic perturbation effects over time are due to time delay of signal transduction. Inspired by dynamic Bayesian networks (Murphy 2002), Fröhlich et al. proposed another framework – dynoNEMs, in which the signalling network structure is unrolled over time (Fröhlich et al. 2011). Failmezger et al. (2013) further extended dynoNEMs for analyzing phenotyping screens of cell morphologies based on time-lapse microscopy. In the next section, we introduce hidden Markov nested effects models (HMNEMs, Wang et al. 2014) to model signalling networks with topological changes over discrete time points.

6.4   Scenario 3b: Capturing rewiring events during network evolution

Modelling rewiring events is particularly important for studies in early organism development. As reviewed in the last section, the existing extensions of NEMs for dealing with time-series phenotying screens have demonstrated their own advantages. However, these methods either model dynamics of signal flow within the static NEM framework (e.g. Anchang et al. 2009) or unroll the static NEM network over time (e.g. Fröhlich et al. 2011).

We approach the challenge from a different perspective: we model the evolving network by a Markov chain on a state space of signalling networks (Wang et al. 2014). The transition probabilities of the Markov process is defined by a geometric distribution, which exploits the topological distance between networks. The underlying assumption is that the more distant two networks are, the less likely the transition is. For the observation model, NEMs provide the formulation of emission probabilities that link the hidden network topologies to the observable perturbation effects. Coupling HMMs with NEMs (HMNEMs) makes it possible to capture the evolving wiring diagram underlying the dynamic developmental process such as stem cell differentiation. In the following sections we describe in detail the model design and inference algorithm for HMNEMs.

6.4.1   Model design

In HMNEMs, the time-varying network is considered as a discrete stochastic process $G_{1:T} = \{G_t\}_1^T$. Let $G_t = (\mathcal{V}, \mathcal{E}_t)$ be the network at time t for $t \in \{1, \ldots, T\}$, in which $\mathcal{V}$ is the set of pathway components, and $\mathcal{E}_t$ is the edge set including all signalling interactions between pathway components at timestep t. Let $D_{1:T} = \{D_t\}_1^T$, where $D_t = [d_{ikt}]_{m \times l}$ is a matrix of observed effects for m effect reporters across l perturbations. Under the first-order Markov assumption the probability of the observation of G_t only depends on its previous network structure G_{t-1} for $t \in \{2, \ldots, T\}$ (the upper layer in Figure 6.6).

A hidden Markov nested effects model (HMNEM) is a hidden Markov model (HMM) with the time-varying network as the transition module and nested effects model as the emission module. Similar to ordinary HMMs, we represent a HMNEM as $\mathcal{H}_{nem} = (\pi, A, B)$, in which π is the initial distribution over states (network structures), and A and B denote the transition probabilities and emission probabilities, respectively. Since the state space of the hidden layer consists of *all* possible network structures, no model selection is needed to determine the number of states, which is fixed but can be very large (e.g. >1 billion states for a network consisting of only six pathway components).

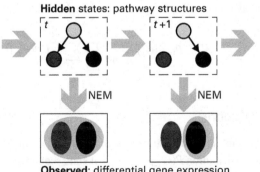

Hidden states: pathway structures

Observed: differential gene expression

Figure 6.6   Hidden Markov nested effects models: the hidden layer represents the evolving pathway over time, the observable layer nested sets of differentially expressed genes after perturbation.

Initial distribution

The initial distribution is set to be a uniform distribution over all the possible network structures. This setting makes the initial distribution take no part in the inference methods, reflecting the fact that prior information about the underlying network is often not available in gene perturbation studies.

Transition probability

We assume here that a signalling network prefers to transit to a state with a similar structure. That is, let P_{uv} be the probability to transit from state u to v, then the more distant the network structure v is from u, the lower the transition probability P_{uv} will be. This assumption is sound in biology, and many other graphical models for modelling time-varying networks also make similar assumptions (e.g. TV-DBN in Song et al. 2009 and TESLA in Ahmed & Xing 2009).

Here we use the geometric distribution to derive the transition probabilities:

$$P_{uv} = P(G_{t+1} = v | G_t = u) = \frac{1}{Z_u}(1 - \lambda)^{s_{uv}}\lambda \ , \tag{6.3}$$

where $Z_u = \sum_u (1 - \lambda)^{s_{uv}}\lambda$ is a normalizing constant; $\lambda \in (0,1)$ is a parameter controlling the 'smoothness'; s_{uv} is the distance between network u and v computed by:

$$s_{uv} = \|A^u - A^v\|_1 := \sum_r \sum_c |a^u_{rc} - a^v_{rc}| \ , \tag{6.4}$$

where A^u and A^v are binary adjacency matrices of networks u and v; precisely, $a_{rc} = 1$ denotes a directed edge from vertice r to c, and 0 otherwise. Each G_t corresponds to a unique adjacency matrix A_t.

Emission probability

The emission probability in a HMNEM is the probability to observe perturbation effects D_t at time t given the current network topology G_t, which can be derived directly from the marginal likelihood of the nested effects model in Eq. (6.2) when the relationships of pathway components to effect reporters are unknown:

$$P(D_t | G_t) = \frac{1}{n^m} \prod_{i=1}^m \sum_{j=1}^n \prod_{k=1}^l P(d_{ikt} | G_t, \theta_i = j). \tag{6.5}$$

6.4.2   MCMC inference

Having established the framework of HMNEMs, our main interest is to infer the joint posterior distribution $P(G_{1:T}, \lambda | D_{1:T})$ of state sequence $G_{1:T}$ and the unknown parameter λ in Eq. (6.3) given the phenotype of gene perturbations over time $D_{1:T}$. To approach this target distribution, we use a Gibbs sampler which draws samples from the two full conditionals: $P(G_{1:T} | \lambda, D_{1:T})$ and $P(\lambda | G_{1:T}, D_{1:T})$.

Sampling $G_{1:T}$

The full conditional of states is obtained by a single-site-update approach, which samples one hidden state at a time. As such, let $G_{-t} := \{G_{t'}; t' \neq t\}$, the target distribution becomes the conditional distribution of each state given all the other states, data and parameter, which can be written as:

$$P(G_t = s | G_{-t}, D_{1:T}, \lambda)$$

$$\propto \begin{cases} P(G_{t+1}|G_t = s, \lambda)P(D_t|G_t = s) & \text{If } t = 1 \\ P(G_t = s|G_{t-1}, \lambda)P(G_{t+1}|G_t = s, \lambda)P(D_t|G_t = s) & \text{If } t = 2, \ldots, T-1 \\ P(G_t = s|G_{t-1}, \lambda)P(D_t|G_t = s) & \text{If } t = T \end{cases}.$$

Direct sampling from this distribution is infeasible. Hence, we resort to the Metropolis-within-Gibbs approach (Geyer 2010) which facilitates sampling by the Metropolis–Hastings (MH) algorithm. To sample networks, we propose a structural MH. By contrast to the method in Madigan et al. (1995), this MH does not restrict the state space to directed acyclic graphs (DAGs). In detail, we use a uniform jumping distribution to propose new graphs: a new graph s' is generated by adding or deleting an edge selected randomly with equal probabilities from all pairs of genes in the current graph s.

Sampling λ

The parameter λ is sampled based on the Metropolis–Hastings algorithm as well. According to Bayes' theorem, the posterior probability of λ can be computed as follows:

$$P(\lambda|G_{1:T}, D_{1:T}) \propto P(D_{1:T}, G_{1:T}|\lambda)P(\lambda)$$

$$= \pi \prod_{t=2}^{T} P(G_t|G_{t-1}, \lambda) \prod_{t=1}^{n} P(D_t|G_t) , \tag{6.6}$$

where the equality statement assumes that the prior probability follows a uniform distribution.

To constrain λ between 0 and 1, we re-parameterize λ by the sigmoid function, such that $\lambda = S(\kappa)$ where $S(\kappa) = \frac{1}{e^{-\kappa}+1}$. Accordingly, the posterior probability of κ is scaled by the determinant of the Jacobian (in this case, the Jacobian is a scalar):

$$P(\kappa|G_{1:T}, D_{1:T}) = P(\lambda|G_{1:T}, D_{1:T})\frac{\partial S(\kappa)}{\partial \kappa}$$

$$= P(\lambda|G_{1:T}, D_{1:T})S(\kappa)(1 - S(\kappa)). \tag{6.7}$$

Expected network

Let $\mathcal{A} := \{A_t\}_1^T$ be the adjacency matrices of $\mathcal{G}$. The expected time-varying network $E[\mathcal{A}] = \{E[A_t]\}_1^T$ is computed by averaging over all adjacency matrices of $\mathcal{G}$ in the estimated posterior distribution obtained from the sampling result:

$$E[A_t] = \sum_{A_t} A_t P(A_t|D_{1:T}) = \frac{1}{N - N_b} \sum_{i=1}^{N-N_b} A_t^{(i)} , \tag{6.8}$$

where N_b is the number of burn-in samples and N is the total number of samples.

In Wang et al. (2014), we demonstrate the potential of HMNEMs by simulations on synthetic time-series perturbation data. We also show the applicability of HMNEMs in two real biological case studies. In one application to neutrophil polarization, HMNEMs capture the transition between initiation, development and maintenance phases during neutrophil polarization. In another application to mouse embryonic stem cells (ESCs), the time-varying network inferred by HMNEMs suggests that underlying early differentiation of ESCs may be the feedback regulations between *Nanog*, *Sox2* and *Oct4*. Our results on these two real biological applications are in part consistent with recent findings in the literature, and generate intriguing hypotheses about the mechanisms of network evolution that can be tested by further experiments.

6.5   Scenario 4: Multi-parametric screen, up to genome-wide

Combining large-scale RNA interference (RNAi) screens with automated imaging generates multi-parametric phenotypes including cell morphology (Fuchs et al. 2010) and tissue architecture (Green et al. 2011). Multi-parametric phenotypes can also result from observing a single phenotype under different biochemical conditions or in different cell lines (Arora et al. 2010, Mulder et al. 2012). Analysing multi-parametric phenotypes is challenging, because they provide richer information than individual reporters but less than high-dimensional phenotypes like microarrays (Markowetz 2010).

Multi-parametric phenotypic data are often represented as association networks clustering together genes with similar phenotypes (Fuchs et al. 2010) and we describe our strategy to infer functional modules in Section 6.5.1. While association networks are a powerful approach to find groups of genes contributing to the same function or pathway, they do not capture directed relationships, like signal flow, between modules. In the second half of the section, we describe our ongoing work to extend NEMs to larger data sets with less rich phenotypes and less targeted experimental design and infer signal flow between gene modules (Section 6.5.2).

6.5.1   Predicting functional modules

In this section, we introduce posterior association networks (*PANs*) to infer networks of functional interactions and enriched modules (Wang et al. 2012).

Synthetic genetic interactions estimated from combinatorial gene perturbation screens provide systematic insights into synergistic interactions of genes in a biological process (Mani et al. 2008, Baryshnikova et al. 2010, Battle et al. 2010, Costanzo et al. 2010). For example, combinatorial drug treatments in bacteria and double mutants in yeast have been implemented to explore their underlying cellular networks (Tong et al. 2004, Collins et al. 2007, Costanzo et al. 2010, Farha & Brown 2010). Very recently, RNAi-based combinatorial gene silencing was applied to *Drosophila* cell culture for signalling pathway reconstruction (Horn et al. 2011).

A major limitation of combinatorial gene silencing, however, lies in its scalability in higher organisms such as humans. Genetic interaction profiling requires double knockdown experiments over all possible combinations of RNAi reagents targeting each pair of genes; thus, the very recent application to *Drosophila* cell culture took more than 70 000 pairwise perturbations between only 93 genes involved in signal transduction (Horn et al. 2011). This explains why genetic interaction profiling for metazoan genes is still limited to a relatively small scale. Moreover, the quality of RNAi screens may suffer from false positives and false negatives due to a lack of efficacy and specificity in silencing reagents (Echeverri & Perrimon 2006, Echeverri et al. 2006). Meta-data analysis or high-quality custom screens are needed to overcome these shortcomings (Echeverri et al. 2006, Booker et al. 2011).

Instead of combinatorial perturbations, we propose to make efficient use of perturbation data on single genes to predict their functional connections. Our motivation is inspired by the fact that genes that genetically associate very often exhibit correlated phenotypes (Costanzo et al. 2010). Only those coherent modules that are highly functionally connected are then subjected to comprehensive biological analysis for deciphering their synergistic functions in a particular process. Thus, our approach starts from building a large-scale landscape of putative functional interactions and results in a condensed core functional module to prioritize further tests for genetic interactions. This strategy makes it possible to integrate publicly available data sets of single gene perturbations performed across multiple cell lines or under different biochemical conditions.

To predict a PAN and functional modules, we developed a unified computational framework (Figure 6.7) involving the following major procedures:

Profiling functional associations

A conventional way to quantify the functional association between two genes is to compute the similarity between their phenotypic profiles based on correlation coefficients (e.g. Bakal et al. 2007). Here, we prefer the uncentred correlation coefficient (also known as *cosine similarity*), because it considers both magnitude and direction and has been very successful in exploring gene expression patterns (Eisen et al. 1998, Dadgostar et al. 2002, de Hoon et al. 2002). Thus, we will focus on cosine similarities throughout this chapter, although other correlation coefficients can be used without changing our methodology.

Beta-mixture modelling

Motivated by the density pattern of association profiles, we propose to model functional associations by a mixture of three components representing positive association (+), negative association (−) and lack of association (×), respectively. We employ a stratification strategy to take into consideration potential prior knowledge for the functional network such as protein–protein interactions. To fit the beta-mixture model, we performed MAP (maximum *a posteriori*) based on the EM algorithm (Dempster et al. 1977).

Network inference

To assess the strength of evidence for having a functional interaction, a model selection step is performed for each pair of genes. We compute signal-to-noise ratios (SNRs),

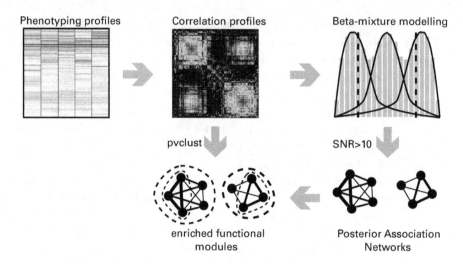

Figure 6.7　Posterior association networks. PAN takes as input various types of phenotyping screens (e.g. gene expression, biochemical signals, imaging data of cell morphologies and tissue architectures) that have already been preprocessed. Two parallel sub-workflows are subsequently performed to predict (i) significant functional interactions between genes by beta-mixture modelling on functional association profiles, and (ii) significant gene clusters by hierarchical clustering on functional association profiles. Superimposing the predicted significant gene clusters onto the predicted posterior association network, we finally obtain modules enriched for functional interactions.

which are posterior odds for edge a_u between a pair of genes $u = (i, j)$ in favour of association to lack of association:

$$K_u = \frac{P(z_{u\times} = 0 \mid a_u, \mathbf{\Pi}, \boldsymbol{\theta}, \mathbf{\Gamma}^*)}{P(z_{u\times} = 1 \mid a_u, \mathbf{\Pi}, \boldsymbol{\theta}, \mathbf{\Gamma}^*)} \ , \tag{6.9}$$

where

- $z_{u\times}$ is a latent variable indicating the affiliation of gene pair u to mixture component $\times$ designating a lack of relationship;
- $\boldsymbol{\theta}$ denotes shape parameters of the three beta distributions;
- $\mathbf{\Pi}$ denotes the set of mixture coefficients affiliated with different partition sets;
- $\mathbf{\Gamma}^*$ is a matrix of hyperparameters of a Dirichlet prior with each row corresponding to a stratum and each column to a mixture component.

A cutoff score K_0 can be set to filter out non-significant edges, guided by the interpretation of Bayes factors by Harold Jeffreys (Jeffreys 1998). The sign of each edge can be simply determined by comparing the posterior probabilities for it belonging to the mixture component representing positive and negative associations.

Searching for modules

We search for coherent functional modules in the inferred PAN by performing hierarchical clustering on functional association profiles, each of which is a vector of cosine similarities between one gene and all genes screened. The method compares functional profiles of genes instead of their individual functions, and it has been demonstrated to be a highly desirable measure to group genes with similar interaction patterns (Costanzo et al. 2010). To assess the uncertainty of the clustering analysis, we computed a p-value for each cluster using multiscale bootstrap resampling (Suzuki & Shimodaira 2006). The clusters derived from hierarchical clustering are projected onto the inferred posterior association network to generate functional modules. Top significant modules enriched for significant functional interactions are selected according to four module filtering steps.

We demonstrate the general applicability of our computational methodology on a publicly available data set of single RNAi perturbations across four cell lines in Ewing's sarcoma (ES) (Arora et al. 2010). Using our approach, we prioritized one module enriched for confirmed and promising potential therapeutic targets for ES and highly associated with signalling pathways that are known to be critical for proliferation of ES cells (Wang et al. 2012). We also applied the computational framework to study self-renewal of epidermal stem cells using RNA interference screening data for 332 known and predicted chromatin modifiers (Wang et al. 2012). We predicted a highly significant module enriched for functional interactions, and confirmed their dense genetic interactions using combinatorial gene perturbation (Mulder et al. 2012).

6.5.2   Inferring signalling flows between modules

Clustering approaches like the one described above rely on *guilt by association* (GBA), a powerful principle underlying most approaches in functional genomics. Roughly it states that similar phenotypes hint at similar function. GBA is the main rationale by which new members of a functional unit (a protein complex, a pathway) are found: if gene X shares many features with genes already known to contribute to functional unit Y, then X most probably also contributes to Y. However, these successes can all too easily hide important limitations of the GBA principle.

The most important one is: while GBA allows inferring membership of a functional unit, it does not allow inferring a more finely resolved structure within the unit. For example, we can learn that gene X may play a role in the WNT pathway, because it looks very similar to other WNT genes in an RNAi screen, but we don't know where and how it contributes to the pathway. This is an important limitation, since it prevents reconstructing directions of signal flow in the cell or a hierarchy between functional units. Thus, the picture provided by GBA while useful for functional classification lacks the details required of a predictive model of the inner workings of the cell.

To address these shortcomings we work on alternatives to GBA that do not implement symmetric similarity measures (like Pearson correlation) but directed relationships as used in NEMs. To make NEMs applicable to larger data sets with less rich phenotypes and less targeted experimental design, we relax the assumptions of NEMs by balancing the resolution of the inferred network against the richness of the observed phenotypes: rich phenotypes result in detailed pathway models, while less rich phenotypes result in models of interactions between gene clusters. We extend NEMs to a more generalized framework (Figure 6.8) and use a four-step inference strategy (Figure 6.9).

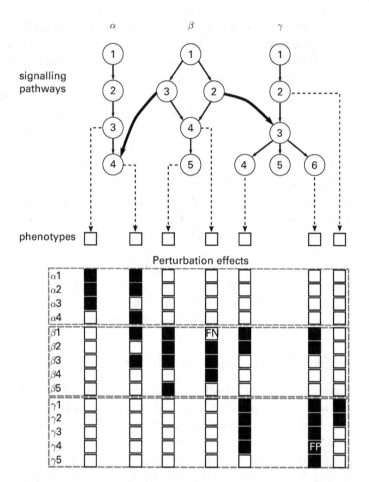

Figure 6.8   Extending NEMs to reconstruct signalling flows between pathways. The upper panel shows a toy example of three signalling pathways (α, β and γ) with cross-talks. In each pathway, only a part of S-genes directly influence distinct downstream phenotypic reporters. The perturbation effects are illustrated in the lower panel, where rows correspond to S-genes and columns to phenotypes. Due to the sparsity of phenotypes, S-genes can exhibit exactly the same or highly similar perturbation effects (e.g. $\alpha 1$ and $\alpha 2$, $\gamma 1$ and $\gamma 2$). Thus, different from NEMs, the limited number of phenotypes is not sufficient to infer the very detailed signalling network structure.

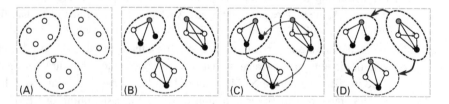

Figure 6.9   The major procedures to infer cross-talks between pathways. (A) Partitioning genes to modules. (B) Searching for boundary nodes. Upper boundary nodes and lower boundary nodes are coloured in grey and black, repectively. (C) Inferring signalling relationships between boundary nodes. (D) Inferring cross-talks between modules.

Partitioning genes to modules

We first partition genes to signalling pathway modules by hierachical clustering (Figure 6.9A). To measure how distant two S-genes are, the posterior probability of two genes (e.g. A and B) being unrelated ($\phi_{AB} = A \cdot \cdot B$) is used as a dissimilarity score:

$$
\begin{aligned}
d(A, B) &= P(\phi_{AB} = A \cdot \cdot B | D) \\
&\propto P(D | \phi_{AB} = A \cdot \cdot B) P(\phi_{AB} = A \cdot \cdot B) \\
&\overset{1}{\propto} P(D | \phi_{AB} = A \cdot \cdot B) \\
&\overset{2}{=} \prod_{i=1}^{m} \sum_{j \in \{A,B\}} \prod_{k=1}^{l} P(d_{ik} | \phi_{AB} = A \cdot \cdot B, \theta_i = j) ,
\end{aligned}
\tag{6.10}
$$

where D represents observed perturbation data; m and l denote the number of E-genes and replicates, respectively. Neglecting the normalizing constant $P(D)$, the above equation is derived based on the following assumptions: (1) a uniform distribution of $P(\phi_{AB})$ over all four possible relationships: $A \rightarrow B, B \rightarrow A, A \leftrightarrow B$ and $A \cdot \cdot B$ in Eq. (6.10); (2) the same assumptions as in Eq. (6.2) for computing the likelihood.

This intriguing score takes the advantage of the nested structure of perturbation effects instead of global dissimilarity for grouping genes to pathway modules. The dissimilarity scores are computed for all pairs of genes and are subsequently used for hierarchical clustering. The cutoff level for clustering can be viewed as a control of the resolution of the final signalling map constituted by pathway modules and signaling flows between them.

Searching for boundary nodes

In each module, we search for representatives – *boundary nodes*, which are genes sitting at the top or bottom of the signalling hierarchy (Figure 6.9B). Boundary nodes encode the information of signalling hierarchies within the minimal number of genes. These representatives enable collapsing the initial gene-level network to a module-level map while retaining the information of signalling cross-talks.

Inferring signalling flows between modules

It can be proved theoretically that the signalling flows between modules can be fully recovered from the relationships between boundary nodes. Thus, in this step we first infer a network of signalling interactions between boundary nodes (Figure 6.9C). As a result of the two steps above, the number of boundary nodes is much lower than the initial number of S-genes. Therefore, depending on the type of gene perturbation data and the number of boundary nodes, various inference methods for NEMs reviewed in Table 6.1 may be applied here. Finally, the interactions between boundary nodes are collapsed to module-level cross-talks (Figure 6.9D).

In ongoing work we have applied the generalized NEMs in several case studies and found that the asymmetric association score computed using NEMs outperforms conventional distance metrics such as Pearson correlation coefficients and Euclidean distance in partitioning genes to functional modules. We found that the signalling flow between functional gene modules can be predicted efficiently by scaling network resolution to the richness of phenotypic information.

6.6   Conclusions

High-throughput genetic screening (HTS), especially RNAi-based, has revolutionized the methodologies of functional genomics and shown the therapeutic promise in personalized medicine. In the enormous HTS applications, computational approaches have been demonstrated to be extremely important for guiding towards a faster, more focused and more cost-efficient experimental design. Beyond characterizing functions of single genes, recent advances in large-scale rich phenotyping screens have opened up the era of systematic studying the inner workings of the cell.

How to bridge the genome and observed phenome relies on the number of genes perturbed and richness of phenotypes. In this chapter, we introduced computational methodologies specialized for screening data derived from various experimental strategies.

One extreme scenario is that a single reporter is used in large-scale screening. We developed HTSanalyzeR, a bioconductor package, to link observed phenotypes to cellular pathways and networks. HTSanalyzeR provides integrated pipelines for an hypothesis-free and unbiased analysis of HTS data and generates enriched subnetworks for experimental follow-up. The other extreme is multi-level phenotyping screening over time for a single gene perturbed. These rich temporal and spatial phenotypes provide new opportunities for a mechanistic study across epigenome, transcriptome and proteome.

High-dimensional (e.g. microarray based) phenotypic screens provide more information about the interplay of perturbed componets in a specific pathway. We discussed and compared two popular statistical frameworks – Bayesian networks and nested effects models (NEMs). Bayesian networks are a classical graphical model for inferring networks from direct effects observed in perturbed components. NEMs, in contrast, reconstruct signalling relationships from *subset relations* of indirect effects in downstream genes. NEMs and extensions have been successfully applied to recovering signalling pathways involved in *Drosophila immune response*, human breast and colon cancer, early development of stem cells, etc. Furthermore, we introduced hidden Markov NEMs, which extend NEMs to capture rewiring diagrams underlying dynamic biological processes.

Many recent RNAi experiments combine large-scale RNAi screening with automated imaging to generate rich phenotyping data. In this senario, we introduced posterior association networks (PANs) to infer networks of functional interactions and enriched modules. The major advantage of this approach over conventional approches is that prior knowledge can be incorporated to enhance predictive power. To capture more finely resolved structures, we also proposed to use directed relationships instead of symmetric similarity measures. The generalized NEM framework we developed can efficiently predict signalling flows between functional modules by scaling network resolution to the richness of phenotypic information.

References

Ahmed, A. & Xing, E. P. (2009), 'Recovering time-varying networks of dependencies in social and biological studies', *Proceedings of the National Academy of Sciences of the USA* **106**(29), 11 878–11 883.

Alexa, A., Rahnenführer, J. & Lengauer, T. (2006), 'Improved scoring of functional groups from gene expression data by decorrelating go graph structure', *Bioinformatics* **22**(13), 1600–1607.

Anchang, B., Sadeh, M., Jacob, J., Tresch, A., Vlad, M. et al. (2009), 'Modeling the temporal interplay of molecular signaling and gene expression by using dynamic nested effects models', *Proceedings of the National Academy of Sciences of the USA* **106**(16), 6447–6452.

Arora, S., Gonzales, I., Hagelstrom, R., Beaudry, C., Choudhary, A. et al. (2010), 'RNAi phenotype profiling of kinases identifies potential therapeutic targets in Ewing's sarcoma', *Molecular Cancer* **9**(1), 218.

Ashburner, M., Ball, C., Blake, J., Botstein, D., Butler, H. et al. (2000), 'Gene Ontology: tool for the unification of biology', *Nature Genetics* **25**(1), 25–29.

Bakal, C., Aach, J., Church, G. & Perrimon, N. (2007), 'Quantitative morphological signatures define local signaling networks regulating cell morphology', *Science* **316**(5832), 1753–1756.

Baryshnikova, A., Costanzo, M., Kim, Y., Ding, H., Koh, J. et al. (2010), 'Quantitative analysis of fitness and genetic interactions in yeast on a genome scale', *Nature Methods* **7**(12), 1017–1024.

Battle, A., Jonikas, M. C., Walter, P., Weissman, J. S. & Koller, D. (2010), 'Automated identification of pathways from quantitative genetic interaction data', *Molecular Systems Biology* **6**, 379.

Bauer, S., Grossmann, S., Vingron, M. & Robinson, P. (2008), 'Ontologizer 2.0: a multifunctional tool for GO term enrichment analysis and data exploration', *Bioinformatics* **24**(14), 1650–1651.

Beißbarth, T. & Speed, T. (2004), 'GOstat: find statistically overrepresented Gene Ontologies within a group of genes', *Bioinformatics* **20**(9), 1464–1465.

Beisser, D., Klau, G., Dandekar, T., Müller, T. & Dittrich, M. (2010), 'BioNet: an R-package for the functional analysis of biological networks', *Bioinformatics* **26**(8), 1129–1130.

Birmingham, A., Selfors, L., Forster, T., Wrobel, D., Kennedy, C. et al. (2009), 'Statistical methods for analysis of high-throughput RNA interference screens', *Nature Methods* **6**(8), 569–575.

Booker, M., Samsonova, A. A., Kwon, Y., Flockhart, I., Mohr, S. E. et al. (2011), 'False negative rates in *Drosophila* cell-based RNAi screens: a case study', *BMC Genomics* **12**, 50.

Boutros, M. & Ahringer, J. (2008), 'The art and design of genetic screens: RNA interference', *Nature Reviews Genetics* **9**(7), 554–566.

Boutros, M., Brás, L. P. & Huber, W. (2006), 'Analysis of cell-based RNAi screens', *Genome Biology* **7**(7), R66.

Boutros, M., Kiger, A. A., Armknecht, S., Kerr, K., Hild, M. et al. (2004), 'Genome-wide RNAi analysis of growth and viability in *Drosophila* cells', *Science* **303**(5659), 832–835.

Breitkreutz, B., Stark, C., Reguly, T., Boucher, L., Breitkreutz, A. et al. (2008), 'The BioGRID interaction database: 2008 update', *Nucleic Acids Research* **36**(Suppl 1), D637–D640.

Brideau, C., Gunter, B., Pikounis, B. & Liaw, A. (2003), 'Improved statistical methods for hit selection in high-throughput screening', *Journal of Biomolecular Screening* **8**(6), 634–647.

Castro, M., Wang, X., Fletcher, M., Meyer, K. & Markowetz, F. (2012), 'RedeR: R/Bioconductor package for representing modular structures, nested networks and multiple levels of hierarchical associations', *Genome Biology* **13**(4), R29.

Cheung, H. W., Cowley, G. S., Weir, B. A., Boehm, J. S., Rusin, S. et al. (2011), 'Systematic investigation of genetic vulnerabilities across cancer cell lines reveals lineage-specific dependencies in ovarian cancer', *Proceedings of the National Academy of Sciences of the USA* **108**(30), 12 372–12 377.

Collins, S., Miller, K., Maas, N., Roguev, A., Fillingham, J. et al. (2007), 'Functional dissection of protein complexes involved in yeast chromosome biology using a genetic interaction map', *Nature* **446**(7137), 806–810.

Costanzo, M., Baryshnikova, A., Bellay, J., Kim, Y., Spear, E. et al. (2010), 'The genetic landscape of a cell', *Science* **327**(5964), 425.

Dadgostar, H., Zarnegar, B., Hoffmann, A., Qin, X., Truong, U. et al. (2002), 'Cooperation of multiple signaling pathways in CD40-regulated gene expression in B lymphocytes', *Proceedings of the National Academy of Sciences of the USA* **99**(3), 1497–1502.

de Hoon, M., Imoto, S. & Miyano, S. (2002), 'A comparison of clustering techniques for gene expression data', *Proceedings of the 10th International Conference on Intelligent Systems for Molecular Biology*, Abstract 33A.

Dempster, A., Laird, N. & Rubin, D. (1977), 'Maximum likelihood from incomplete data via the EM algorithm', *Journal of the Royal Statistical Society, Series B (Methodological)* **39**(1), 1–38.

Echeverri, C. & Perrimon, N. (2006), 'High-throughput RNAi screening in cultured cells: a user's guide', *Nature Reviews Genetics* **7**(5), 373–384.

Echeverri, C., Beachy, P., Baum, B., Boutros, M., Buchholz, F. et al. (2006), 'Minimizing the risk of reporting false positives in large-scale RNAi screens', *Nature Methods* **3**(10), 777–779.

Eisen, M., Spellman, P., Brown, P. & Botstein, D. (1998), 'Cluster analysis and display of genome-wide expression patterns', *Proceedings of the National Academy of Sciences of the USA* **95**(25), 14 863–14 868.

Failmezger, H., Praveen, P., Tresch, A. & Fröhlich, H. (2013), 'Learning gene network structure from time lapse cell imaging in RNAi knockdowns', *Bioinformatics* **29**(12), 1534–1540.

Falcon, S. & Gentleman, R. (2007), 'Using GOstats to test gene lists for GO term association', *Bioinformatics* **23**(2), 257–258.

Farha, M. & Brown, E. (2010), 'Chemical probes of *Escherichia coli* uncovered through chemical–chemical interaction profiling with compounds of known biological activity', *Chemistry & Biology* **17**(8), 852–862.

Friedman, N. (2004), 'Inferring cellular networks using probabilistic graphical models', *Science* **303**(5659), 799–805.

Friedman, N., Linial, M., Nachman, I. & Pe'er, D. (2000), 'Using Bayesian networks to analyze expression data', *Journal of Computational Biology* **7**(3–4), 601–620.

Fröhlich, H., Beißbarth, T., Tresch, A., Kostka, D., Jacob, J. et al. (2008*a*), 'Analyzing gene perturbation screens with nested effects models in R and Bioconductor', *Bioinformatics* **24**(21), 2549–2550.

Fröhlich, H., Fellmann, M., Sueltmann, H., Poustka, A. & Beissbarth, T. (2007), 'Large scale statistical inference of signaling pathways from RNAi and microarray data', *BMC Bioinformatics* **8**, 386.

Fröhlich, H., Fellmann, M., Sueltmann, H., Poustka, A. & Beissbarth, T. (2008*b*), 'Estimating large-scale signaling networks through nested effect models with intervention effects from microarray data', *Bioinformatics* **24**(22), 2650–2656.

Fröhlich, H., Praveen, P. & Tresch, A. (2011), 'Fast and efficient dynamic nested effects models', *Bioinformatics* **27**(2), 238–244.

Fuchs, F., Pau, G., Kranz, D., Sklyar, O., Budjan, C. et al. (2010), 'Clustering phenotype populations by genome-wide RNAi and multiparametric imaging', *Molecular Systems Biology* **6**, 370.

Geyer, C. (2010), 'Introduction to Markov chain Monte Carlo', *in* S. Brooks, A. Gelman, G. Jones & X.-L. Meng, eds., *Handbook of Markov chain Monte Carlo*, CRC Press, Boca Raton, FL, pp. 3–48.

Green, R., Kao, H., Audhya, A., Arur, S., Mayers, J. et al. (2011), 'A high-resolution *C. elegans* essential gene network based on phenotypic profiling of a complex tissue', *Cell* **145**(3), 470–482.

Hahne, F., Arlt, D., Sauermann, M., Majety, M., Poustka, A. et al. (2006), 'Statistical methods and software for the analysis of high-throughput reverse genetic assays using flow cytometry readouts', *Genome Biology* **7**(8), R77.

Horn, T., Sandmann, T., Fischer, B., Axelsson, E., Huber, W. et al. (2011), 'Mapping of signaling networks through synthetic genetic interaction analysis by RNAi', *Nature Methods* **8**(4), 341–346.

House, C. D., Vaske, C. J., Schwartz, A. M., Obias, V., Frank, B. et al. (2010), 'Voltage-gated Na^+ channel SCN5A is a key regulator of a gene transcriptional network that controls colon cancer invasion', *Cancer Research* **70**(17), 6957–6967.

Jeffreys, H. (1998), *Theory of probability*, 3rd edn, Oxford University Press.

Jensen, L. J., Kuhn, M., Stark, M., Chaffron, S., Creevey, C. et al. (2009), 'String 8: a global view on proteins and their functional interactions in 630 organisms', *Nucleic Acids Research* **37**(Database issue), D412–D416.

Kaderali, L., Dazert, E., Zeuge, U., Frese, M. & Bartenschlager, R. (2009), 'Reconstructing signaling pathways from RNAi data using probabilistic Boolean threshold networks', *Bioinformatics* **25**(17), 2229–2235.

Kessler, J., Kahle, K., Sun, T., Meerbrey, K., Schlabach, M. et al. (2012), 'A SUMOylation-dependent transcriptional subprogram is required for Myc-driven tumorigenesis', *Science* **335**(6066), 348–353.

Li, C. & Wong, W. (2001), 'Model-based analysis of oligonucleotide arrays: model validation, design issues and standard error application', *Genome Biology* **2**(8), 0032.

Liberzon, A., Subramanian, A., Pinchback, R., Thorvaldsdóttir, H., Tamayo, P. et al. (2011), 'Molecular signatures database (MSigDB) 3.0', *Bioinformatics* **27**(12), 1739–1740.

Lu, R., Markowetz, F., Unwin, R. D., Leek, J. T., Airoldi, E. M. et al. (2009), 'Systems-level dynamic analyses of fate change in murine embryonic stem cells', *Nature* **462**(7271), 358–362.

Maathuis, M. H., Colombo, D., Kalisch, M. & Bhlmann, P. (2010), 'Predicting causal effects in large-scale systems from observational data', *Nature Methods* **7**(4), 247–248.

Madigan, D., York, J. & Allard, D. (1995), 'Bayesian graphical models for discrete data', *International Statistical Review/Revue Internationale de Statistique* **63**(2), 215–232.

Malo, N., Hanley, J., Cerquozzi, S., Pelletier, J. & Nadon, R. (2006), 'Statistical practice in high-throughput screening data analysis', *Nature Biotechnology* **24**(2), 167–175.

Mani, R., St Onge, R., Hartman, J., Giaever, G. & Roth, F. (2008), 'Defining genetic interaction', *Proceedings of the National Academy of Sciences of the USA* **105**(9), 3461–3466.

Markowetz, F. (2010), 'How to understand the cell by breaking it: network analysis of gene perturbation screens', *PLoS Computational Biology* **6**(2), e1000655.

Markowetz, F. & Spang, R. (2007), 'Inferring cellular networks – a review', *BMC Bioinformatics* **8**(Suppl 6), S5.

Markowetz, F., Bloch, J. & Spang, R. (2005*a*), 'Non-transcriptional pathway features reconstructed from secondary effects of RNA interference', *Bioinformatics* **21**(21), 4026–4032.

Markowetz, F., Grossmann, S. & Spang, R. (2005*b*), 'Probabilistic soft interventions in conditional Gaussian networks', *Proceedings of 10th International Workshop on Artificial Intelligence and Statistics*.

Markowetz, F., Kostka, D., Troyanskaya, O. G. & Spang, R. (2007), 'Nested effects models for high-dimensional phenotyping screens', *Bioinformatics* **23**(13), i305–i312.

Markowetz, F., Mulder, K. W., Airoldi, E. M., Lemischka, I. R. & Troyanskaya, O. G. (2010), 'Mapping dynamic histone acetylation patterns to gene expression in *nanog*-depleted murine embryonic stem cells', *PLoS Computational Biology* **6**(12), e1001034.

Merico, D., Isserlin, R., Stueker, O., Emili, A. & Bader, G. (2010), 'Enrichment map: a network-based method for gene-set enrichment visualization and interpretation', *PLoS One* **5**(11), e13984.

Mulder, K. W., Wang, X., Escriu, C., Ito, Y., Schwarz, R. F. et al. (2012), 'Diverse epigenetic strategies interact to control epidermal differentiation', *Nature Cell Biology* **14**(7), 753–763.

Müller, P., Kuttenkeuler, D., Gesellchen, V., Zeidler, M. P. & Boutros, M. (2005), 'Identification of JAK/STAT signalling components by genome-wide RNA interference', *Nature* **436**(7052), 871–875.

Murphy, K. (2002), 'Dynamic Bayesian networks: representation, inference and learning', PhD thesis, University of California – Berkeley.

Niederberger, T., Etzold, S., Lidschreiber, M., Maier, K., Martin, D. et al. (2012), 'MC EMiNEM maps the interaction landscape of the mediator', *PLoS Computational Biology* **8**(6), e1002568.

Ogata, H., Goto, S., Sato, K., Fujibuchi, W., Bono, H. et al. (1999), 'KEGG: Kyoto encyclopedia of genes and genomes', *Nucleic Acids Research* **27**(1), 29–34.

Orvedahl, A., Sumpter Jr, R., Xiao, G., Ng, A., Zou, Z. et al. (2011), 'Image-based genome-wide siRNA screen identifies selective autophagy factors', *Nature* **480**(7375), 113–117.

Pearl, J. (1988), *Probabilistic reasoning in intelligent systems: networks of plausible inference*, Morgan Kaufmann, San Mateo, CA.

Pearl, J. (2000), *Causality: models, reasoning, and inference*, Cambridge University Press.

Pe'er, D. (2005), 'Bayesian network analysis of signaling networks: a primer', *Science STKE* **2005**(281), l4.

Pe'er, D., Regev, A., Elidan, G. & Friedman, N. (2001), 'Inferring subnetworks from perturbed expression profiles', *Bioinformatics* **17**(Suppl 1), S215–S224.

Pelz, O., Gilsdorf, M. & Boutros, M. (2010), 'web-cellHTS2: a web-application for the analysis of high-throughput screening data', *BMC Bioinformatics* **11**(1), 185.

Rung, J., Schlitt, T., Brazma, A., Freivalds, K. & Vilo, J. (2002), 'Building and analysing genome-wide gene disruption networks', *Bioinformatics* **18**(Suppl 2), S202–S210.

Sachs, K., Perez, O., Pe'er, D., Lauffenburger, D. A. & Nolan, G. P. (2005), 'Causal protein-signaling networks derived from multiparameter single-cell data', *Science* **308**(5721), 523–529.

Shimoni, Y., Fink, M. Y., Choi, S.-G. & Sealfon, S. C. (2010), 'Plato's cave algorithm: inferring functional signaling networks from early gene expression shadows', *PLoS Computational Biology* **6**(6), e1000828.

Smyth, G. K. (2005), 'Limma: linear models for microarray data', *in* R. Gentleman, V. Carey, S. Dudoit, R. Irizarry & W. Huber, eds., *Bioinformatics and computational biology solutions using R and Bioconductor*, Springer, New York, pp. 397–420.

Song, L., Kolar, M. & Xing, E. P. (2009), 'Time-varying dynamic Bayesian networks', *Advances in Neural Information Processing Systems* **22**, 1732–1740.

Subramanian, A., Tamayo, P., Mootha, V. K., Mukherjee, S., Ebert, B. L. et al. (2005), 'Gene set enrichment analysis: a knowledge-based approach for interpreting genome-wide expression profiles', *Proceedings of the National Academy of Sciences of the USA* **102**(43), 15 545–15 550.

Suzuki, R. & Shimodaira, H. (2006), 'Pvclust: an R package for assessing the uncertainty in hierarchical clustering', *Bioinformatics* **22**(12), 1540.

Tong, A., Lesage, G., Bader, G., Ding, H., Xu, H. et al. (2004), 'Global mapping of the yeast genetic interaction network', *Science* **303**(5659), 808.

Tresch, A. & Markowetz, F. (2008), 'Structure learning in nested effects models', *Statistical Applications in Genetics and Molecular Biology* **7**(1), 9.

Vaske, C. J., House, C., Luu, T., Frank, B., Yeang, C.-H. et al. (2009), 'A factor graph nested effects model to identify networks from genetic perturbations', *PLoS Computational Biology* **5**(1), e1000274.

Wagner, A. (2001), 'How to reconstruct a large genetic network from n gene perturbations in fewer than n^2 easy steps', *Bioinformatics* **17**(12), 1183–1197.

Wang, X., Castro, M. A., Mulder, K. W. & Markowetz, F. (2012), 'Posterior association networks and functional modules inferred from rich phenotypes of gene perturbations', *PLoS Computational Biology* **8**(6), e1002566.

Wang, X., Terfve, C., Rose, J. C. & Markowetz, F. (2011), 'HTSanalyzeR: an R/Bioconductor package for integrated network analysis of high-throughput screens', *Bioinformatics* **27**(6), 879–880.

Wang, X., Yuan, K., Hellmayr, C., Liu, W. & Markowetz, F. (2014), 'Reconstructing evolving signalling networks by hidden Markov nested effects models', *Annals of Applied Statistics* **8**(1), 448–480.

Zhang, J., Chung, T. & Oldenburg, K. (1999), 'A simple statistical parameter for use in evaluation and validation of high throughput screening assays', *Journal of Biomolecular Screening* **4**(2), 67–73.

Zhang, X., Yang, X., Chung, N., Gates, A., Stec, E. et al. (2006), 'Robust statistical methods for hit selection in RNA interference high-throughput screening experiments', *Pharmacogenomics* **7**(3), 299–309.

7 High-content screening in infectious diseases: new drugs against bugs

André P. Mäurer, Peter R. Braun, Kate Holden-Dye, and Thomas F. Meyer

Despite immense achievements in the past century in hygiene control, and the development of vaccines and antibiotics, infectious diseases continue to pose a tremendous threat to public health globally. There are still devastating infections for which there are no effective vaccines or antimicrobial therapies. Moreover, the problem of drug resistance in bacteria and viral populations and the increasing appreciation that pathologies resulting from infections are responsible for a number of chronic conditions, are creating an ever-growing need for novel preventive and therapeutic approaches. In line with this, a new host-targeted approach has been suggested for antimicrobial drug research that exploits the central role of the host cell during infection. Decades of research have taught us that infections are supported by host cell functions, and that infection pathology is frequently host dependent. Accordingly, the pharmacological targeting of host cell factors promises novel opportunities to prevent and treat infectious disease. Such an approach may be anticipated to expand the number of druggable targets, produce broad-spectrum compounds and impede the generation of resistance. The discovery of RNA interference (RNAi) has created opportunities to explore gene functions in cellular systems in a targeted manner. RNAi loss-of-function approaches have proved invaluable for the identification of host proteins important for pathogen viability. These approaches can be applied on a high-throughput scale, which demands sophisticated liquid handling and high-content image analysis. Here, we provide an overview of the current status of high-content screening (HCS) in loss-of-function analyses in infectious disease research and discuss how these powerful techniques can be applied to identify host factors with previously unknown roles in infection and its pathology.

7.1 The challenge of fighting infectious diseases

Infections by pathogenic species of bacteria, viruses, fungi and protozoa have had considerable impact on mankind throughout history. Advances in our understanding of the importance of hygiene control, and later, improvements in diagnostics and the development and successful employment of vaccines and antimicrobial drugs, have substantially benefited human health, and provided social and economic benefits. However, infectious

Systems Genetics: Linking Genotypes and Phenotypes, ed. F. Markowetz and M. Boutros. Published by Cambridge University Press. © Cambridge University Press 2015.

diseases remain a major burden to global health in the twenty-first century (Figure 7.1). Approximately 25 per cent of total annual deaths worldwide are estimated to result directly from infection by pathogenic agents, a conservative figure that does not include the additional deaths that result from the pathological consequences of prior infections or chronic infections (Morens et al. 2004). Moreover, it was recently reported that in 2010, approximately 64 per cent of deaths worldwide in children under the age of 5 were the result of infection, a figure that will challenge efforts aimed at achieving reductions in global child mortality rates (Liu et al. 2012).

Infections have also been associated with various forms of cancer in humans. Initial breakthroughs in establishing causal links between oncogenesis and pathogen infection were supported by the intrinsic properties of these agents, i.e. tumour viruses tend to deposit some of their genes in the host cells, thus leaving unmistakable markers in the resulting tumours. In the case of *Helicobacter pylori*, the only bacterial agent to date that is widely accepted to be oncogenic, it was the unique epidemiological association of this pathogenic bacterial agent in the acidic environment of the human stomach that founded its aetiologic connection with gastric cancer. As a consequence, we can now say that infections are a leading cause of human cancers globally (Pisani et al. 1997), particularly in less-developed countries (de Martel et al. 2012). The latest calculations of global cancer burdens attributed 23 per cent and 7 per cent of cancers to infectious aetiologies in less-developed and more-developed countries, respectively. It is estimated that, overall, approximately 1 in 6 cancer deaths worldwide are caused by an infectious agent (de Martel et al. 2012). And in the current climate, in which HIV-1 is endemic in large parts of the globe, HIV-1 co-infection is a significant compounding factor in the progression of cancers associated with chronic infections, including hepatocellular carcinoma associated with hepatitis C virus infection (Puoti et al. 2004) and cancers associated with human papillomavirus (HPV) infection (Frisch et al. 2000). It is still possible that infectious agents could be the cause of other human cancers, for which we have not yet discovered the aetiologic link. Strikingly, there are strong links between inflammation and oncogenesis in humans, it is feasible that the origin of inflammatory events involved in cell transformation could well derive from pathogen infection. Thus, there is still a need for substantial efforts aimed at increasing our appreciation of links between chronic pathologies and infectious aetiologies. Indeed, so far we may only have an understanding of the 'tip of the iceberg'.

7.2   Classic strategies for antimicrobial drug development and their limitations

In 1928, Alexander Fleming made a momentous breakthrough in medicine upon his discovery that a fungal extract, penicillin, inhibited the growth of the pathogen *Staphylococcus aureus*. Penicillin compounds would go on to become the first targeted antimicrobials that were broadly active against some of the most devastating bacterial diseases in history; this began the modern era of antimicrobial research. By the 1950s the widespread use of penicillin and the development of vaccines for diseases such as

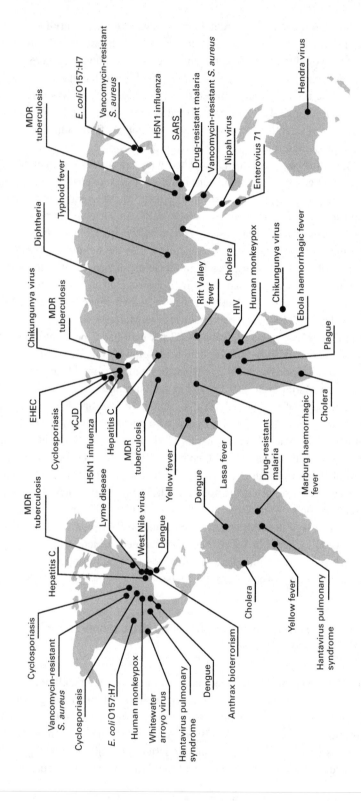

Figure 7.1 Global regions associated with the origin of selected emerging diseases (adapted from Morens et al. 2004).

polio triggered a somewhat optimistic statement from the US Surgeon General declaring that 'the war against infectious diseases has been won' (Fauci 2001). Historically, antimicrobial drug development has involved the identification of compounds that selectively inhibit the growth of pathogens by directly targeting determinants of pathogen biology crucial for viability and fitness, e.g. bacterial cell wall synthesis (penicillin), viral budding (the neuraminidase inhibitors oseltamivir and zanamivir) or viral genome replication (reverse transcriptase inhibitors, including zidovudine and tenofovir); it is on this paradigm that almost all currently available antimicrobial therapies are based, and this approach has vastly improved our abilities to both treat and prevent many infectious diseases.

7.2.1   Limitations to the classical antimicrobial strategy

As well as heralding improved diagnostics and access to antimicrobial prophylactic and therapeutic drugs, the twenty-first century presents a specific set of new challenges for antimicrobial research (Cohen 2000). In particular, the emergence of new infections, the re-emergence of old infections and dwindling efficacy of established treatment protocols in the face of antimicrobial resistance require the expedited development of novel approaches.

Emerging and re-emerging infectious diseases

Emerging infectious diseases are those diseases that were previously unknown in man or known diseases whose incidence in humans has significantly increased in the past two decades. These diseases are frequently caused by zoonotic pathogens, i.e. those that originated in animals and then transferred to humans; examples include Nipah virus (Field & Kung 2011) and West Nile virus (Reed et al. 2003). In the last 50 years alone more than 335 emerging infectious diseases have been reported (Morens et al. 2004) (Figure 7.1).

Re-emerging infectious diseases comprise previously known diseases that have since reappeared after a significant decline in incidence and include newly evolved strains such as multi-drug-resistant tuberculosis (MDR-TB) or methicillin-resistant *S. aureus* (MRSA). Re-emerging pathogens can theoretically develop as a corollary of any number of environmental (e.g. climate change and deforestation) (Vittor et al. 2009), biological (overuse of antibiotics) (Blossom & McDonald 2007, Davies & Davies 2010) or societal (for example, disruption to disease control programmes) (Gayer et al. 2007) factors, which create novel biological niches and opportunities that can be rapidly exploited by pathogens (Cutler et al. 2010).

The increased 'globalization' of modern society, particularly increased global travel and trade, facilitates the spread of emerging and re-emerging infectious diseases and has been widely implicated as increasing the threat of epidemic disease affecting human, animal and plant populations (Hufnagel et al. 2004). A prominent example to illustrate the speed with which an apparently new pathogen can now spread across the globe is that of the severe adult respiratory syndrome (SARS) outbreak between November 2002 and July 2003. The SARS outbreak was caused by an emerging coronavirus, SARS-CoV.

The outbreak originated in Hong Kong, but within weeks SARS had disseminated globally, to 37 countries (Smith 2006). The epidemic resulted in 8422 cases and 916 deaths worldwide and an overall estimated case fatality rate of approximately 11 per cent (http://www.who.int/csr/sars/archive/2003_05_07a/en). For many of the diseases caused by emerging or re-emerging pathogens we currently have no drugs or vaccines available, or the drugs we do have are losing clinical efficacy. These limitations highlight the need for new antimicrobial strategies to fight infectious diseases.

Pathogen escape and drug resistance

Besides these challenges, the clinical utility of current antimicrobials can be limited by pathogen 'escape'. Pathogens can escape the activity of antimicrobials through the acquisition of characteristics that confer resistance, i.e. loss of drug efficacy, which provides them with a survival benefit in the presence of the drug. This can occur by a number of different mechanisms including mutation of drug-binding sites (e.g. mutations in penicillin-binding proteins and resistance to penicillins) or mutations at sites outside of drug-binding sites that nevertheless introduce changes in drug-binding sites (e.g. HIV-1 resistance to protease inhibitors), enzymatic degradation of drugs (e.g. beta-lactamases and degradation of penicillins), drug efflux, in which transporter proteins present in bacterial membranes export drugs out of pathogens' cytoplasm (e.g. resistance to tetracyclines), and bypassing of metabolic pathways targeted by antibiotics (e.g. resistance to sulfonamides). Resistance to currently available antimicrobials has been widely reported for a number of bacterial (including MRSA, *Mycobacterium tuberculosis*, *Pseudomonas aeruginosa* and *Chlamydia* spp.) and viral pathogens (HIV-1, influenza A virus) and hampers the development of future antimicrobials, especially for viral pathogens with RNA genomes that replicate rapidly or have low-fidelity polymerases to introduce mutations during replication (Goldberg et al. 2012, Theuretzbacher 2011). In some cases, pathogen strains are circulating that are now resistant to all our currently available most effective antimicrobial therapies. This was recently reported for the sexually transmitted pathogen *Neisseria gonorrhoeae* (Stoltey & Barry 2012), and is well known for extensively drug-resistant tuberculosis (Gandhi et al. 2010); restricted therapeutic options and treatment failures threaten to become more common. Moreover, we are now faced with the alarming prospect that resistance has evolved to every antibiotic ever used clinically, irrespective of chemical class or molecular target of the drug.

Silent populations and pathway redundancy

Pathogens can also escape the activity of antimicrobial drugs through the existence of metabolically inactive or 'persistent' sub-populations (Steuernagel & Polani 2010), which can be refractory to standard drug treatments and represent hard to diagnose reservoirs of infection that can later be reactivated (for example tuberculosis) (Koul et al. 2011). The herpes simplex 1 and 2 viruses, responsible predominantly for oral and genital herpes, respectively, adopt latent forms that can persist within cells for prolonged periods and which can be reactivated by certain stimuli. Reactivation, which cannot be prevented by currently available drugs, can cause devastating disease in immunocompromised individuals (Piret & Boivin 2011, Wilson et al. 2009). Also, HIV-1 integrates into the host genomic DNA in CD4+ immune cells, creating a replication competent drug-resistant viral reservoir that can undermine attempts to achieve an HIV-1 cure with current antiretrovirals (Smith et al. 2012).

Finally, the development of novel antimicrobials using the established method of targeting products of bacterial or viral genomes is also impeded by redundancies in pathogen systems. To illustrate this, it was recently reported that out of 700 salmonella enzymes, over 400 are non-essential for salmonella bacteria virulence, which reflects extensive metabolic redundancies and an ability to utilize surprisingly diverse host nutrients. The authors stated that this suggests a shortage of new metabolic targets for broad-spectrum antibiotics (Becker et al. 2006). These limitations highlight the need for new strategies in fighting infectious diseases.

7.2.2   Targeting host cell processes as a new antimicrobial strategy

A promising new strategy for antimicrobial drug development is based on the largely neglected role of the host cell in both the initiation and the progression of infection. From the initial contact of pathogens with their host target cells, to the successful coloniza-tion and completion of replication cycles, pathogens are, to varying degrees, dependent on host cells. Co-evolution of host and microbe has led to a high degree of pathogen specialization, ensuring maximal exploitation of beneficial host functions by pathogens and, at the same time, evasion of host immune responses (Woolhouse et al. 2002). This close interaction makes pathogens vulnerable to anything that inhibits the host func-tions on which they depend (Figure 7.2). The overwhelming importance of the host cell during infection provides ample novel opportunities to prevent and treat infections by targeting host cell factors and functions that support pathogens or are responsible for pathology upon infection.

There are several potential advantages to a drug development strategy that exploits the dependency of pathogens on their hosts by targeting host factors. First is the extension of possible drug targets that such an approach provides. For example, Karlas et al. (2010) identified in excess of 150 host factors that are required for influenza infection, which contrasts with the 11 proteins in total that are encoded by the influenza A virus genome. The majority of the factors identified in that study were essential for replication of two different H1N1 influenza viruses, including the pandemic swine-origin influenza A virus strain. A subset of these common factors was also required for efficient replication of the highly pathogenic avian H5N1 virus, indicating that these host factors have potential as targets for broad-spectrum antivirals.

Second, targeting host factors for antimicrobial drug development has an additional, particularly important advantage, over classical antimicrobial drug strategies. In con-trast to 'classical' antimicrobials, the development of resistance to drugs targeting host factors could reasonably be expected to be much slower. Whereas resistance to 'classi-cal' antimicrobials can be mediated by a single point mutation, or other comparatively simple mechanisms (Bright et al. 2005), resistance to a drug that targets a host fac-tor involved in a pathway that is essential for pathogen survival, and devoid of cellular redundancy, may require the pathogen to functionally supplement a complete pathway.

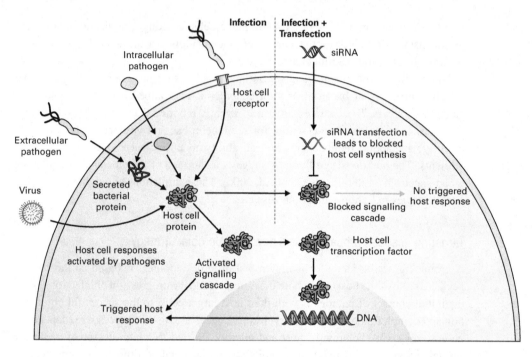

Figure 7.2   Host–pathogen interactions. Pathogens interact with host cells in different ways; they can trigger host cell signalling cascades via interaction with host cell receptors or via the action of secreted bacterial effector proteins. Activated host cell pathways lead to altered host responses, either directly or indirectly, by changing the gene expression profile of the host cell. This close interaction makes pathogens dependent on host cells, and renders them more vulnerable to anything that blocks or blocks access to the host functions on which they depend. Small interfering RNA (siRNA) can be used to block the synthesis of host proteins in host cells to study host–pathogen interactions.

There are already a limited number of antimicrobial drugs in the clinic that adopt a 'host-targeted' mechanism of action. A prominent example is the antiretroviral drug maraviroc. Maraviroc is an antagonist of the HIV-1 C-C chemokine receptor type 5 (CCR5) co-receptor. It blocks the interaction of the viral protein gp120 with CCR5 and consequently inhibits entry of the virus into cells. However, resistance to maraviroc has been observed both in vitro and in vivo and is suggested to be the result of HIV resistance mutations that permit virus binding to alternative sites on CCR5 (Berro et al. 2012). Novel CCR5-inhibitors (PRO 140, TNX-355) and inhibitors of another HIV-1 co-receptor, CXCR4, are currently being investigated (http://aidsinfo.nih.gov/drugs/), which illustrates the potential of this approach, even for the treatment of a pathogen that can rapidly evolve resistance, such as HIV-1.

However, this example also demonstrates that to promote and expedite the 'host-targeted' approach to antimicrobial development demands a complex understanding of host–pathogen systems (Figure 7.2) at the molecular level, which will lead to the identification of a broad range of host proteins that are essential for pathogen propagation but are dispensable for the host (or that can at least be targeted with a favourable therapeutic index). Modern post-genomic approaches, such as gene expression analysis or loss-of-function screening in high throughput, coupled with advances in high-content imaging technologies, hold real promise for achieving this on a large scale.

7.3   Post-genomic approaches for investigating host–pathogen interactions

Since the completion of the first human genome sequence in 2004 (Collins et al. 2004), whole-genome sequencing technologies have improved and vastly accelerated the collection of genomic information from a number of hosts and their pathogens. In this 'post-genomic era', and with the advances in high-throughput technologies, this information has been applied in high throughput for whole-genome expression profiling studies (Kellam et al. 2000).

Genome-wide microarray analyses of gene expression profiles in pathogens (Conway & Gary 2003, Leroy & Raoult 2010, Maeurer et al. 2007) and host responses to pathogens (Belcher et al. 2000, Rappuoli 2000) can provide quantitative and qualitative information concerning the role of pathogen virulence factors and host proteins in infection and disease (Brzuszkiewicz et al. 2011, Grimes et al. 2005). Analysis of the data can identify groups of genes that are important for the infection process and can implicate the activation or repression of key host regulatory pathways, some of which are common to infections with different pathogens (Pennings et al. 2008). More recently, the power of simultaneously analysing the host transcriptome and microbiome has been demonstrated by combining metagenomics analysis of genome sequences obtained directly from natural sources, e.g. microbial genomes from infected tissues, with microarray expression studies (Schwartz et al. 2012). Proteomic technologies have also advanced and now permit the quantitative global analysis of infection-induced changes to proteomes in host–pathogen systems (Holland et al. 2011, Shui et al. 2008, Zhang et al. 2005). In this way, the functions of previously uncharacterized genes and proteins can be predicted, virulence-associated genes identified and the effects of drug treatments investigated.

Following the discovery of RNA interference (RNAi), experimental techniques became available that allowed researchers to systematically shut down gene expression in cultured human cells – a powerful analytical loss-of-function strategy. In contrast to routine transcriptional profiling (microarray) and proteomics technologies (Garbis et al. 2005), RNAi loss-of-function approaches are more readily able to generate direct causal links between gene function and phenotype. Moreover, RNAi technologies were themselves rapidly adapted for use in high-throughput methods, to facilitate the analysis of genetic determinants on genome-wide scales.

7.3.1   High-throughput liquid handling for RNAi screening of infectious diseases

High-throughput screening (HTS) requires the precondition of well-established, reliable and highly reproducible liquid-handling procedures. For genome-wide RNAi screens, with multiple siRNAs per gene and several technical replicates, sample numbers easily exceed 200 000, making a high level of automation for experimental handling indispensable. However, while the establishment of fully automated procedures is labour intensive, semi-automated or manual approaches can be considered for small-scale screens. High-throughput screens of host–pathogen systems typically require many more liquid dispensing, transfer, washing and pipetting steps compared with a 'typical' cellular assay, i.e. in the absence of infectious agent.

Figure 7.3 shows two different examples of a HTS experimental set-up performed in our facility. The first involves high-throughput siRNA screening to identify host factors that have a role in the replication of a species, here, from the genus *Chlamydia*. These intracellular bacterial pathogens are highly dependent on host cell signalling pathways and host nutrients for the successful completion of their life cycles (Braun et al. 2008). *Chlamydia trachomatis* is a medically relevant pathogen responsible for the highest burden of sexually transmitted diseases caused by bacterial species and, as the cause of the devastating ocular infection trachoma, is a leading cause of infectious blindness. We established an experimental set-up and were able to identify 59 host factors involved in *C. trachomatis* survival and infectivity, using several quantifiable parameters simultaneously in a loss-of-function screen (Gurumurthy et al. 2010). We measured (1) cell number, to determine siRNA effects on cell viability and proliferation; (2a) number of *Chlamydia*-containing vacuoles in the primary infection; (2b) size of *Chlamydia*-containing vacuoles in the primary infection and (3) number of infectious progeny produced during primary infection (Figure 7.3A and B). Each experimental run was performed over 8 days, required triplicate plates for the different readouts and comprised more than nine liquid transfer steps. It is thus quite evident that such protocols demand high-precision liquid-handling capabilities. Also, even a comparatively simple approach, such as screening putative antiviral compounds for their ability to interfere with a viral replication cycle, can require complex liquid-handling protocols. We have investigated the influence of small molecules on influenza virus replication in cell culture (Figure 7.3C and D). In our example, cells are pretreated with drug candidates at eight different concentrations. After 2 h, cells are infected with influenza and the drug treatment is continued for a total of 36 h, at which point viral progeny particles are quantified. However, influenza virus infection of host cells requires special conditions, which demand changing of the cell culture medium and several wash steps prior to every step of the infection protocol. The final protocol comprises several liquid transfer steps, some of which require particular pipetting techniques; for example, thorough washing of cells without detaching them.

Typically, the standard microtitre formats used in HTS are 96- or 384-well plates. These have to be used in assays that measure pathogen progeny production as an output parameter. In such a procedure, progeny from the first cellular infection are transferred to cells in a second plate to measure infectivity. Where there is no requirement for measuring progeny production, cell arrays, where siRNAs are printed in single spots on glass slides, can be used. This new approach, with several thousands of siRNA spots per slide, has been successfully applied for the identification of novel host factors involved in *Trypanosoma cruzi* infection in a genome-wide, high-content siRNA screen (Genovesio et al. 2011). Since a genome-wide library with one data point per gene can be covered

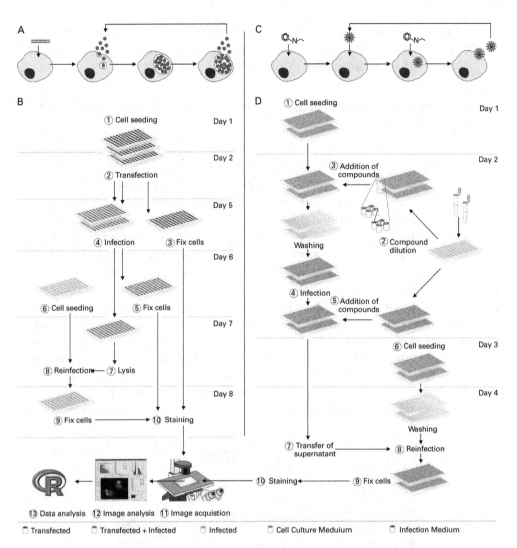

Figure 7.3 HTS/HCS design. (A) RNAi screening *Chlamydia trachomatis* (*Ctr*) infection in vitro: after siRNA transfection, host cells are infected with *Ctr* elementary bodies (EBs) (green), which differentiate into replicative reticulate bodies (RBs) (red). RBs multiply and re-differentiate back into infectious EBs. (B) Screening set-up design reflects the biphasic life cycle: (Steps 1–5) Cells are seeded and transfected in triplicate. 72 h after transfection, one plate is fixed to monitor effects of siRNAs on cell growth. Remaining plates are infected with *Ctr*, and 24 h later one plate is fixed to evaluate bacterial inclusion number and size. (Steps 6–9) 48 h later fresh cells are infected with lysates from unfixed plates and fixed after 24 h. (Steps 10–13) Host cell nuclei and *Chlamydia* are stained for detection and analysis (adapted from Gurumurthy et al. 2010). (C) Drug screening in influenza infection: host cells are pretreated with chemical compound candidates, infected and supernatants are used to infect fresh cells for detection of viral particles. (D) Screening set-up: (Steps 1–5) Compounds are serially diluted and cells are pretreated prior to infection followed by virus replication in compound presence. (Steps 6–8) After completion of viral replication, supernatants from treated, infected cells are used to reinfect fresh cells. (Steps 9–13) Cells are fixed; nuclei and virus are immunofluorescently stained for automated microscopy (11), and computationally analysed (12 and 13). A black and white version of this figure will appear in some formats. For the color version, please refer to the plate section.

using only a few slides, instead of using approximately 70 multi-titre plates, the advantages of cell arrays are reduced material and automation requirements and increased homogeneity of the derived data (Vanhecke & Janitz 2004).

7.3.2   Biosafety considerations for RNAi screening of infectious diseases: BSL2, BSL3 and BSL4

Experimental work with pathogenic microorganisms must be performed in accordance with strict biosafety procedures and levels of biocontainment in order to protect the working scientist and to exclude the possibility of unwanted pathogen release (Chosewood & Wilson 2007). There are four international standard biosafety levels that govern the degree of containment deemed appropriate for such work, with biosafety level BSL1 ranked as the lowest level of containment, up to BSL4, which is applicable to those pathogens that cause severe human disease for which there are no available treatments or vaccines. Adherence to BSL requirements is a crucial consideration when designing HTS protocols using BSL2, BSL3 or BSL4 pathogens. As it is very time-consuming and cost intensive to run robots for HTS in high-containment laboratories there are only a few fully automated screening facilities that are operational under BSL3 or BSL4 conditions (Figure 7.4) worldwide. Furthermore, there are additional challenges for high-content analysis using pathogens that require high levels of biocontainment. There are very few examples of high-content imaging facilities that are located within BSL3 and BSL4 laboratories, and pathogen-infected samples are typically inactivated before leaving high-containment laboratories for analysis. Pathogens can be inactivated

Figure 7.4   A high-throughput laboratory fulfilling BSL3 requirements. The liquid handler with two 384-channel pipetting heads (Beckman Coulter) and its periphery devices are placed in a custom-designed class II enclosure inside a BSL3 laboratory.

by fixing of infected cells, for example using formaldehyde, paraformaldehyde or methanol, and subsequent fluorescence staining and automated microscopy can be performed under less restrictive BSL1 or BSL2 conditions, presuming the inactivation procedure itself does not interfere with the assay. However, such procedures must be rigorously and exhaustively investigated for each individual pathogen to ensure no breach of biocontainment (Gene 2011). In particular, in addition to ascertainment of efficient inactivation of the experimental sample itself, one must also determine that decontamination of all materials leaving the high-security laboratory, such as plate lids or the outer surfaces of multi-titer plates, has been achieved. When established, however, such protocols can be safely and successfully employed for high-content endpoint analysis of BSL3 pathogens. Such a procedure was used by Karlas et al. (2010) for hit validation from a screen using a highly pathogenic influenza virus strain and others have used similar strategies of low-level laboratory automation for the analysis of fixed samples for the high-content analysis of high-biocontainment pathogens (Boerner et al. 2010, Brodin et al. 2010, Sessions et al. 2009).

An alternative to pathogen inactivation is to substitute the BSL3 or BSL4 pathogen of interest for a related pathogen that can be safely worked with at lower levels of biocontainment; for example, *M. tuberculosis* has been substituted with the Bacillus Calmette–Guerin (BCG) vaccine strain (Chosewood & Wilson 2007). There are also several reports of the successful use of pseudotyped viruses in HTS for high-containment pathogens such as the *Filoviridae* sp., ebolavirus. Pseudotyped viruses contain viral envelope glycoproteins – responsible for binding to the surface of host cells – from high-containment pathogens, but these are packaged along with the replicative machinery of a lower-containment, retroviral pathogen. These viruses can be safely used in lower-containment facilities in cellular assays (Basu et al. 2011). This approach does, however, have obvious limitations as it can be used only to identify host factors and screen compounds that interfere with the attachment or entry of the high-containment pathogen; moreover, pseudotyped viruses may exhibit altered morphology from the wild-type pathogen from which the viral glycoprotein derives (Saeed et al. 2010).

All these approaches have caveats – every measure of compromise that is introduced to simplify and reduce the cost of the experimental set-up can be assumed to shift the experimental system further away from the biological ideal and may adversely affect the outcome.

7.3.3   State-of-the-art high-throughput loss-of-function RNAi screening in host–pathogen systems

The first loss-of-function screens investigating the role of host factors during infection used *Drosophila melanogaster* cell lines as simple model systems to monitor the growth and development of intracellular bacteria following RNAi gene silencing of cellular genes (Cherry 2008). In 2005, two genome-wide studies applied high-content RNAi screening in *Drosophila* cells to identify host factors involved in the intracellular propagation of *Listeria monocytogenes*, but with different approaches. Agaisse et al. (2005) used high-throughput microscopy and automated image analysis to group

Drosophila SL2 cells into five different phenotypes after infection with a GFP-labelled *L. monocytogenes* strain; by contrast, Cheng et al. (2005) infected *Drosophila* S2 cells with *L. monocytogenes* wild-type and mutant strains and relied on visual inspection of samples and the manual flagging of samples with observable phenotypes after indirect immunofluorescence labelling of bacteria. Similar approaches in *Drosophila* S2 cells led to the identification of a CD36 family member as a host factor specifically involved in the uptake of *Mycobacterium fortuitum* (Philips et al. 2005), and to the finding that the mitochondrial proteins Tom40 and Tom22 are involved in the intracellular growth of *Chlamydia cavia* (Derré et al. 2007). The exceptional sensitivity of *Drosophila* cells to RNAi and their development as tractable models of infection for select mammalian pathogens made them valuable tools for early high-content infectious disease RNAi screens; however, they are not ideal biological models of mammalian host–pathogen systems in all respects. Thus, to complement and improve upon the *Drosophila* system for RNAi high-content screening, mammalian cell systems were developed, which promised more biologically relevant screening conditions. High-content RNAi screening of human cell infection was soon used to identify host factors crucial for viral replication and has contributed to our understanding of the infection, and pathogenesis, of West Nile virus, HIV, hepatitis C virus, influenza viruses and others (Brass et al. 2009, Coller et al. 2012, Karlas et al. 2010, Krishnan et al. 2008, Li et al. 2009, Zhou et al. 2008).

Viruses, by their very definition, are intrinsically highly dependent on their hosts; thus RNAi screening for host factors essential for viral propagation would be predicted to identify numerous putative drug targets. However, as was first shown in *Drosophila* cells, intracellular bacterial and protozoal parasitic pathogens – with varying degrees of metabolic independence from their hosts – are also dependent on host cell functions. Pathogens such as species of the protist genus *Plasmodium* which cause malaria, have complex life cycles that frequently include intracellular as well as extracellular stages, and different vector and host species. Whereas some parasitic life stages can be monitored comparatively easily to identify putative drug candidates, e.g. by measuring the quantity of parasitic DNA in *Plasmodium*-infected erythrocytes (Guiguemde et al. 2010), high-content analysis proved to be more precise for monitoring the role of host genes in the less tractable intracellular *Plasmodium berghei* liver stage (Prudêncio et al. 2008, Rodrigues et al. 2008), or *T. cruzi* infection (Genovesio et al. 2011). In addition to using determinants of pathogen function/viability as readouts for loss-of-function HCS, it is possible to focus entirely on aspects of host physiology, in particular those that are known to be important for infection pathogenesis; for example, we derived a time-resolved high-content cellular assay to monitor activation of the nuclear factor κB (NF-κB) inflammatory pathway during infection with the gastric pathogen *H. pylori* (Bartfeld et al. 2010); and we also used loss-of-function screening to investigate the inhibition of host cell apoptosis by *C. trachomatis* (Sharma et al. 2011), since apoptosis is a host response that may be important in determining infection pathology, especially in persistent chlamydial infections (Ying et al. 2007). In recent years, an abundance of high-content RNAi screens have contributed to our understanding of pathogens' infection strategies, in general terms and, in addition, specifically regarding facets such as pathogen survival, growth, dissemination and immune evasion.

7.4   Advanced high-content screening in pathogen research

Whereas some early high-throughput RNAi screens used homogeneous assay systems such as infection-dependent beta-galactosidase activity of a reporter cell line (Nguyen et al. 2006), quantifying luciferase reporter gene expression in cell lines or analysis of microscopy images soon became the method of choice in this field.

There currently exists a variety of commercially available assays that can be used to quantify typical cellular loss-of-function output parameters, such as cell proliferation, cell toxicity, DNA damage and apoptosis. However, HTS in infectious disease research often calls for the de novo establishment of assays for pathogen replication or specific host signalling pathways that are triggered after infection. Possible approaches that can be adapted for use in HTS cellular assays include the use of luciferase reporter constructs, fluorescently tagged proteins and fluorescent antibody staining. Luciferase assays are simpler to process, since they provide one value per well of a plate assay, which can be used in downstream statistical analyses; however, data from HCS automated imaging using fluorescent and other readouts can be exploited in a manner that provides information for an increased number of image features that more completely describes the biological phenotypes; thus this approach often is worthy of the extra time and effort that must be devoted towards establishing new protocols (Swedlow et al. 2009). Which technique is most suitable to the particular system being investigated depends upon the scientific question that is being addressed, and also on the amount of personnel and financial investment possible; generating luciferase reporter constructs or tagged proteins can be laborious, whilst antibody staining is more expensive and antibody batch-to-batch variation has to be taken into consideration. Depending on assay requirements, however, antibody staining may be the only option, e.g. for genetically intractable pathogens.

Modern automated high-throughput microscopy in combination with a broad variety of different fluorescent staining approaches are currently most frequently used in screening, and screens based on non-fluorescent reporter constructs, such as luciferase expression, tend to be in the minority. Only a few of the early publications in this field used manual inspection of microscopy images to evaluate data, and commercial or open-source automated image analysis software bundles, such as CellProfiler, Fiji, KNIME, ScanꞰ or Matlab, which are generally less costly in terms of man–hours and less subject to human bias, are current standards (Conrad & Gerlich 2010). Whilst a typical image analysis pipeline for an adherent cell assay mostly consists of image background correction followed by object detection (nucleus, cell) and object quantification steps, image analysis for HCS of cell infection adds another level of complexity. Such complexity derives from the variation of recovered infectious progeny (infectivity) even within a seemingly homogeneous cell population, strain-to-strain and batch-to-batch variability in the pathogen sample and, more importantly, the often complicated life cycles of pathogens for which HCS assays must be adapted such that they can accurately probe very complex biological processes, e.g. cell-to-cell spread of *L. monocytogenes* (Chong et al. 2011).

Cellular HCS assays quantifying the effects of gene loss-of-function typically use complex phenotypic endpoint measurements, often on a multi-parameter level, which places high demands on image acquisition and image bioinformatics. A genome-wide siRNA screen for 20 000 genes covered by two to four siRNAs, for which several images and fluorescent channels might be taken, will easily exceed 200 000 images, which must be stored and analysed automatically using image analysis software. Thus, to realize the full potential of HCS, specialist knowledge and skills are still required.

When comparing host gene hit lists identified by different groups in cellular-infection RNAi screens there is often limited consensus between published 'hit lists'. This can be seen, for example, in screens for HIV 1 (Bushman et al. 2009) and for influenza virus replication (Min & Subbarao 2010). The lack of consensus can be attributed to differences in individual screening protocols, such as the use of alternative cell lines and pathogen strains, choice of siRNA libraries etc., but are also likely to derive from differences in chosen readouts, e.g. imaging features and data analysis procedures. HCS can contribute to improving hit consensus in screening data by providing a more complete phenotypic description of the gene knockdown. As an example of what can be achieved with high-content multi-parameter output, Fuchs et al. (2010) recently published the use of multi-parametric phenotyping in a genome-wide RNAi screen applying automated image analysis to assign cell morphological changes at the level of individual cells. The authors assigned 51 numerical descriptors of targeted genes to functional modules. Even though not every one of these descriptors represented a classic 'biological' feature, together they formed a vector precisely characterizing the state of cells, which could be used to predict biological function.

7.4.1   Analysing influenza viral replication

The identification of host factors and signalling networks that are potential bottlenecks for the propagation of viral pathogens provides the basis for a new strategy towards development of future antiviral drugs that are effective against medically important viruses, such as seasonal and pandemic influenza viruses. Several groups have now published RNAi screens analysing the influence of host cell factors on influenza virus replication, using a variety of technical approaches.

Initially, the effects of host factor knockdown on influenza replication were investigated using *Drosophila* cells (Hao et al. 2008). However, the biological relevance of the factors identified in such studies for infection of mammalian cells is not always immediately apparent; for example, flies are not natural hosts of influenza virus, and a modified virus that was able to infect *Drosophila* cells had to be used in this experimental system. Moreover, flies do not express all the proteins required for influenza virion assembly and infectivity. Nevertheless, studies such as this can be used to demonstrate feasibility of an approach and can provide initial insights – the authors of this study also confirmed the relevance of some of the targets identified in *Drosophila* cell for influenza virus infection of mammalian cells.

Recently, the results of several genome-wide RNAi screens, using different approaches, in human cells have generated target information for factors that modify the influenza virus life cycle (Brass et al. 2009, Karlas et al. 2010, König et al. 2009). Using fluorescence microscopy Brass et al. (2009) quantified the influenza envelope glycoprotein, hemagglutinin, as a marker of influenza A virus infection in siRNA-treated osteosarcoma cells, and identified 120 host-dependency factors, implicating various cellular functions including endosomal acidification and RNA splicing in influenza virus replication. In an alternative approach, König et al. (2009) infected siRNA-transfected A549 cells, a human lung epithelial cell line, with a recombinant influenza A virus in which the coding region of haemagglutinin was replaced by a luciferase reporter gene. Using this method they identified 295 host factors with a role in the viral life cycle; however, the absence of functional haemagglutinin protein limited this study to the identification of those genes involved in the early stages of infection, and viral RNA transcription and translation, but could not identify factors involved in virus assembly, budding and release. Karlas et al. (2010) used an approach that enabled the identification of 287 mammalian cell host factors that are important in the early and late stages of the influenza A virus life cycle. The authors applied a method in which, first, siRNA-treated A549 human lung carcinoma cells were infected with influenza A viral strains and cells were immunostained for viral nucleoprotein as a marker of infection; then, second, the culture supernatants from infected A549 cells were transferred to 293T human embryonic kidney reporter cells containing an inducible influenza A virus-specific luciferase construct. Figure 7.5 shows the microscopic image analysis workflow Karlas et al. (2010) used for monitoring influenza A virus infection in A549 cells during siRNA screening. This protocol consisted of two steps: first, cell nuclei were defined using DNA Hoechst staining, followed by detection of the Cy3 signal (corresponding to a Cy3-conjugated secondary antibody detecting viral nucleoprotein) in the nucleus and cytoplasm. If the Cy3 signal exceeded a certain level in a cell, then the cell was counted as influenza infected.

7.4.2   Detecting intracellular bacterial replication using antibody staining

Intracellular bacterial pathogens are etiologies of many clinically important diseases including tuberculosis (*M. tuberculosis*), legionellosis (*Legionellosis pneumophila*) and enteritis (salmonellae). Typically, these pathogens have evolved to maximally exploit host cells for the establishment, support and propagation of infection, whilst simultaneously evading host immune responses. Due to the intimate relationships they assume with host cells, intracellular bacterial pathogens are ideal candidates for loss-of-function screening. *Chlamydia trachomatis* is an intracellular bacterial pathogen that has an extraordinary biphasic life cycle, alternating between extracellular metabolically inactive 'elementary bodies' and intracellular metabolically active reticulate bodies (RB). Chlamydial RBs reside in an intracellular vacuole termed the 'inclusion'. Depending on the chlamydial strain, inclusions can last up to three days in cells and can reach considerable size (Figure 7.6). Several challenges, therefore, face automated image analysis for such screens; in particular, for image analysis software it is almost impossible to differentiate between the nucleus of a cell and a late phase inclusion as both are detected using DNA stains such as 4',6-diamidino-2-phenylindole (DAPI). Workarounds are possible to enable the software to differentiate between a true cell nucleus and a bacterial inclusion. An example of such a workaround is to use chlamydial-specific staining, e.g. an

a   Hoechst staining (DNA)

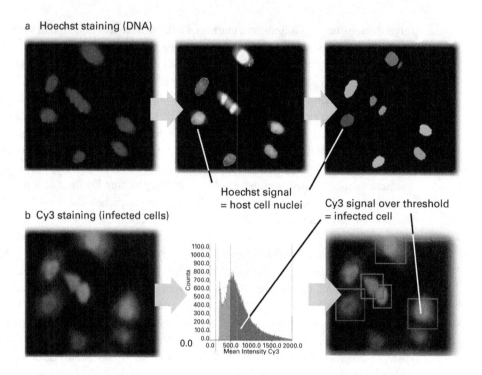

Hoechst signal
= host cell nuclei

Cy3 signal over threshold
= infected cell

b   Cy3 staining (infected cells)

Figure 7.5   Detecting influenza-infected cells. (a) Nuclei are detected using Hoechst staining. (b) Influenza virus infected cells are detected based on the cytoplasmic Cy3 signal (corresponding to a Cy3-conjugated secondary antibody detecting viral nucleoprotein). A black and white version of this figure will appear in some formats. For the color version, please refer to the plate section.

antibody against the major outer membrane protein, after initial DNA (DAPI) staining. From the DAPI and the pathogen-specific channels it is then possible to construct a so-called 'virtual channel' in which the inclusions are represented as 'black holes'. Cell nuclei can be discriminated from inclusions on this 'virtual channel' (Figure7.6). The additional level of complexity added by using intracellular pathogens in a common cell-based system often calls for additional image and data analysis steps to clearly separate pathogen and host cell image information.

7.4.3   Analysing translocation of host-cell signalling factors triggered by infection

Besides identifying individual host factors that influence pathogen replication, loss-of-function HCS can be used to investigate the role of entire host cell signalling pathways in cellular infection. A prominent example is the NF-κB signalling pathway, which is triggered by a variety of pathogens (Neish & Naumann 2011). This is a transcription-ally active protein complex that is a key mediator of both innate and adaptive immune responses to pathogens. The mammalian NF-κB protein family includes the five sub-units p65, p50, p52, RelB and c-Rel. In the inactivated state, these subunits reside in the cytoplasm sequestered by their inhibitors, such as the inhibitor of κBα (IκBα). Upon

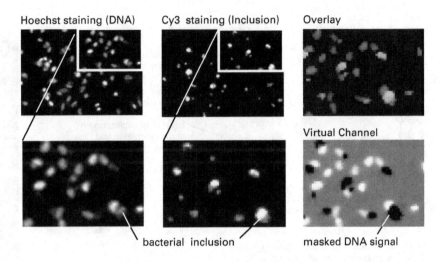

Hoechst staining (DNA)   Cy3 staining (Inclusion)   Overlay

Virtual Channel

bacterial inclusion   masked DNA signal

Figure 7.6   Definition of a virtual channel. High-content assays can include so-called 'virtual channels' in which two distinct channels are combined into a new channel. To avoid the erroneous detection of inclusions as nuclei, the channel associated with DNA Hoechst staining, which includes both cellular nuclei and *C. trachomatis* inclusions, is overlayed with an inclusion-specific channel that is derived from a fluorescently labelled bacteria-specific antibody (e.g. Cy3 labelled). The latter signal is subtracted from the Hoechst signal, leading to a masked DNA signal where inclusions are located, therefore avoiding false nuclei detection.

activation of the NF-κB pathway, for example via the activation of toll-like receptors (TLRs) by pathogen-associated molecular patterns (PAMPs), the inhibitor is degraded to enable the NF-κB subunits to translocate into the nucleus where they activate their target genes (Hoffmann & Baltimore 2006).

High-content screening assays can be used to analyse the cellular localization or translocation of proteins from the cytoplasm to the nucleus upon infection. To these ends, stably integrated genetic constructs can be used to couple the protein of interest to a fluorescent protein such as green fluorescent protein (GFP), or alternatively, protein localization and phosphorylation levels can be monitored using specific fluorescently labelled antibodies (Figure 7.7b). As an example of an automated image analysis assay measuring the NF-κB subunit p65, cells are seeded into multi-titer plates, stimulated, fixed and nuclei stained with Dapi/Hoechst. In the example shown in Figure 7.6, objects were detected using image analysis software based on the DAPI/Hoechst staining, defined as 'nuclei' and the surrounding area based on a virtual masking ring was then defined as 'cytoplasmic'. In a second step the detected objects were filtered based on the measured features 'area' and 'mean object intenstity' allowing the differentiation between smaller staining artifacts and real nuclei objects. For each real nuclei object the log2 ratio for the quantified GFP signal in the nuclei versus cytoplasmic area was calculated. This ratio was then used to define whether a cell is 'activated' or not. The percentage of activated cells was finally used as the basis for the analysis of the screening data for the translocation of p65 to cellular nuclei.

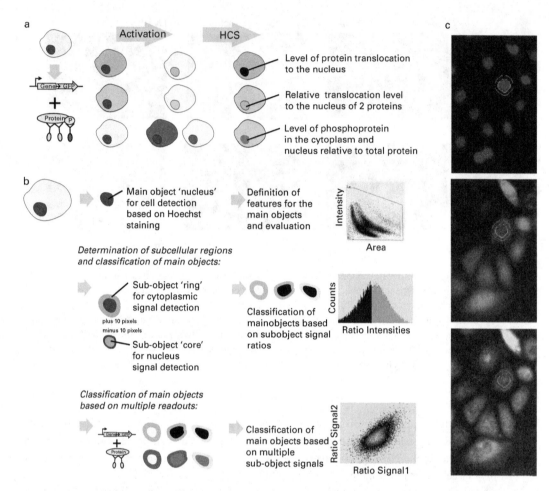

Figure 7.7   Analysing host-cell protein translocation and phosphorylation events. High-content screening assays can be used to analyse the translocation of host-cell factors, such as NF-κB or MAPKs, from the cytoplasm to the nucleus upon activation. (a) To monitor translocation events cell lines used for screening can be stably genetically modified using a fluorescent protein, such as GFP, coupled to the protein of interest. Alternatively, antibodies can be used to monitor changes to the total, and phosphorylation, levels of target proteins. Following activation of cells with a selected stimulus, it is possible to quantify protein nuclear translocation events, as well as phosphorylation events in the nucleus or cytoplasm. In addition, high-content analyses can be used to determine ratios of, for example, the phosphorylated protein species versus its total protein level. (b) An example workflow for automated analysis of microscopy images. First, Hoechst staining is used to define cellular nuclei, the 'main object'. Based on certain features, including size, intensity and area, cell nuclei staining can be evaluated and erroneously detected objects can be excluded from subsequent analyses. For the analysis of host-cell events, subcellular objects, such as 'core' and 'ring' objects, are defined based on the main object. Classification of the main object (nuclei, cell) can then be performed using signal ratios for the defined sub-objects. (c) Cells stained with Hoechst dye (top image) and a NF-κB–GFP reporter construct (middle image) and antibody staining for total- and phospho-MAPK (bottom image). A black and white version of this figure will appear in some formats. For the color version, please refer to the plate section.

There are published examples describing the use of HTS to study the role of host factors during infection that contribute to NF-κB activation. Kim et al. (2010) used RNAi screening to investigate the role of host factors required for nuclear translocation of NF-κB p65 during infection with the intracellular bacterial pathogen *Shigella flexneri*, a cause of bacillary dysentery. This pathogen enters epithelial cells thereby activating an extensive inflammatory response. *Shigella flexneri* infection activates the NOD-1-dependent nuclear translocation of NF-κB p65 and upregulation of inflammatory genes, including IL-8. The authors infected epithelial cells (pretreated with siRNAs) with DsRed *S. flexneri*, stained with an anti-p65 antibody and Hoechst and quantified p65 nuclear translocation automatically by defining the cytoplasmic area and generating a nuclear/cytoplasmic p65 ratio. Combined with enzyme-linked immunosorbent assay (ELISA) analysis of IL-8 secretion, they could assign a role for IKKα as a host factor required for NF-κB p65 nuclear translocation and IL-8 secretion during *S. flexneri* infection. Gewurz et al. (2012) performed a genome-wide screen for human cellular factors that regulate the activation of the NF-κB pathway by the Epstein–Barr virus latent membrane protein (LMP1) – a known oncoprotein that activates NF-κB. They screened HEK293 cells containing an inducible LMP1 construct transfected with siRNAs. Activation of NF-κB was quantified using a NF-κB GFP reporter and fluorescent-activated cell sorting analysis. The authors of this study identified 155 cellular factors implicated in NF-κB activation by LMP1, many of which were also required for NF-κB pathway activation by the cytokines IL-1β and tumour necrosis factor α (TNFα). Bartfeld et al. (2010) developed a methodology that can be used in HTS applications to investigate NF-κB pathway activation dynamics. Using lentivirally transduced p65–GFP monoclonal cell lines (human lung, A549; human gastric, AGS; murine fibroblast, L929) the authors investigated NF-κB activation dynamics, by automated microscopy and image analysis of p65 nuclear translocation, in response to different stimuli, including infection with the gastric pathogen *H. pylori*, which was observed to induce oscillations in p65 nuclear translocation.

7.4.4 Analysing phosphorylation of host-cell signalling factors triggered by infection

Besides the translocation of host-cell factors such as NF-κB into the nucleus, the triggering of cell signalling pathways can also be detected by monitoring phosphorylation events. A prominent example is the mitogen-activated protein kinase/extracellular signal-related kinases (MAPK/ERK) pathway, which has important roles in oncogenesis and infection (Chuluunbaatar et al. 2012, Feng et al. 1999, Gurumurthy et al. 2010).

Activation of all MAPKs is regulated by a central three-tiered core signalling module, comprised of an apical MAPK kinase kinase (MAP3K), a MAPK kinase (MEK or MKK) and a downstream MAPK. Phosphorylation of ERK1/2 by MEK1/2 results in its translocation to the nucleus where it phosphorylates transcriptional factors such as those that drive the early response of cells to epidermal growth factor (EGF) (Lenormand et al. 1993). Interestingly, activated ERK1/2 regulates growth-factor-responsive targets in the cytosol, but also is able to translocate to the nucleus where it phosphorylates a number of transcription factors regulating gene expression. This double role of ERK1/2 presents a challenge for screening image analysis as, besides its nuclear translocation, the phosphorylation level of ERK1/2 in the cytoplasm and in the nucleus has to be monitored. To achieve this, we have developed a double antibody staining protocol. While a total ERK antibody is used to monitor the ERK level itself and its translocation to the nucleus, the use of a phosphorylation-specific antibody facilitates monitoring of the ERK1/2 phosphorylation level in the cytoplasm and the nucleus (Figure 7.7a and c). This enabled us not only to monitor the translocation of total and phosphorylated ERK1/2 into nuclei, but also the phosphorylation level in each cell in the cytoplasm and the nucleus, thus providing deeper insights into the potential functionality of the involved host cell factors.

7.4.5   Measuring complex cellular phenotypes during infection

High-content analysis can be used to analyse any number of visual output parameters in host–pathogen systems that have biological significance. For example, the size of a chlamydial vacuole directly corresponds to the growth of this intracellular pathogen, whereas the number of vacuoles or the level of secondary infection are representative of other biological facets, i.e. invasion and productivity of infection, respectively (Gurumurthy et al. 2010). In another example, the number and average size of *Plasmodium merozoites* have been regarded as reflecting invasion and intracellular growth, respectively (Rodrigues et al. 2008).

A challenging example for imaging analysis is the infection of epithelial cells by *Neisseria* spp., which are facultative intracellular bacterial pathogens of medical importance – *N. gonorrhoeae* is the cause of the sexually transmitted infection gonorrhoea. Piliated *Neisseria* bacteria attach to epithelial cells and form aggregates, or microcolonies, on their surface. Cellular factors, such as actin, are recruited to the area under the microcolonies. Only a very small proportion of these bacteria is actually invasive, the majority only attach to cells and do not invade (Boettcher et al. 2010). Multi-channel fluorescence microscopy can be used to monitor several biological processes involved in *Neisseria* infection simultaneously (Figure 7.8). First, cellular DNA content is stained with Hoechst to monitor cell number. This parameter provides information on siRNA cytotoxicities and the modulation of the cell cycle, and is needed for normalization of other parameters. Since microcolonies can resemble cellular nuclei in terms of size, shape and Hoechst signal intensity, certain informatics methods have to be applied to distinguish microcolonies from nuclei (refer to Section 7.4.2). In a second fluorescence channel, GFP-expressing *N. gonorrhoeae* are used to monitor the attachment of bacteria to the cells, as well as the number and size of microcolonies. Labelling exclusively extracellular bacteria with an additional fluorescent marker serves to distinguish between extra- and intracellular bacteria. The fourth fluorescence detection channel is used to monitor cellular actin recruitment to microcolonies. With this rather complicated assay, four different fluorescent markers are applied in parallel to monitor six different parameters, each one of them a true representative of a distinct biological process. Therefore, many aspects of host–pathogen interactions are elucidated within the same experiment (Figure 7.8).

Actin (Cy3) Nuclei (Hoechst) Overlay (Cy3, GFP and Hoechst)

All bacteria (GFP) Extracellular bacteria (Cy5)

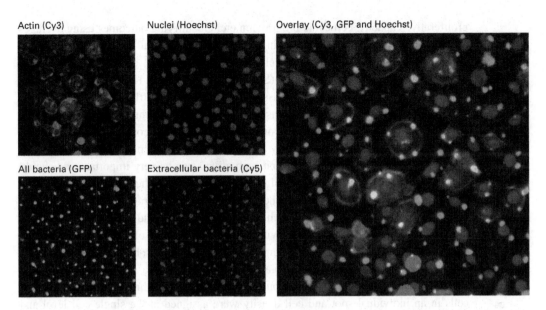

Figure 7.8 Parallel detection of several independent parameters. Several fluorescent channels can be used in parallel for simultaneous analysis of many biological features. In our example, four fluorescence channels are used to monitor different facets of *N. gonorrhoeae* epithelial cell monolayer infection: GFP-expressing bacteria are used to quantify bacterial attachment to the host cells, and formation of aggregates (microcolonies); additionally, external bacteria are labelled with Cy5 to distinguish intra- from extracellular bacteria. Cy3 labelling is used to detect the host actin cytoskeleton, and Hoechst staining to detect cellular nuclei. The overlay gives an impression of the overall complexity (courtesy C. Lange and C. Rechner). A black and white version of this figure will appear in some formats. For the color version, please refer to the plate section.

7.5 Single-cell population analyses in high-content screening

In general, sample-based or control-based normalization methods of output parameters in HCS use the mean or median output values (e.g. fluorescence intensity) from a measured cell population (i.e. a plate well) for each treated sample as the most straightforward normalization method. However, the application of mean values can result in a loss of biological information and can adversely affect statistical power and functional analysis as it fails to account for individual cell-to-cell variability (Snijder & Pelkmans 2011). Alternatively, one can perform functional analyses at the single-cell population level, thus taking account of intrapopulation cell-to-cell variability. Single-cell analysis in high-content loss-of-function screening is a technically challenging but advancing method.

Imaging and computational methods at the single-cell level can be used to account for population-context effects in cellular assays. Population-context effects are representative of the phenotypic heterogeneity in populations of cells that results from responses to intracellular and extrinsic stimuli; for example, cell density will vary throughout a population of cells, which will have consequences for cell size and proliferative capacity, cell cycle progression etc. Quantifying outcomes using averages of entire cell populations, i.e. averaging out cell-to-cell variability, rather than collecting data from multiple, individual cells within a population, disregards biologically significant population-context effects (Snijder & Pelkmans 2011), which may be particularly relevant for cellular infection assays (Suratanee et al. 2010), and may adversely affect data reproducibility and consensus of hit list targets identified between research groups. The relevance of population-context effects for the outcomes of functional analyses has, thus far, been addressed by a limited number of reports. However, Snijder et al. (2009) recently demonstrated that such effects can be very important and should be accounted for during experimental design and analysis. The authors of that report investigated how cell intrapopulation heterogeneity affected viral infection. Population-context-dependent cell properties, such as cell location, size and density, were found to be important factors for determining viral infectivity. In terms of loss-of-function analyses, Knapp et al. (2011) recently demonstrated population-context effects were very important in an RNAi screen investigating host factors with a role in hepatitis C and dengue virus infection. Population-context cell features including cell size, number of cells in an individual spot and cell density were assigned at the single-cell level and were shown to influence efficiency of viral infectivity, and hence be important for functional analysis of RNAi data. They report that such a method improves sensitivity and reproducibility, and propose a novel method to analyse high-throughput, high-content cellular assays, based on single-cell measurements including normalization of the measurements from each individual cell against its population context. Most recently, in a very comprehensive study, Snijder et al. (2012) and co-workers convincingly showed that accounting for population-context effects during RNAi loss-of-function experiments improved hit list consensus between replicate screens and screens performed by different laboratories. The authors investigated the contribution of population-context effects to infection phenotypes during RNAi experiments using small-scale (49 human kinases) and large-scale (approximately 7000 genes) experimental systems with three different HeLa cell lines and 17 different mammalian viruses (consisting of DNA/RNA and enveloped/non-enveloped viruses). In total, they calculated 200 quantitative image-based parameters for 2.4×10^9 individual cells, from which, using supervised machine learning, each cell could be assigned to corresponding phenotypic classes. They found that in many instances, infection phenotypes (e.g. increased or decreased infectivities) were due, in large part, to cell population-context effects. Since RNAi itself perturbs cell population-context-dependent properties, siRNA treatment might increase cell growth, which would increase population size and local density, a large proportion of the infection phenotypes resulting from siRNA treatment could actually be traced to population-context effects (indirect effects) rather than direct effects of gene knockdown on infection. A particular complication for true hit identification in siRNA screening observed by the authors was masking, in which a direct effect (most often the experimentally desired effect) of RNAi knockdown on virus infectivity was counteracted by an opposite population context-dependent (indirect) effect on infectivity. For example, BR serine/threonine kinase (BRSK1) knockdown inhibited human rotavirus 2 infection; however this effect was masked in the overall outcome of infection by an indirect effect of one siRNA used to target the BRSK1 gene as this siRNA increased the relative number of cells at islet edges, which human rotavirus 2 preferentially infects. They reasoned that indirect effects of RNAi knockdown on infection could be predicted, based on single-cell analysis of population-context-dependent properties, and showed that taking indirect effects of RNAi knockdown into consideration during the analysis of screening data improved consensus/compatibility in screening data, i.e. hit lists containing only direct hits compared favourably, whereas total (direct and indirect hits) hit lists correlated poorly. Such approaches are likely to be particularly useful for selecting which hits should be used for further validation and investigation, which frequently involves time-consuming in vitro and in vivo investigations; for example, accounting for population-context effects in RNAi screening of genes involved in SV40 virus infection improved functional annotation of hits and consensus between screening data and reduced the target list to genes that have a known role in propagation of this virus, excluding annotations that had a more general role in cell physiology, e.g. in this system, alternative splicing.

Based on recent reports, therefore, single-cell analyses in HTS for loss-of-function investigations of host–pathogen systems is likely to develop as an important method in the future, where it will offer new opportunities for cell-population-dependent normalization methods and hit definition.

7.6   Future directions

Although HCS in infectious disease research is still in its infancy, the discussion we have presented here demonstrates that these early explorations have provided remarkable new insights into the role that host cells, and host factors, play during infection with pathogens. However, if the preliminary successes of these new technologies are to be improved upon and sustained in academic and pharmaceutical research, several challenges must be addressed. A particularly important obstacle that must be navigated for future applications of such technologies is the seeming lack of hit consensus of HTS, HCS – optimizing hit list consensus between different laboratories is a major task. Improvements in this regard could be achieved by the introduction of infrastructures that enable researchers to deposit protocols, reporter cells, imaging raw data, data pipelines and analysed data; moreover, there is a need for the establishment of a set of minimum screening standards, which should include common siRNA control sequences and standard protocols, for example.

To date, siRNA loss-of-function screening on a genome-wide scale has been limited to a small number of screening centres, predominantly because of the requirements for sophisticated liquid-handling techniques, expensive microscopy units and expert knowledge for the statistical analysis of raw images and data that are collected. Performing a whole-genome siRNA screen is also a huge financial investment for academic research groups; for a primary screen up to 80 000 different sequences must be analysed, in multiple biological replicates. The subsequent requirement for a second, extensive hit validation round further increases screening costs. Therefore, to increase the accessibility of such technologies for non-screening laboratories, in addition to more stringent statistical analytical tools, there is high demand for technologies that enhance the miniaturization of cellular assays that can be applied in high-throughput, high-content protocols. Besides the use of 1536-well plates, the use and development of cell-based arrays should be promoted, if technically possible. Decreased screening costs would enable the broader application of these technologies. An alternative approach target identification in drug development is to computationally predict activities of drug candidates, i.e. their toxicity, mutagenicity and solubility profiles, in virtual screening (Seidel et al. 2011). This approach can lead to a significant reduction in the number of chemical compounds to be initially screened experimentally and can thus reduce man–hours and financial costs. So far, several screening approaches exist, namely siRNA-based loss-of-function, small compound and virtual screening. However, as siRNA-based screening enables identification of proteins involved in biological processes directly, combining high-throughput RNAi-based loss-of-function with the latter two could enable stringent lead discovery pipelines, preserving financial input as well as manpower. Still, this approach would need the combination of high-tech approaches and in-depth, specialist knowledge from different areas. This drug discovery pipeline would use the benefits of these three high-content approaches and could unleash the development of new lead drug candidates, not only for infectious disease research, and will likely feature more often as this field develops.

Currently, most HTS in cell-based assays uses established cell lines, typically of human origin; this applies for both RNAi loss-of-function as well as chemical small-compound screens. Most of the cell lines used, however, are derived from tumours or have been immortalized by transforming genes, thus allowing permanent growth characteristics in vitro. Cell lines are tractable in vitro models that have, amongst other features, the added advantages of typically homogeneous growth and ease of transfection and/or transformation by stable genetic modification, e.g. using lentiviral constructs. However, such in vitro adapted permanent lines almost always exhibit altered cellular morphology and disrupted signalling pathways. In particular, besides the expression of oncogenes and the depletion of tumour suppressor genes, redundant parallel signalling routes have been shut off, leading to an artificially enhanced sensitivity of cells towards specific siRNA or chemical inhibitors. Moreover, genetic anomalies in transformed cell lines can be numerous (approximately 30 000 mutations in tumour cells), which can involve large numbers of pathways. As a result, part of the downstream validation of siRNA screening hits involves investigating target function in different cellular systems, to exclude cell-line-specific artifacts. One of the great challenges for the future of these technologies is the establishment of high-throughput assays that are compatible with primary human cells. Future solutions may involve the use of adult human stem cells, which may be effectively and permanently propagated in culture (Barker et al. 2010). However, applying such systems to high-throughput RNAi will also require optimization of transformation protocols, since primary cells are often refractory to transformation with nucleic acids.

Current HCS protocols largely rely on fixed endpoint analysis, and averaging across cell populations. Another challenge, therefore, is the development of image and data analysis tools that can be used to follow changes over time at the single-cell level in whole-cell populations for a huge number of samples. Such techniques will provide more in-depth insights into cell signalling and temporally regulated processes during infection.

RNAi has, thus far, not proven to be directly suitable for treatment of human disease in vivo; however, it has proven to be an invaluable tool in biological research for its possibility to systematically identify host proteins that are involved in specific biological processes. Combining high-throughput loss-of-function screening approaches with virtual in silico analysis and small-compound screening will generate cost-effective drug development pipelines that can be used to identify small compounds that target host factors identified by RNAi screening. Taken together, HTS has the potential to be one of the major research tools in the next decade.

References

Agaisse, H., Burrack, L., Philips, J., Rubin, E., Perrimon, N. et al. (2005), 'Genome-wide RNAi screen for host factors required for intracellular bacterial infection', *Science* **309**(5738), 1248–1251.

Barker, N., Bartfeld, S. & Clevers, H. (2010), 'Tissue-resident adult stem cell populations of rapidly self-renewing organs', *Cell Stem Cell* **7**(6), 656–670.

Bartfeld, S., Hess, S., Bauer, B., Machuy, N., Ogilvie, L. et al. (2010), 'High-throughput and single-cell imaging of NF-kappaB oscillations using monoclonal cell lines', *BMC Cell Biology* **11**(1), 21.

Basu, A., Li, B., Mills, D., Panchal, R., Cardinale, S. et al. (2011), 'Identification of a small-molecule entry inhibitor for filoviruses', *Journal of Virology* **85**(7), 3106–3119.

Becker, D., Selbach, M., Rollenhagen, C., Ballmaier, M., Meyer, T. et al. (2006), 'Robust salmonella metabolism limits possibilities for new antimicrobials', *Nature* **440**(7082), 303–307.

Belcher, C., Drenkow, J., Kehoe, B., Gingeras, T., McNamara, N. et al. (2000), 'The transcriptional responses of respiratory epithelial cells to *Bordetella pertussis* reveal host defensive and pathogen counter-defensive strategies', *Proceedings of the National Academy of Sciences of the USA* **97**(25), 13 847–13 852.

Berro, R., Klasse, P., Moore, J. & Sanders, R. (2012), 'V3 determinants of HIV-1 escape from the CCR5 inhibitors Maraviroc and Vicriviroc', *Virology* **427**(2), 158–165.

Blossom, D. & McDonald, L. (2007), 'The challenges posed by reemerging *Clostridium difficile* infection', *Clinical Infectious Diseases* **45**(2), 222–227.

Boerner, K., Hermle, J., Sommer, C., Brown, N., Knapp, B. et al. (2010), 'From experimental setup to bioinformatics: an RNAi screening platform to identify host factors involved in HIV-1 replication', *Biotechnology Journal* **5**(1), 39–49.

Boettcher, J., Kirchner, M., Churin, Y., Kaushansky, A., Pompaiah, M. et al. (2010), 'Tyrosine-phosphorylated caveolin-1 blocks bacterial uptake by inducing Vav2-RhoA-mediated cytoskeletal rearrangements', *PLoS Biology* **8**(8), e1000457.

Brass, A., Huang, I., Benita, Y., John, S., Krishnan, M. et al. (2009), 'The IFITM proteins mediate cellular resistance to influenza A H1N1 virus, West Nile virus, and dengue virus', *Cell* **139**(7), 1243–1254.

Braun, P., Al-Younes, H., Gussmann, J., Klein, J., Schneider, E. et al. (2008), 'Competitive inhibition of amino acid uptake suppresses chlamydial growth: involvement of the chlamydial amino acid transporter BrnQ', *Journal of Bacteriology* **190**(5), 1822–1830.

Bright, R., Medina, M., Xu, X., Perez-Oronoz, G., Wallis, T. et al. (2005), 'Incidence of adamantane resistance among influenza A (H3N2) viruses isolated worldwide from 1994 to 2005: a cause for concern', *The Lancet* **366**(9492), 1175–1181.

Brodin, P., Poquet, Y., Levillain, F., Peguillet, I., Larrouy-Maumus, G. et al. (2010), 'High-content phenotypic cell-based visual screen identifies *Mycobacterium tuberculosis* acyltrehalose-containing glycolipids involved in phagosome remodeling', *PLoS Pathogens* **6**(9), e1001100.

Brzuszkiewicz, E., Weiner, J., Wollherr, A., Thuermer, A., Huepeden, J. et al. (2011), 'Comparative genomics and transcriptomics of *Propionibacterium acnes*', *PLoS One* **6**(6), e21581.

Bushman, F., Malani, N., Fernandes, J., D'Orso, I., Cagney, G. et al. (2009), 'Host cell factors in HIV replication: meta-analysis of genome-wide studies', *PLoS Pathogens* **5**(5), e1000437.

Cheng, L., Viala, J., Stuurman, N., Wiedemann, U., Vale, R. et al. (2005), 'Use of RNA interference in *Drosophila* S2 cells to identify host pathways controlling compartmentalization of an intracellular pathogen', *Proceedings of the National Academy of Sciences of the USA* **102**(38), 13 646–13 651.

Cherry, S. (2008), 'Genomic RNAi screening in *Drosophila* S2 cells: what have we learned about host–pathogen interactions?', *Current Opinion in Microbiology* **11**(3), 262–270.

Chong, R., Squires, R., Swiss, R. & Agaisse, H. (2011), 'RNAi screen reveals host cell kinases specifically involved in *Listeria monocytogenes* spread from cell to cell', *PLoS One* **6**(8), e23399.

Chosewood, L. & Wilson, D. (2007), *Biosafety in microbiological and biomedical laboratories*, DIANE Publishing, Darby, PA.

Chuluunbaatar, U., Roller, R. & Mohr, I. (2012), 'Suppression of extracellular signal-regulated kinase activity in herpes simplex virus 1-infected cells by the Us3 protein kinase', *Journal of Virology* **86**(15), 7771–7776.

Cohen, M. (2000), 'Changing patterns of infectious disease', *Nature* **406**(6797), 762–767.

Coller, K., Heaton, N., Berger, K., Cooper, J., Saunders, J. et al. (2012), 'Molecular determinants and dynamics of hepatitis C virus secretion', *PLoS Pathogens* **8**(1), e1002466.

Collins, F., Lander, E., Rogers, J., Waterston, R. & Conso, I. (2004), 'Finishing the euchromatic sequence of the human genome', *Nature* **431**(7011), 931–945.

Conrad, C. & Gerlich, D. (2010), 'Automated microscopy for high-content RNAi screening', *Journal of Cell Biology* **188**(4), 453–461.

Conway, T. & Gary, K. (2003), 'Microarray expression profiling: capturing a genome-wide portrait of the transcriptome', *Molecular Microbiology* **47**(4), 879–889.

Cutler, S., Fooks, A. & Van Der Poel, W. (2010), 'Public health threat of new, reemerging, and neglected zoonoses in the industrialized world', *Emerging Infectious Diseases* **16**(1), 1–17.

Davies, J. & Davies, D. (2010), 'Origins and evolution of antibiotic resistance', *Microbiology and Molecular Biology Reviews* **74**(3), 417–433.

de Martel, C., Ferlay, J., Franceschi, S., Vignat, J., Bray, F. et al. (2012), 'Global burden of cancers attributable to infections in 2008: a review and synthetic analysis', *The Lancet Oncology* **13**(6), 607–615.

Derré, I., Pypaert, M., Dautry-Varsat, A. & Agaisse, H. (2007), 'RNAi screen in *Drosophila* cells reveals the involvement of the TOM complex in *Chlamydia* infection', *PLoS Pathogens* **3**(10), e155.

Fauci, A. (2001), 'Infectious diseases: considerations for the 21st century', *Clinical Infectious Diseases* **32**(5), 675–685.

Feng, G., Goodridge, H., Harnett, M., Wei, X., Nikolaev, A. et al. (1999), 'Extracellular signal-related kinase (ERK) and p38 mitogen-activated protein (MAP) kinases differentially regulate the lipopolysaccharide-mediated induction of inducible nitric oxide synthase and IL-12 in macrophages: *Leishmania* phosphoglycans subvert macrophage IL-12 production by targeting ERK MAP kinase', *Journal of Immunology* **163**(12), 6403–6412.

Field, H. & Kung, N. (2011), 'Henipaviruses: unanswered questions of lethal zoonoses', *Current Opinion in Virology* **1**(6), 658–661.

Friedrich, B.M., Scully, C.E., Brannan, J.M., Ogg, M.M., Johnston, S.C. et al. (2011), 'Assessment of high-throughput screening (HTS) methods for high-consequence pathogens', *Journal of Bioterrorism & Biodefense* **3**, 005.

Frisch, M., Biggar, R. & Goedert, J. (2000), 'Human papillomavirus-associated cancers in patients with human immunodeficiency virus infection and acquired immunodeficiency syndrome', *Journal of the National Cancer Institute* **92**(18), 1500–1510.

Fuchs, F., Pau, G., Kranz, D., Sklyar, O., Budjan, C. et al. (2010), 'Clustering phenotype populations by genome-wide RNAi and multiparametric imaging', *Molecular Systems Biology* **6**, 370.

Gandhi, N., Nunn, P., Dheda, K., Schaaf, H., Zignol, M. et al. (2010), 'Multidrug-resistant and extensively drug-resistant tuberculosis: a threat to global control of tuberculosis', *The Lancet* **375**(9728), 1830–1843.

Garbis, S., Lubec, G. & Fountoulakis, M. (2005), 'Limitations of current proteomics technologies', *Journal of Chromatography A* **1077**(1), 1–18.

Gayer, M., Legros, D., Formenty, P. & Connolly, M. (2007), 'Conflict and emerging infectious diseases', *Emerging Infectious Diseases* **13**(11), 1625–1631.

Genovesio, A., Giardini, M., Kwon, Y., de Macedo Dossin, F., Choi, S. et al. (2011), 'Visual genome-wide RNAi screening to identify human host factors required for *Trypanosoma cruzi* infection', *PloS One* **6**(5), e19733.

Gewurz, B., Towfic, F., Mar, J., Shinners, N., Takasaki, K. et al. (2012), 'Genome-wide siRNA screen for mediators of NF-κB activation', *Proceedings of the National Academy of Sciences of the USA* **109**(7), 2467–2472.

Goldberg, D., Siliciano, R. & Jacobs, W. (2012), 'Outwitting evolution: fighting drug-resistant TB, malaria, and HIV', *Cell* **148**(6), 1271–1283.

Grimes, G., Moodie, S., Beattie, J., Craigon, M., Dickinson, P. et al. (2005), 'GPX-macrophage expression atlas: a database for expression profiles of macrophages challenged with a variety of pro-inflammatory, anti-inflammatory, benign and pathogen insults', *BMC Genomics* **6**(1), 178.

Guiguemde, W., Shelat, A., Bouck, D., Duffy, S., Crowther, G. et al. (2010), 'Chemical genetics of *Plasmodium falciparum*', *Nature* **465**(7296), 311–315.

Gurumurthy, R., Maurer, A., Machuy, N., Hess, S., Pleissner, K. et al. (2010), 'A loss-of-function screen reveals Ras-and Raf-independent MEK-ERK signaling during *Chlamydia trachomatis* infection', *Science's STKE* **3**(113), ra21.

Hao, L., Sakurai, A., Watanabe, T., Sorensen, E., Nidom, C. et al. (2008), '*Drosophila* RNAi screen identifies host genes important for influenza virus replication', *Nature* **454**(7206), 890–893.

Hoffmann, A. & Baltimore, D. (2006), 'Circuitry of nuclear factor κb signaling', *Immunological Reviews* **210**(1), 171–186.

Holland, C., Schmid, M., Zimny-Arndt, U., Rohloff, J., Stein, R. et al. (2011), 'Quantitative phosphoproteomics reveals link between *Helicobacter pylori* infection and RNA splicing modulation in host cells', *Proteomics* **11**(14), 2798–2811.

Hufnagel, L., Brockmann, D. & Geisel, T. (2004), 'Forecast and control of epidemics in a globalized world', *Proceedings of the National Academy of Sciences of the USA* **101**(42), 15 124–15 129.

Karlas, A., Machuy, N., Shin, Y., Pleissner, K., Artarini, A. et al. (2010), 'Genome-wide RNAi screen identifies human host factors crucial for influenza virus replication', *Nature* **463**(7282), 818–822.

Kellam, P. (2000), 'Host–pathogen studies in the post-genomic era', *Genome Biology* **1**(2), 1009–1011.

Kim, M., Jeong, H., Kasper, C. & Arrieumerlou, C. (2010), 'IKKα contributes to canonical NF-κB activation downstream of Nod1-mediated peptidoglycan recognition', *PLoS One* **5**(10), e15371.

Knapp, B., Rebhan, I., Kumar, A., Matula, P., Kiani, N. et al. (2011), 'Normalizing for individual cell population context in the analysis of high-content cellular screens', *BMC Bioinformatics* **12**(1), 485.

König, R., Stertz, S., Zhou, Y., Inoue, A., Hoffmann, H. et al. (2009), 'Human host factors required for influenza virus replication', *Nature* **463**(7282), 813–817.

Koul, A., Arnoult, E., Lounis, N., Guillemont, J. & Andries, K. (2011), 'The challenge of new drug discovery for tuberculosis', *Nature* **469**(7331), 483–490.

Krishnan, M., Ng, A., Sukumaran, B., Gilfoy, F., Uchil, P. et al. (2008), 'RNA interference screen for human genes associated with West Nile virus infection', *Nature* **455**(7210), 242–245.

Lenormand, P., Pages, G., Sardet, C., L'Allemain, G., Meloche, S. et al. (1993), 'MAP kinases: activation, subcellular localization and role in the control of cell proliferation', *Advances in Second Messenger and Phosphoprotein Research* **28**, 237–244.

Leroy, Q. & Raoult, D. (2010), 'Review of microarray studies for host–intracellular pathogen interactions', *Journal of Microbiological Methods* **81**(2), 81–95.

Li, Q., Brass, A., Ng, A., Hu, Z., Xavier, R. et al. (2009), 'A genome-wide genetic screen for host factors required for hepatitis C virus propagation', *Proceedings of the National Academy of Sciences of the USA* **106**(38), 16 410–16 415.

Liu, L., Johnson, H., Cousens, S., Perin, J., Scott, S. et al. (2012), 'Global, regional, and national causes of child mortality: an updated systematic analysis for 2010 with time trends since 2000', *The Lancet* **379**(9832), 2151–2161.

Maeurer, A., Mehlitz, A., Mollenkopf, H. & Meyer, T. (2007), 'Gene expression profiles of *Chlamydophila pneumoniae* during the developmental cycle and iron depletion-mediated persistence', *PLoS Pathogens* **3**(6), e83.

Min, J. & Subbarao, K. (2010), 'Cellular targets for influenza drugs', *Nature Biotechnology* **28**(3), 239–240.

Morens, D., Folkers, G. & Fauci, A. (2004), 'The challenge of emerging and re-emerging infectious diseases', *Nature* **430**(6996), 242–249.

Neish, A. & Naumann, M. (2011), 'Microbial-induced immunomodulation by targeting the NF-κB system', *Trends in Microbiology* **19**(12), 596–605.

Nguyen, D., Wolff, K., Yin, H., Caldwell, J. & Kuhen, K. (2006), 'UnPAKing human immunodeficiency virus (HIV) replication: using small interfering RNA screening to identify novel cofactors and elucidate the role of group I PAKs in HIV infection', *Journal of Virology* **80**(1), 130–137.

Pennings, J., Kimman, T. & Janssen, R. (2008), 'Identification of a common gene expression response in different lung inflammatory diseases in rodents and macaques', *PLoS One* **3**(7), e2596.

Philips, J., Rubin, E. & Perrimon, N. (2005), '*Drosophila* RNAi screen reveals CD36 family member required for mycobacterial infection', *Science* **309**(5738), 1251–1253.

Piret, J. & Boivin, G. (2011), 'Resistance of herpes simplex viruses to nucleoside analogues: mechanisms, prevalence, and management', *Antimicrobial Agents and Chemotherapy* **55**(2), 459–472.

Pisani, P., Parkin, D., Munoz, N. & Ferlay, J. (1997), 'Cancer and infection: estimates of the attributable fraction in 1990', *Cancer Epidemiology Biomarkers & Prevention* **6**(6), 387–400.

Prudêncio, M., Rodrigues, C., Hannus, M., Martin, C., Real, E. et al. (2008), 'Kinome-wide RNAi screen implicates at least 5 host hepatocyte kinases in *Plasmodium* sporozoite infection', *PLoS Pathogens* **4**(11), e1000201.

Puoti, M., Bruno, R., Soriano, V., Donato, F., Gaeta, G. et al. (2004), 'Hepatocellular carcinoma in HIV-infected patients: epidemiological features, clinical presentation and outcome', *Aids* **18**(17), 2285–2293.

Rappuoli, R. (2000), 'Pushing the limits of cellular microbiology: microarrays to study bacteria–host cell intimate contacts', *Proceedings of the National Academy of Sciences of the USA* **97**(25), 13 467–13 469.

Reed, K., Meece, J., Henkel, J. & Shukla, S. (2003), 'Birds, migration and emerging zoonoses: West Nile virus, Lyme disease, influenza A and enteropathogens', *Clinical Medicine & Research* **1**(1), 5–12.

Rodrigues, C., Hannus, M., Prudêncio, M., Martin, C., Gonçalves, L. et al. (2008), 'Host scavenger receptor SR-BI plays a dual role in the establishment of malaria parasite liver infection', *Cell Host & Microbe* **4**(3), 271–282.

Saeed, M., Kolokoltsov, A., Albrecht, T. & Davey, R. (2010), 'Cellular entry of Ebola virus involves uptake by a macropinocytosis-like mechanism and subsequent trafficking through early and late endosomes', *PLoS Pathogens* **6**(9), e1001110.

Schwartz, S., Friedberg, I., Ivanov, I., Davidson, L., Goldsby, J. et al. (2012), 'A metagenomic study of diet-dependent interaction between gut microbiota and host in infants reveals differences in immune response', *Genome Biology* **13**(4), R32.

Seidel, T., Ibis, G., Bendix, F. & Wolber, G. (2011), 'Strategies for 3D pharmacophore-based virtual screening', *Drug Discovery Today: Technologies* **7**(4), e221–e228.

Sessions, O., Barrows, N., Souza-Neto, J., Robinson, T., Hershey, C. et al. (2009), 'Discovery of insect and human dengue virus host factors', *Nature* **458**(7241), 1047–1050.

Sharma, M., Machuy, N., Boehme, L., Karunakaran, K., Maeurer, A. et al. (2011), 'HIF-1α is involved in mediating apoptosis resistance to *Chlamydia trachomatis*-infected cells', *Cellular Microbiology* **13**(10), 1573–1585.

Shui, W., Gilmore, S., Sheu, L., Liu, J., Keasling, J. et al. (2008), 'Quantitative proteomic profiling of hostpathogen interactions: the macrophage response to *Mycobacterium tuberculosis* lipids', *Journal of Proteome Research* **8**(1), 282–289.

Smith, M., Wightman, F. & Lewin, S. (2012), 'HIV reservoirs and strategies for eradication', *Current HIV/AIDS Reports* **9**(1), 5–15.

Smith, R. (2006), 'Responding to global infectious disease outbreaks: lessons from SARS on the role of risk perception, communication and management', *Social Science & Medicine* **63**(12), 3113–3123.

Snijder, B. & Pelkmans, L. (2011), 'Origins of regulated cell-to-cell variability', *Nature Reviews Molecular Cell Biology* **12**(2), 119–125.

Snijder, B., Sacher, R., Rämö, P., Damm, E., Liberali, P. et al. (2009), 'Population context determines cell-to-cell variability in endocytosis and virus infection', *Nature* **461**(7263), 520–523.

Snijder, B., Sacher, R., Rämö, P., Liberali, P., Mench, K. et al. (2012), 'Single-cell analysis of population context advances RNAi screening at multiple levels', *Molecular Systems Biology* **8**(1), 579.

Steuernagel, O. & Polani, D. (2010), 'Multiobjective optimization applied to the eradication of persistent pathogens', *IEEE Transactions on Evolutionary Computation* **14**(5), 759–765.

Stoltey, J. E. & Barry, P. M. (2012), 'The use of cephalosporins for gonorrhea: an update on the rising problem of resistance', *Expert Opinion in Pharmacotherapy* **13**(10), 1411–1420.

Suratanee, A., Rebhan, I., Matula, P., Kumar, A., Kaderali, L. et al. (2010), 'Detecting host factors involved in virus infection by observing the clustering of infected cells in siRNA screening images', *Bioinformatics* **26**(18), i653–i658.

Swedlow, J., Goldberg, I. & Eliceiri, K. (2009), 'Bioimage informatics for experimental biology', *Annual Review of Biophysics* **38**, 327–346.

Theuretzbacher, U. (2011), 'Resistance drives antibacterial drug development', *Current Opinion in Pharmacology* **11**(5), 433–438.

Vanhecke, D. & Janitz, M. (2004), 'High-throughput gene silencing using cell arrays', *Oncogene* **23**(51), 8353–8358.

Vittor, A., Pan, W., Gilman, R., Tielsch, J., Glass, G. et al. (2009), 'Linking deforestation to malaria in the Amazon: characterization of the breeding habitat of the principal malaria vector, *Anopheles darlingi*', *American Journal of Tropical Medicine and Hygiene* **81**(1), 5–12.

Wilson, S., Fakioglu, E. & Herold, B. (2009), 'Novel approaches in fighting herpes simplex virus infections', *Expert Review of Anti-Infective Therapy* **7**(5), 559–568.

Woolhouse, M., Webster, J., Domingo, E., Charlesworth, B., Levin, B. et al. (2002), 'Biological and biomedical implications of the co-evolution of pathogens and their hosts', *Nature Genetics* **32**(4), 569–577.

Ying, S., Pettengill, M., Ojcius, D. & Haecker, G. (2007), 'Host-cell survival and death during *Chlamydia* infection', *Current Immunology Reviews* **3**(1), 31–40.

Zhang, C., Chromy, B. & McCutchen-Maloney, S. (2005), 'Host–pathogen interactions: a proteomic view', *Expert Review of Proteomics* **2**(2), 187–202.

Zhou, H., Xu, M., Huang, Q., Gates, A., Zhang, X. et al. (2008), 'Genome-scale RNAi screen for host factors required for HIV replication', *Cell Host & Microbe* **4**(5), 495–504.

8 Inferring genetic architecture from systems genetics studies

Xiaoyun Sun, Stephanie Mohr, Arunachalam Vinayagam, Pengyu Hong, and Norbert Perrimon

In recent years many efforts have been invested in comprehensively evaluating the behavior and relationships of all genes/proteins in a particular biological system and at a particular state. Here, we review how genome-wide RNAi screens together with mass spectrometry can be integrated to generate high-confidence functional interactome networks. Next we review the mathematical modeling methods available today that allow the computational reconstruction of such networks. Network modeling will play an important role in generating hypotheses, driving further experimentation and thus novel insights into network structure and behavior.

8.1 Introduction

Most biologists study a specific biological problem by investigating the activities of a limited number of genes or proteins involved in a particular biological process. This traditional approach is critical and has proven to be extremely successful to reveal the detailed molecular functions of individual genes and proteins. For example, genetic studies of embryonic patterning in *Drosophila* identified about 40 genes with striking segmentation defects that fell into distinct phenotypic classes: gap genes, pair rule genes, segment polarity genes, and homeotic genes (Nusslein-Volhard & Wieschaus 1980). Detailed analyses of the mutant phenotypes and functions of even this relatively small set of genes led to a comprehensive molecular framework of the process of embryonic patterning (St Johnston & Nusslein-Volhard 1992). Reductionist approaches, however, are not sufficient for generating the big picture of how a biological system, including multiple levels of many different gene products and the interactions among them, works at different physiological states or developmental stages (Friedman & Perrimon 2007). Thus, as our knowledge of individual genes and proteins accumulates, there is a need to comprehensively evaluate the behavior and relationships of all genes/proteins in a particular biological system and at a particular state. In recent years, progress has been made in multicellular organisms towards this goal mostly in tissue culture, a platform that allows a sufficient amount of homogeneous material to be easily obtained.

Profiling the parameters involved in various biological processes at genome scale has become a promising strategy to address such a systems biology problem. This

Systems Genetics: Linking Genotypes and Phenotypes, ed. F. Markowetz and M. Boutros. Published by Cambridge University Press. © Cambridge University Press 2015.

approach is now possible due to advances in RNA interference (RNAi), whereby the functions of all annotated genes in a genome can be systematically interrogated (Mohr et al. 2010). Furthermore, major technical advances in proteomics, transcriptomics, and cellular imaging now provide sophisticated means to measure biological parameters quantitatively and at high-throughput scale. Altogether, these approaches allow the generation of phenotypic signatures for all genes expressed in a cell of interest that describe their roles in the biological process under scrutiny. The goal of applying these methods is not only to provide functional information on the activities of many genes/proteins, but also to enable the construction of networks that faithfully reflect the dynamics of biological activities in a particular system. This approach is challenging as both biological and technical noise can affect the quality of the data sets generated, and requires in particular robust cellular assays, careful consideration of the reproducibility of the data generated, integration of orthogonal data sets, and rigorous computational analyses.

Three types of experimental data sets are most frequently integrated in network construction: transcriptomics, proteomics, and interactomics. Transcriptomics provides information about both the presence/absence and relative abundance of RNA transcripts, thereby indicating the active components within the cell. Transcriptome data measured by genome-wide microarray or RNA-seq (transcriptome profiling that uses deep-sequencing technologies) are widely used for network construction, as RNA molecules are easily accessible in comparison to proteins and metabolites. Proteomics describes the entire population of expressed proteins in a cell or tissue. It aims to identify and quantify the cellular levels of genome-wide protein expression in a specific biological system. Interactomics include protein–DNA, protein–RNA, and protein–protein interactions (PPIs). Protein–DNA interactions mainly occur between transcription factors and their target DNA, whereas protein–RNA interactions depict potential regulatory roles of specific proteins to target RNAs. PPIs define the fundamental genetic regulatory network of the cell. They are extremely valuable for network construction, as with this approach the relationships among interacting proteins are clearly established, in contrast to the often indirect and sometimes complicated regulation of components within genetic networks. Finally, in addition to transcriptomics, proteomics, and interactomics, the analysis of phenotypic signatures, based on cellular features extracted from image analyses, has emerged as a powerful method that provides rich phenotypic information on dynamic and more complex cellular processes, such as nuclear translocation, cytokinesis, and cell migration (Perlman et al. 2004, Bakal et al. 2007).

Here, we review how some of these methods can be applied today to *Drosophila* cells to gain insights on the organization of biological networks. Specifically, we describe how genome-wide RNAi screens are used to identify gene activities in the cell that affect the output of a network, then we describe how PPIs, generated from mass spectrometry, can be easily integrated with RNAi data sets to generate a high-confidence functional interactome. Finally, we review the mathematical modeling methods available which, when applied to the integrated data sets generated from RNAi and PPI, allow the computational reconstruction of the network – the goal of which being to generate a number of hypotheses ultimately driving further experimentations leading to novel insights.

8.2   Identification of network components by RNAi

In *Drosophila* cells, RNAi knockdown is easily achieved using in vitro synthesized long dsRNAs (typically 150 to 500 bp), and readily adaptable to screening of cultured or primary cells in miniaturized platforms (e.g. 384-well plates) (Boutros et al. 2004). Thus, *Drosophila* cell-based RNAi screening can be done in high-throughput mode, providing a platform for genome-scale functional analysis of cellular processes (DasGupta & Gonsalves 2008, Bakal & Perrimon 2010, Falschlehner et al. 2010, Mohr et al. 2010). Information about reagents and results from *Drosophila* cell-based RNAi screens is available at a number of databases, including FLIGHT (http://flight.icr.ac.uk/) (Sims et al. 2006), GenomeRNAi (http://genomernai.de/GenomeRNAi/) (Gilsdorf et al. 2010), and FlyRNAi, the *Drosophila* RNAi Screening Center (http://www.flyrnai.org) (Flockhart et al. 2006). To date, a large number of screens have been performed in *Drosophila* cells, yielding insights into a number of biological processes and systems (Mohr et al. 2010). Researchers often screen grouped subsets of genes, e.g. all genes encoding kinases or genes identified using another high-throughput method or bioinformatics analysis. However, full-genome screening remains the most unbiased and comprehensive approach. An important aspect of RNAi screening distinguishing it from some other high-throughput methods is that RNAi results not only implicate genes in a given pathway but can also indicate the direction of action (i.e. a positive or negative regulator in a given pathway).

A wide variety of high-throughput screening methodologies, instruments, and assays are available for RNAi screening ((Shumate & Hoffman 2009, Mohr et al. 2010) (Figure 8.1). Among the most straightforward to perform and analyze are total-well luciferase or fluorescence readouts, which are collected using a luminometer or fluorimeter (plate-reader). These outputs are typically expressed as a ratio (e.g. of values obtained with a transcriptional reporter versus a constitutive expressed control). From these numerical outputs, positive hits are typically identified after calculating Z-scores and choosing an appropriate Z-score cut-off value. Researchers often rely on prior knowledge of components of a process or system in order to select an appropriate Z-score value. This is often done empirically (i.e. I know gene X is involved and reagents targeting X gave a Z-score of n in the screen; therefore, I will use n as my cut-off). However, it can also be done more systematically, such as using RNAiCut, which is based on the assumption that gene products corresponding to true positives are more interconnected to one another at the level of PPIs (Kaplow et al. 2009). A number of other types of assays, supported by specialized plate-readers or laser-scanning cytometers, are similar in that all cells in the well are measured and the data are in the form of one or more numerical outputs that can be analyzed based on Z-scores. Among these, the in-cell Western approach, in which immunofluorescence labeling of a phospho-protein or protein is compared to a total protein control, has proved particularly useful for interrogation of signal transduction in *Drosophila* cells (Friedman & Perrimon 2006, Friedman & Perrimon 2007, Kockel et al. 2010, Friedman et al. 2011).

Although relatively simple outputs continue to be informative for screening, researchers are increasingly turning to high-content image-based screens in order

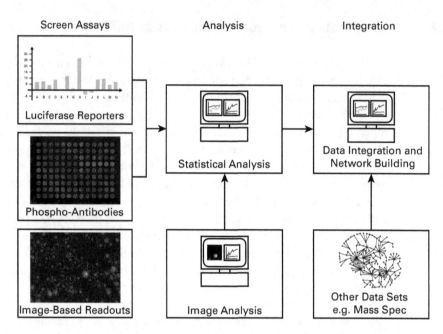

Figure 8.1   Cell-based assays used to describe cellular phenotypes. Several types of assays and corresponding instruments are available for RNAi screening and other high-throughput *Drosophila* cell-based methods. Luciferase levels provide a relatively simple way to monitor overall induction or suppression of a transcriptional reporter. Methods such as the in-cell Western approach allow for monitoring of endogenous proteins. With this approach, immunofluorescence levels of a phospho-protein or protein are compared with levels detected using non-phospho-specific antibody or total protein dye. High-content imaging allows for detection of subcellular protein distributions, organelles, and other features. For high-content imaging, epifluorescence or confocal microscopy is used to detect one or several cellular and subcellular readouts, followed by single- or multi-parametric image analysis of simple or complex phenotypes. The complexity of image data requires specialized analysis. In all cases, phenotypes are reduced through analysis to numerical values such as Z-scores. These results can then be combined with results from other high-throughput assays, e.g. mass spectrometry, or with information from the literature in order to build high-confidence gene networks.

to obtain high-quality results relevant to complex phenotypes (Bakal 2011). Instruments developed for acquisition of high-content screen image data include automated epifluorescence, fluorescence confocal, and laser-scanning microscopes (Shumate & Hoffman 2009, Zanella et al. 2010). Most of these instruments image a subregion of the well and multiple images per well must be acquired in order to image enough cells for statistically meaningful results. Through the use of one or more fluorescent dye or antibody, as well as the introduction of fluorescence protein-tagged fusion proteins or reporters, several different readouts can be simultaneously collected and measured. Even a single image-based readout such as the DNA dye 4′,6-diamidino-2-phenylindole (DAPI) can be used to count cells, define the nucleus (e.g. as a reference for detecting nuclear versus cytoplasmic localization), measure nuclear area, monitor cell cycle stages, and more. As screen-imaging instruments facilitate collection of data for several different fluorescent tags, the number of features that can be evaluated singly or in relationship to one another can be very large. Thus, high-content screening facilitates detection of complex cellular and subcellular phenotypes, such as changes in the subcellular distribution of a protein, or in the size, shape, or number of cells or organelles (see for example Bakal et al. 2007). Importantly, analysis of multiple parameters can be used to improve the quality of RNAi screen results. Assessment of the Z factor for analyzed images during assay development, for example, can be used as a measure of robustness prior to the screen, guiding optimization of the screen assay (Kummel et al. 2010). Moreover, following image data acquisition, some parameters prove more informative than others in identifying on-target and relevant cellular responses (Collinet et al. 2010, Kummel et al. 2010). Development in the area of screen image analysis is growing rapidly. Various academic and commercial groups have developed software tools useful for analysis of high-content screen data sets, including machine-learning approaches (reviewed in Ljosa & Carpenter 2009, Niederlein et al. 2009). Following analysis, the screen output is reduced to one or more numerical values, which can then be evaluated using Z-scores or another approach.

Despite the power of the RNAi approach, a variety of caveats apply that are relevant to the analysis, interpretation, and integration of RNAi screen results in the context of system-wide analyses. Perhaps the most common problems are systematic errors or bias in the assay; stochastic effects or noise inherent in high-throughput data sets; and reagent-specific off-target effects (Falschlehner et al. 2010, Mohr et al. 2010, Booker et al. 2011, Seinen et al. 2011). Most systematic errors are easily addressed prior to primary data acquisition (e.g. through correction of instrument-derived dispensing errors) but others persist and can affect interpretation of results. For example, some cell-based assays show bias, such as favoring identification of hits in one direction or the other relative to controls, e.g. favoring identification of positive-acting or negative-acting factors in a pathway (DasGupta et al. 2007). When such a bias exists and cannot be fully addressed through assay optimization, robust interrogation of a given network might require screening with more than one assay, i.e. by performing related assays with opposite biases. In general, screening of multiple time-points or conditions, as well as screens that combine more than one dsRNA reagent to look at additive or synergistic effects of double knockdown, can help reduce false negative discovery. For example, genes whose knockdown results in weak phenotypes not detected above noise in a single-knockdown screen might be picked up as significant positives when combined in a double-knockdown screen (Bakal et al. 2008).

Regarding the identification of false positives in RNAi screens, a number of approaches are available (Figure 8.2). The most rigorous approach to validation of screen hits is a rescue test, in which a construct that confers gene activity but can evade the RNAi reagent is tested for the ability to reverse the observed phenotype. For many screens, initial positive hits are retested with two or more unique reagents per gene in order to filter out potential false positive results. An additional filter that can be applied is to remove initial positive hits for which there is no evidence of gene expression in the cell line that was tested, with the underlying assumption that reagents targeting genes for which there is no evidence of expression are more likely to be exerting their effects

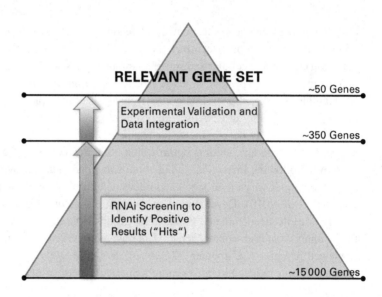

Figure 8.2   Validation and integration of RNAi screen data can help identify high-confidence gene networks. Full-genome screening allows researchers to identify screen hits (positive hits), reducing the candidate gene list to a more manageable size. Subsequent to the screen, experimental validation, as well as integration and filtering with other data sets, can be used to identify a high-confidence set of genes likely to be involved in a given process or pathway. Commonly used methods for experimental validation include testing with more than one unique dsRNA reagent per gene, testing for concordance between quantitative reverse transcriptase PCR (qPCR) analysis of mRNA knockdown and observed phenotypes, and RNAi rescue tests. Commonly used computational approaches include comparison of the set of screen hits to evidence for gene expression (with the assumption that reagents targeting genes known to be expressed are more likely to be exerting on-target effects), and integration with the results of other high-throughput studies, such as mass spectrometry analyses, or information culled from the published literature.

through off-targets (Booker et al. 2011). Informatics-based analysis of reagent quality (e.g. number of predicted off-targets) can also be taken into account in assessing primary screen data. However, the experimental approaches are impractical to apply at large scale, and the systematic approaches can limit but not eliminate false discovery. As a result, systematic and robust detection of false positive and negative results is simply not practical to do for all genes tested in a large-scale RNAi study. Thus, in general, a researcher's curated list of positive results from an RNAi screen is likely to be based upon a combination of statistical analysis, experimental verification, and/or prior knowledge of the process or pathway under study, and genome-wide information is usually only available in the form of analyzed but unverified primary screen data (e.g. Z-scores for all primary hits). In addition to these methods, overlapping orthogonal data sets with RNAi results, as described below, provides a powerful filter to identify high-confidence network components.

8.3   Identification of network components using proteomics

Protein–protein interactions play critical roles in many cellular processes, such as signaling cascades and regulatory complex formation. In addition, information acquired from PPI data are definitive (i.e. proteins A and B interact with each other) and can contain quantitative features (i.e. the strength of interactions), making PPI data a core resource for network construction. Hence identifying all functional PPIs is not only important for understanding the structure and function of biological systems, but also for the construction of reliable networks.

Although classic biochemical experiments based on co-immunoprecipitation can readily identify interactions between specific proteins, they lack the ability to explore interactions at whole-proteome scale. In recent years, several techniques measuring proteome-wide PPIs have been developed. Two methods in particular facilitate high-throughput studies, the yeast two-hybrid approach (Fields 2005) and tandem affinity purification coupled with mass spectrometry (TAP/MS) (Gavin et al. 2006, Krogan et al. 2006, Jeronimo et al. 2007). The yeast two-hybrid approach is a useful approach for high-throughput identification of putative interaction partners, but it can be prone to false positive identification and interactions are detected in a heterologous context. Thus, TAP/MS analysis has been increasingly used to identify novel and large-scale PPIs under physiologically relevant conditions (Gavin et al. 2006, Krogan et al. 2006, Jeronimo et al. 2007).

TAP and MS are two essential components of the TAP/MS technique: TAP efficiently isolates native protein complexes from cells for proteomics analysis. It is followed by MS analysis, a powerful analytical technique used to determine the molecular structures of peptides. The advantage of MS is that it identifies multi-subunit protein complexes isolated from the cell lysate with extremely high sensitivity and accuracy. The TAP/MS approach has been used successfully to characterize protein complexes from various cells and multicellular organisms. In addition, this technique can be combined with quantitative proteomics approaches to better understand the dynamics of protein complex assembly. As will be discussed below, TAP/MS can also be integrated with RNAi data, so that high-confidence and even dynamic networks can be reconstructed.

To study pathway-specific interactions, special cell lines need to be generated first, with each cell line stably expressing a TAP-tagged version of a starting protein of interest (the bait protein), such as a major signaling component. The reason for tagging the components is to facilitate isolation of those components later. Along with the special cell lines, a negative control cell line (i.e. not expressing any TAP-tagged proteins) is recommended for subtracting non-specific interactors. Both types of cells are treated with specific conditions to provide protein lysates. The lysates are incubated with affinity purification beads, where the TAP-tagged protein is pulled down via its tag, together with associated proteins (the prey proteins) and other proteins retained through non-specific binding. The protein samples collected are then broken down into peptides with proteases and analyzed by MS, where a list of peptide sequences from each sample is reported as the results. A necessary data preprocessing step is to identify the source proteins of the peptide sequences and calculate the number of peptides for each prey protein identified in each sample. To increase confidence in PPI identification, multiple replicates are recommended for each cell line and condition.

Early TAP/MS analytic methods identify PPIs by binary mode (i.e. indicating the presence or absence of a specific protein) (Zhu et al. 2007). Newer methods take into account quantitative information such as the label-free quantitative spectral count (SC), which is the number of peptides detected in MS. The challenge for TAP/MS data analysis is to minimize false positive interactions and increase the sensitivity to identify true interactions. Currently, there are three popular computational tools for TAP/MS data analysis: NSAF (Normalized Spectral Abundance Factor) (Sardiu et al. 2008); CompPASS (Comparative Proteomic Analysis Software Suite) (Sowa et al. 2009); and SAINT (Significance Analysis of Interactome) (Choi et al. 2011).

With NSAF, the relative abundance of proteins is estimated based on the total number of peptides (i.e. SC) identified in the sample. In general, larger proteins are expected to generate more peptides and hence a larger SC than smaller proteins. To account for the variation of protein size, the SC for each protein is divided by the protein length, which is defined as the spectral abundance factor (SAF). Individual SAFs also need to be normalized by the sum of all SAFs for proteins in the sample to account accurately for run-to-run variation (Eq. 8.1).

$$NSAF(i) = \frac{\left(\frac{SC_i}{L_i}\right)}{\sum_{i=1}^{N}\left(\frac{SC_i}{L_i}\right)}. \tag{8.1}$$

The NSAF for a protein i is the SC of a protein divided by the protein's length (L), divided by the sum of SC/L for all N proteins in the experiment. NSAF is simple, easy to compute, and has been demonstrated to be effective in detecting significant PPIs. However, NSAF is an empirical transformation of SCs, which does not incorporate any information from negative controls. Moreover, it does not add weight to interactions that are detected in all biological replicates (most likely true interactions) and does not penalize interactors detected in all purifications (e.g. sticky proteins that interact with all bait proteins). Thus, although NSAF is useful to some extent, it clearly needs further improvement.

The CompPASS method computes PPI scores by adjusting observed SCs relative to the reproducibility of detection across biological replicates, as well as the frequency of observing the prey protein in purifications with different baits. The first step in Comp-PASS is the generation of a stats table (Table 8.1). In the table, each row is the unique protein identified from the TAP/MS experiments (interactor) and each column is the bait protein used in those experiments. Each element of the table is the SC of an interacting protein from the particular baits TAP/MS experiment. After the stats table is created from all experiment runs in the project, CompPASS calculates a mean value of the SC (M) for each interactor, then calculates a Z-score and D-score for each interaction pair.

The first score is the Z-score, which is specific for a particular interaction; the mean is subtracted from the SC, and is divided by the standard deviation (Eqs. 8.2 and 8.3). X is the SC, i is the bait number, j is the index of interactor, n is the total number of

Table 8.1 Stats table in ComPASS analysis

	Bait 1	Bait 2	Bait 3	Bait 4	Bait k	Mean
Interactor 1	$X_{1,1}$	$X_{1,2}$	$X_{1,3}$	$X_{1,1}$	$X_{1,k}$	M_1
Interactor 2	$X_{2,1}$	$X_{2,2}$	$X_{2,3}$	$X_{2,4}$	$X_{2,k}$	M_2
Interactor 3	$X_{3,1}$	$X_{3,2}$	$X_{3,3}$	$X_{3,4}$	$X_{3,k}$	M_3
Interactor n	$X_{n,1}$	$X_{n,2}$	$X_{n,3}$	$X_{n,4}$	$X_{n,k}$	M_n

interactors, k is the total number of baits, M is the mean of the SC, and σ is the standard deviation of the SC for each interactor. Although the Z-score can identify interactors for which the SC is significantly different from the mean, it fails to discriminate two interactors with dramatically different SCs if the experiment has only one replicate. For example, if in a single experiment, the SCs for A and B SC are 2 and 20, respectively, then the two proteins will have the same Z-score, as the mean and standard deviation are the same for a single data point.

$$Z_{ij} = \frac{X_{ij} - M_i}{\sigma_i}, \text{ where } M_i = \frac{1}{k}\sum_{j=1}^{k} X_{ij}. \tag{8.2}$$

The second is the D-score (Eq. 8.3), which takes into account both the reproducibility of detection across biological replicates and the frequency of observing prey protein in purifications of different baits. The variables are the same as for Eqs. (8.2 and 8.3). Here, f is a term which is 0 or 1 depending on whether or not the interactor was found a given particular bait. $\sum f$ is the summation across all baits. $k/\sum f$ is the frequency of this particular interactor across all baits. P is the number of replicate runs in which the interaction is present. The reproducibility term allows for better discrimination between a likely false positive (i.e. an interactor found in one run but not in any of the other multiple replicates) and a likely true positive (i.e. an interactor with a low SC yet found in all replicates).

$$D_{i,j} = \sqrt{X_{ij}\left(\frac{k}{\sum_{i=1}^{k} f_{ij}}\right)^P} \text{ where } f_{ij} = \begin{cases} 1 & \text{if } X_{ij} > 0 \\ 0 & \text{else} \end{cases}. \tag{8.3}$$

CompPASS is easy to compute and takes into consideration two important factors: reproducibility and the frequency at which each interactor is detected in multiple replicates. It uses a different approach to distinguish the background and real interactors rather than directly utilizing the negative control data sets. Further, CompPASS takes a maximum of two replicates which might not be enough for some experiments with large variance in their biological replicates.

The SAINT approach assigns a confidence score to a PPI by converting the normalized SC into the probability of a true interaction between the two proteins. The parameters for true and false distributions, $P(X_{ij}|\text{true})$ and $P(X_{ij}|\text{false})$, and the prior probability of interactions in the data set, $P(\text{true})$ and $P(\text{false})$, are inferred from the normalized SCs for all interactions that involve prey i and the bait j. The posterior probability of a true interaction, $P(\text{true}|X_{ij})$, can be calculated from parameters using Bayes' rule (Eq. 8.4).

$$P(\text{true} \mid X_{ij}) = \frac{P(X_{ij}|\text{true})P(\text{true})}{P(X_{ij}|\text{true})P(\text{true}) + P(X_{ij}|\text{false})P(\text{false})}. \tag{8.4}$$

SAINT modeling can be performed with or without negative control data. When negative controls are not available, the distribution of false interactions can be estimated in reference to the quantitative information for the same interactor across purifications of all other baits in the data set. When TAP/MS data contain negative controls, SAINT estimates the SC distribution for false interactions directly from the negative controls. The incorporation of negative control data improves the robustness of modeling, especially for small data sets.

The SAINT model is based on label-free quantification using the SC. It constructs separate distributions for true and false interactions to derive the probability of a bona fide PPI. The probability model can also be used to estimate the false discovery rate (FDR), and can be extended to model other types of quantitative parameters such as peptide ion intensity. However, SAINT specifically excludes proteins with one or two SCs. The necessity of this arbitrary step is questionable. Moreover, the complicated reference procedure in SAINT demands high computational costs.

Overall these approaches can effectively analyze TAP/MS data sets, but they also have room for improvement. For example, NSAF and CompPASS compute scores based on the transformation of SC, and SAINT demands high computational costs largely due to its complicated reference procedure. New algorithms should be investigated in the future that eliminate false positives more effectively and that require lower computational costs.

8.4   Integration of RNAi and proteomic data sets

RNAi provides information about which genes affect the activity of a network. However, many gene activities can affect the activity of a network indirectly (e.g. general maintenance activities such as those related to overall protein translation or stability). In addition, as discussed above, although methods are available to ensure that RNAi effects are on target, such as genomic rescue, the most rigorous validation approaches can be tedious and are not commonly used large-scale for validation of RNAi results. Proteomic data sets, which frequently reflect the interactions among different proteins at a genome-wide scale, can help address both of these issues, as the integration of RNAi and proteomic data sets can facilitate validation, leading relatively quickly to a high-confidence functional interactome.

Different data sets are usually ranked using different types of scores (e.g. Z-score is used to evaluate the result of RNAi screens; a probability value or p-value is generated from TAP/MS data sets by various analytic methods). Thus, the first challenge in integrating RNAi and PPI results is to combine different data sets with different scoring functions. One common approach is to choose an appropriate cut-off value (threshold) for each data set and integrate them using the voting system (Zhong & Sternberg 2007), in which a simple statistical model is used to integrate multiple data sets in the absence of data training. Basically, with this approach, a gene/protein that appears in each data set gets one vote, and the total number of votes are computed and used to determine either inclusion or exclusion from the network. When the threshold vote number equals the total number of data sets (i.e. scored in all data sets), the system becomes a filtering model and only the intersection of all data sets is selected. When the threshold is set to 1, the system selects the union of all data sets. The threshold directly affects both the false positive and false negative rates in the final data integration results, and should be set according to different analytic purposes. A small value of the threshold will give rise to a relatively complete network, but more errors might be associated. On the other hand, high threshold value can generate a high-confidence network, but it might also eliminate some useful information.

Because of the simplicity of the voting system, and because there is no requirement for a training data set, it has been extensively used in a variety of investigations (Walhout et al. 2002, Gunsalus et al. 2005). For example, in a recent study of the canonical receptor tyrosine kinase (RTK)/RAS/extracellular signal-regulated kinase (ERK) pathway in *Drosophila* (Friedman et al. 2011), a comprehensive network was integrated by combining unbiased ERK activation genome-wide RNAi screens with TAP/MS network structural data. In this study, RNAi screen results were filtered using interactors identified in the PPI network in order to achieve significant enrichment of pathway regulators. The results showed that about 50% of the filtered PPI network scored in the RNAi screens.

The integration of multiple data sources improves the specificity and reliability of individual high-throughput data sets. It can also be an effective approach to reduce the level of false negative discovery (i.e. protein complexes identified in the network reconstruction can guide experimental validations of some interactions not scored in the original TAP/MS data sets). Furthermore, by combining RNAi and TAP/MS data sets with time-course measurements, aspects of the dynamic regulation of the network can be revealed.

8.5   Network modeling: the next step

Following the construction of a network involved in a particular biological process (e.g. the *Drosophila* RTK/ERK network; Friedman et al. 2011) involves network reconstruction using mathematical modeling. Network reconstruction aims to build a mathematical model through a learning algorithm, so that the output of the model fits with provided biological data, and the relationships of the network components (genes/proteins) are clearly defined. Essential in this network reconstruction process is a solid computational analysis, which involves data preparation, network architecture selection, and structure and parameter learning. Data preparation is the fundamental step and largely determines the quality of the analysis outcome. Appropriate network model selection depends on both the available data type and the aims of the computational analysis. The final

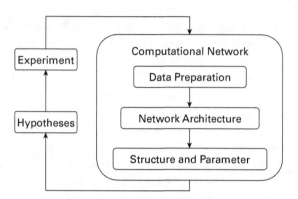

Figure 8.3   Biological experiment and computational network reconstruction cycle. Hypothesis-driven biological experiments are analyzed by computational approaches, aiming to reconstruct an underlying network. This involves data preparation, network architecture selection, structure and parameter learning steps. The inferred network improves our understanding of biological systems and further aids the guidance for future experimental designs. A new round of experiments enables further improvement of subsequent network construction.

network can be built through repetitive structure and parameter learning and refining processes (illustrated in Figure 8.3). A good network not only depicts the detailed regulation of its components but also provides high confidence and promising directions for future experimental design.

8.5.1   Data preparation

Good data preparation is key to network reconstruction and a balancing act. On the one hand, in order to minimize experimental efforts and costs, the number of experiments conducted should be minimal; on the other hand, accurate reconstruction of biological network demands a considerable quantity of reliable data. In determining the amount of data required for network inference, the complexity of the system and the quality of the network are integral and related factors (Hecker et al. 2009). Generally, the quality of a network largely depends on the quality of the given data. Large variation and high levels of measurement noise in the experiment data will impair the quality of the constructed network. Thus, it is important to carry out multiple replicates to minimize the effects of variation and noise. Incorporation of a larger number of parameters allows a network to more accurately represent the complexity of a system; however, this requires collection of a larger amount of experimental data and also adds to the total computational time.

Data preprocessing is an important step in data preparation. It directly affects both the performance of the network inference algorithms and the inference results. Methods for data preprocessing need to be applied selectively according to different data types, experimental designs, and network inference methods. For instance, certain methods only allow for input of binary numbers, so measured expression levels have to be converted into two discrete expression values. Other methods require time-series data, so the appropriate interpolation of experiment data at different time-points has to be conducted during data preprocessing.

In general, to construct a reliable network while also limiting network complexity and computation time, the following strategies should be considered in data preparation (Hecker et al. 2009). First, the amount of data should be increased either by increasing the number of measurements or through data integration. Second, the number of network components should be reduced by grouping together genes/proteins with similar functions. Third, the number of network parameters should be restricted so that the dimensionality of the network search space can be reduced. And finally, specific prior knowledge from various sources should be incorporated to reduce the number of parameters.

8.5.2   Network model selection

To reconstruct a network, it is important to start with an appropriate type of network model. A network model adopts the mathematical function to depict the general behavior of the network components. Most network models can be represented by a graph containing both nodes and edges. The nodes represent network components (e.g. genes, proteins, or protein complexes), and edges between nodes represent the interactions between network components. Edges are either directed (indicating the directionality of the interaction, for example, if we have $A \rightarrow B$, node B is regulated by node A) or undirected (indicating presence/absence of the interaction, for example, if we have A—B, nodes A and B interact). Once the network model is defined, details of the model will be learned from the experimental data: the network structure illustrates the interactions among all the components in the system and the model parameters characterize different aspects of the interactions, e.g. their types/strength.

Several network models have been proposed over the past few years. These models make distinct assumptions about the underlying molecular mechanisms with varying degrees of simplification. In these network models, the activity of a component can be represented by Boolean (0 or 1), discrete (e.g. 1, 2, 3), or continuous (real) values; the type of relationships between the variables (A and B) can be directed ($A \rightarrow B$) or undirected (A—B), linear (e.g. $A = \alpha_1 B + \alpha_0$) or non-linear (e.g. $A = B^2$). The type of model can be deterministic or stochastic, static or dynamic. A deterministic model always predicts the same outcome when the initial conditions are the same, whereas a stochastic model characterizes the probability distribution of possible outcomes. Dynamic models generally define a parametric model of interactions and try to estimate the parameters from different time-points (e.g. time-course gene expression data). Static models characterize causal interactions that are consistent across the measurement (van Someren et al. 2002).

Currently there are five distinct and widely used network models (Figures 8.4 and 8.5): information theory model, Boolean network, differential equation model, Bayesian network (BN), and dynamic Bayesian network (DBN). The strengths and weaknesses of these network models will be addressed below (summary in Table 8.2).

The information theory model is one of the simplest network models (Stuart et al. 2003). It represents the regulatory system with an undirected graph, in which nodes represent components of the system and edges are interactions between components.

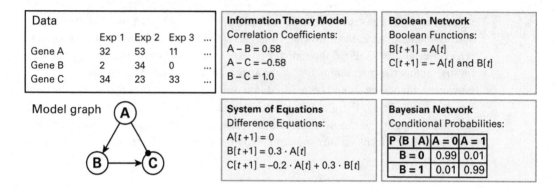

Figure 8.4   Overview of network models.

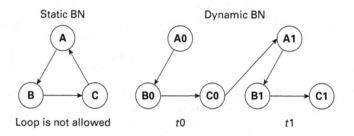

Figure 8.5   Differences between static and dynamic Bayesian networks (BNs). A feedback loop from gene A to gene B to gene C and back to gene A is not allowed in static BNs. However, this feedback loop can be represented in a dynamic BN by separating the feedback edges in two time slices.

Simplicity and low computational costs are the major advantages of the information theory model. It has been widely applied to study global properties of large-scale regulatory systems. However, a drawback of this model is that it is static and cannot adequately account for complex regulation involving multiple gene/protein components.

A Boolean network is a discrete dynamical network (Kauffman 1969). It can be represented as a directed graph, in which nodes represent components of the system and take one of two discrete values (true or false). Edges between nodes can be represented by Boolean functions made up of simple Boolean operations, e.g. AND, OR, NOT. The Boolean network allows efficient analysis of large regulatory networks. It is relatively easy to interpret, has directed edges, and allows multiple genes to participate in the network, and more importantly, it is dynamic. Boolean networks require the transformation of continuous gene expression signals to binary data. This can be performed, for instance, by clustering and thresholding using support vector regression (Martin et al. 2007). Despite these features, a Boolean network is generally criticized because it only allows for two discrete expression levels, clearly an oversimplification of biological processes.

The differential equation model represents changes in gene or protein expression as a function of the expression level of other molecules and environmental factors, and has been widely used to analyze genetic regulatory systems. This model can adequately

Table 8.2 Advantages and disadvantages of network models

	Information theory model	Boolean network	Differential equations model	Bayesian network	Dynamic Bayesian network
Simplicity	Yes	Yes	No	No	No
Low computational cost	Yes	Yes	No	No	No
Multiple genes participate in one function	No	Yes	Yes	Yes	Yes
Directed/ Undirected	Undirected	Directed	Directed	Directed	Directed
Large data set needed	No	No	Yes	Yes	Yes
Deterministic/ Stochastic	Deterministic	Deterministic	Deterministic	Stochastic	Stochastic
Handle incomplete data	No	No	No	Yes	Yes
Handle feedback loops	No	No	No	No	Yes

account for the dynamic behavior of networks by incorporating time-dependent variables, ranging within the set of non-negative real numbers. There are two types of differential equations: linear and non-linear. Linear differential equations can be simply represented as linear algebraic equations. However, the simplification obtained by linearization is not sufficient to identify large-scale networks, and complex dynamic behaviors such as stable oscillatory states cannot be explained using simple linear systems. In contrast, non-linear differential equations can well explain the complicated cellular regulation systems. However, non-linear functions present two major challenges. First, mathematical difficulties are associated with non-linear functions for parameter identification. Second, reliable identification of non-linear interactions normally requires a very large data size. Thus, inference of non-linear systems usually employs predefined functions that reflect prior knowledge to decrease the computational effort needed. But still, the problem of data insufficiency limits the practical relevance of non-linear models. Nevertheless, differential equations have directed edges, allow multiple genes to participate in the regulation, and are dynamic, such that they are good candidates for simulating gene regulatory events.

A Bayesian network (BN) represents a set of random variables and their conditional dependencies via a directed acyclic graph (DAG), which is a directed graph without feedback loops. The nodes of the graph represent molecular components and its edges represent the causal relationships between molecular components. The relationships are quantitatively encoded in the parameters representing the conditional probabilities (e.g. the probability of a gene being up/downregulated given the status of other components connecting to the gene). Unconnected nodes represent variables that are conditionally independent of one another. Bayesian networks can handle different types of data (e.g. discrete and continuous expression data) and their inference does not require discretization of the data. Nodes in BNs can have multiple parents, such that multiple gene participation is allowed. The approach makes use of the Bayes rule and can be used to reflect the stochastic nature of gene regulation (Werhli & Husmeier 2007). However, the BN is static, the learning process needs relatively large data sets, and the computational cost of the approach is relatively high. Moreover, similar to other network models mentioned above, BNs cannot handle feedback loops, which are an intrinsic feature of many biological systems.

The dynamic Bayesian network (DBN) is similar to BN except DBN is able to model dynamic behavior of networks and feedback loops, which occur frequently and are an essential property of many biological systems. The dynamic Bayesian network adopts the hidden Markov model (HMM), a stochastic probability model with hidden variables (Churchill 1989, Rabiner 1989) to model feedback loops by breaking them down into multiple time slices (Figure 8.5). It can also handle heterogeneous, incomplete, or noisy data (Sun & Hong 2007). Its probability function fits well with the stochastic nature of gene regulation. The drawback of DBN is that the learning process needs large time-series data sets and the computational complexity is very high.

8.5.3   Network inference

Network inference is achieved through both parameter learning and structure learning. Learning starts with a candidate graph of relationships (a good start will be a graph bearing prior knowledge), followed by parameter learning and structure learning. In parameter learning, the best parameters for each node need to be determined from a given graph and experimental data. And in structure learning, each candidate model is scored according to the graph and the learned parameters. The higher the score, the better the network structure fits the provided data. The final network structure inference result is usually represented by the graph with the highest score, a Bayesian average of multiple graphs, or a distribution of graphs.

8.5.4   Challenges in network reconstruction

There are considerable challenges in computational network reconstruction from biological data. First, the large scale of data from these experiments has inherent variability, as reflected by systematic errors (bias) and stochastic effects (noise) (Hecker et al. 2009). Systematic errors can be nearly eliminated by data normalization. Stochastic effects cannot be completely corrected by data processing, but can be minimized by the application of repeated measurements. Second, many data from biological experiments are incomplete. For example, proteomics data do not contain gene expression information; and vice versa. For most biological systems, it is impossible to collect a complete set of data covering every possible measurement. Thus, data integration should be applied to make maximal use of the available data, and the appropriate network models capable of handling incomplete data sets (e.g. DBN) need to be adopted. Third, even for a simple organism, the functional regulation network is complex, as the activity of gene products is regulated by many factors, including transcription factors (TFs) and co-factors that influence transcription, processing of proteins and transcripts, and/or post-translational modification or turnover of proteins. Moreover, positive and negative feedback add further complexity to the regulation of the network. Finally, the inclusion of large data sets and high degree of network complexity inevitably drive up computational costs. Therefore, depending on the model quality and complexity, the available data, and the intended application of identified networks, the suitable model architecture should be carefully chosen in order to efficiently achieve the best results.

8.6 Applications of network reconstruction

We expect mathematical modeling of networks to play an important role in generating hypotheses, driving further experimentation, and providing novel insights. Some instructive examples come from studies in other organisms. Bonneau et al. were able to reconstruct a significant portion of the regulatory network of the archaeon *Halobacterium* NRC-1 by integrating genome annotation and gene expression profiles (Bonneau et al. 2006). Several predictions made by the learned network were experimentally tested and verified. Lorenz et al. demonstrated the value of using automatic network inference to identify the regulators of complex phenotypes such as aging (Lorenz et al. 2009). They applied their method to reconstruct interactions in a 10-gene network from the Snf1 signaling pathway, which is required for expression of glucose-repressed genes upon caloric restriction. They also experimentally validated a few predicted interactions, including the demonstration that Snf1 and its transcriptional regulators Hxk2 and Mig1 act as modulators of lifespan. Kaderali et al. developed a Bayesian learning approach that infer pathway topologies from gene knockdown data using Bayesian networks with probabilistic Boolean threshold functions (Kaderali et al. 2009). They demonstrated the power of their results using RNAi data from the JAK/STAT pathway in a human hepatoma cell line. Hong and colleagues reached beyond network reconstruction in single cells and developed a theoretical framework for automatic inference of multicellular regulatory networks (in this case, for *Caenorhabditis elegans* vulval development) by integrating heterogeneous biological data (such as PPIs and gene knockout/knockdown phenotype data) (Sun & Hong 2007, Sun & Hong 2009). The reconstructed model was capable of simulating stochastic *C. elegans* vulval induction under many different genetic conditions, and hence allowed researchers to gain a systematic view about how animal development is dynamically regulated by interacting cells through complex networks of proteins and genes.

To date, most studies have applied modeling to relatively small networks, centered around one or two pathways. A comparison of two-pathway-centered networks can be used to help identify the main routes of pathway cross talk (see review by Hughey et al. 2010). Simple flow-based models have been used to analyze larger networks. For instance, modeling signal propagation within mammalian hippocampal CA1 neurons revealed global properties of regulation, such as point of signal branching and positions of positive and negative feedback loops within the network (Ma'ayan et al. 2005). An application of network modeling is to go beyond direct observations that can be made from the data and uncover novel components of a cellular response, providing new insights into the biological processes under study. For example, a recent study in yeast integrates genetic perturbation data with protein–protein and protein–DNA interaction networks to predict probable signal flow paths (Huang & Fraenkel 2009, Yeger-Lotem et al. 2009). The model characterized the highest probable flow paths in the PPI network by connecting genetic hits identified from perturbation screens to the corresponding expression changes, revealing novel components within such flow paths. More recently, flow-based network modeling was applied to identification of novel human phospho-ERK modulators (Vinayagam et al. 2011). The flow model used known pERK modulators as source nodes. Hidden nodes downstream of multiple source nodes were predicted to be novel pERK modulators, a prediction that was subsequently validated in a cell-based assay. In the context of the *Drosophila* screens described above these approaches now need to be implemented to gain further insights into the structure of the signaling networks.

Predicting novel drug targets is another key application of network models. Network-centered drug-discovery platforms are still in their infancy but some progress has been made. Biological networks are robust in response to removal of most nodes due to redundancy. However, non-redundant nodes appear to be more vulnerable. Network models may facilitate prediction of robust and vulnerable targets based on the network structure. A drug might be effective if it hits a point of fragility in the network; however, targeting an unexpected or extreme point of fragility might lead to more troublesome drug side-effects or toxicity (Figure 8.6). Thus, the goal for network modeling is to find

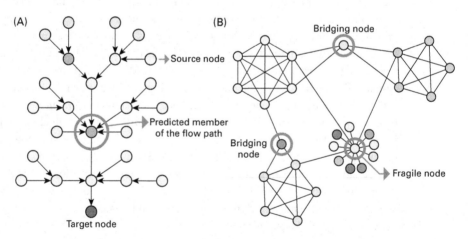

Figure 8.6   Application of network models. (A) Schematic representation of a signal flow path predicted by network modeling (from a directed network). Light gray nodes are source nodes, which are known experimentally to be part of the signal flow. Black nodes are target nodes, where signals converge. The dark gray nodes are hidden nodes, which are predicted to be part of the signal flow. (B) Schematic representation of bridging nodes, which connect two modules. Here, the flow model is applied to an undirected network to distinguish between bridging nodes and fragile nodes (i.e. nodes that have high-betweenness).

a set of nodes that are critical in the network structure but at the same time, not so critical that targeting them is likely to lead to global functional impairment (Kitano 2007, Fliri et al. 2009, Schadt et al. 2009). Flow-based models have been proposed to identify bridging nodes (Figure 8.6), which link two modules. Targeting such nodes only prevents information flow between the modules of interest, not global impairment (Hwang et al. 2008). Recent advancements in developing tools to control complex networks (Liu et al. 2011) will offer a radically new way to develop network-based drug targets. It will be exciting to see how increasingly sophisticated and accurate models contribute to our ability to design new avenues of research and gain novel insights into biology.

Acknowledgements

S.E.M. is supported by R01 GM067761 from NIGMS. N.P. is an investigator of the Howard Hughes Medical Institute.

References

Bakal, C. (2011), '*Drosophila* RNAi screening in a postgenomic world', *Briefings in Functional Genomics* **10**(4), 197–205.

Bakal, C. & Perrimon, N. (2010), 'Realizing the promise of RNAi high throughput screening', *Developmental Cell* **18**(4), 506–7.

Bakal, C., Aach, J., Church, G. & Perrimon, N. (2007), 'Quantitative morphological signatures define local signaling networks regulating cell morphology', *Science* **316**(5832), 1753–6.

Bakal, C., Linding, R., Llense, F., Heffern, E., Martin-Blanco, E. et al. (2008), 'Phosphorylation networks regulating JNK activity in diverse genetic backgrounds', *Science* **322**(5900), 453–6.

Bonneau, R., Reiss, D. J., Shannon, P., Facciotti, M., Hood, L. et al. (2006), 'The Inferelator: an algorithm for learning parsimonious regulatory networks from systems-biology data sets de novo', *Genome Biology* **7**(5), R36.

Booker, M., Samsonova, A. A., Kwon, Y., Flockhart, I., Mohr, S. E. et al. (2011), 'False negative rates in *Drosophila* cell-based RNAi screens: a case study', *BMC Genomics* **12**, 50.

Boutros, M., Kiger, A. A., Armknecht, S., Kerr, K., Hild, M. et al. (2004), 'Genome-wide RNAi analysis of growth and viability in *Drosophila* cells', *Science* **303**(5659), 832–5.

Choi, H., Larsen, B., Lin, Z.-Y., Breitkreutz, A., Mellacheruvu, D. et al. (2011), 'SAINT: probabilistic scoring of affinity purification-mass spectrometry data', *Nature Methods* **8**(1), 70–3.

Churchill, G. A. (1989), 'Stochastic models for heterogeneous DNA sequences', *Bulletin of Mathematical Biology* **51**(1), 79–94.

Collinet, C., Stoter, M., Bradshaw, C. R., Samusik, N., Rink, J. C. et al. (2010), 'Systems survey of endocytosis by multiparametric image analysis', *Nature* **464**(7286), 243–9.

DasGupta, R. & Gonsalves, F. C. (2008), 'High-throughput RNAi screen in *Drosophila*', *Methods in Molecular Biology* **469**, 163–84.

DasGupta, R., Nybakken, K., Booker, M., Mathey-Prevot, B., Gonsalves, F. et al. (2007), 'A case study of the reproducibility of transcriptional reporter cell-based RNAi screens in *Drosophila*', *Genome Biology* **8**(9), R203.

Falschlehner, C., Steinbrink, S., Erdmann, G. & Boutros, M. (2010), 'High-throughput RNAi screening to dissect cellular pathways: a how-to guide', *Biotechnology Journal* **5**(4), 368–76.

Fliri, A. F., Loging, W. T. & Volkmann, R. A. (2009), 'Drug effects viewed from a signal transduction network perspective', *Journal of Medicinal Chemistry* **52**(24), 8038–46.

Flockhart, I., Booker, M., Kiger, A., Boutros, M., Armknecht, S. et al. (2006), 'FlyRNAi: the *Drosophila* RNAi screening center database', *Nucleic Acids Research* **34**(Database issue), D489–94.

Friedman, A. & Perrimon, N. (2006), 'High-throughput approaches to dissecting MAPK signaling pathways', *Methods* **40**(3), 262–71.

Friedman, A. & Perrimon, N. (2007), 'Genetic screening for signal transduction in the era of network biology', *Cell* **128**(2), 225–31.

Friedman, A. A., Tucker, G., Singh, R., Yan, E., Vinayagam, A. et al. (2011), 'Proteomic and functional genomic landscape of receptor tyrosine kinase and Ras/ERK signaling', *Science Signaling* **4**(196), rs10.

Gavin, A.-C., Aloy, P., Grandi, P., Krause, R., Boesche, M. et al. (2006), 'Proteome survey reveals modularity of the yeast cell machinery', *Nature* **440**(7084), 631–6.

Gilsdorf, M., Horn, T., Arziman, Z., Pelz, O., Kiner, E. et al. (2010), 'GenomeRNAi: a database for cell-based RNAi phenotypes – 2009 update', *Nucleic Acids Research* **38**(Database issue), D448–52.

Gunsalus, K. C., Ge, H., Schetter, A. J., Goldberg, D. S., Han, J.-D. J. et al. (2005), 'Predictive models of molecular machines involved in *Caenorhabditis elegans* early embryogenesis', *Nature* **436**(7052), 861–5.

Hecker, M., Lambeck, S., Toepfer, S., van Someren, E. & Guthke, R. (2009), 'Gene regulatory network inference: data integration in dynamic models – a review', *Biosystems* **96**(1), 86–103.

Huang, S. S. & Fraenkel, E. (2009), 'Integrating proteomic, transcriptional, and interactome data reveals hidden components of signaling and regulatory networks', *Science Signaling* **2**(81), ra40.

Hughey, J. J., Lee, T. K. & Covert, M. W. (2010), 'Computational modeling of mammalian signaling networks', *Wiley Interdisciplinary Reviews. Systems Biology and Medicine* **2**(2), 194–209.

Hwang, W. C., Zhang, A. & Ramanathan, M. (2008), 'Identification of information flow-modulating drug targets: a novel bridging paradigm for drug discovery', *Clinical Pharmacology and Therapeutics* **84**(5), 563–72.

Jeronimo, C., Forget, D., Bouchard, A., Li, Q., Chua, G. et al. (2007), 'Systematic analysis of the protein interaction network for the human transcription machinery reveals the identity of the 7SK capping enzyme', *Molecular Cell* **27**(2), 262–4.

Kaderali, L., Dazert, E., Zeuge, U., Frese, M. & Bartenschlager, R. (2009), 'Reconstructing signaling pathways from RNAi data using probabilistic Boolean threshold networks', *Bioinformatics* **25**(17), 2229–35.

Kaplow, I. M., Singh, R., Friedman, A., Bakal, C., Perrimon, N. et al. (2009), 'RNAiCut: automated detection of significant genes from functional genomic screens', *Nature Methods* **6**(7), 476–7.

Kauffman, S. A. (1969), 'Metabolic stability and epigenesis in randomly constructed genetic nets', *Journal of Theoretical Biology* **22**(3), 437–67.

Kitano, H. (2007), 'Biological robustness in complex host–pathogen systems', *Progress in Drug Research* **64**, 239, 241–63.

Kockel, L., Kerr, K. S., Melnick, M., Bruckner, K., Hebrok, M. et al. (2010), 'Dynamic switch of negative feedback regulation in *Drosophila* Akt-TOR signaling', *PLoS Genetics* **6**(6), e1000990.

Krogan, N. J., Cagney, G., Yu, H., Zhong, G., Guo, X. et al. (2006), 'Global landscape of protein complexes in the yeast *Saccharomyces cerevisiae*', *Nature* **440** (7084), 637–43.

Kummel, A., Gubler, H., Gehin, P., Beibel, M., Gabriel, D. et al. (2010), 'Integration of multiple readouts into the Z-factor for assay quality assessment', *Journal of Biomolecular Screening* **15**(1), 95–101.

Liu, Y. Y., Slotine, J. J. & Barabasi, A. L. (2011), 'Controllability of complex networks', *Nature* **473**(7346), 167–73.

Ljosa, V. & Carpenter, A. E. (2009), 'Introduction to the quantitative analysis of two-dimensional fluorescence microscopy images for cell-based screening', *PLoS Computational Biology* **5**(12), e1000603.

Lorenz, D. R., Cantor, C. R. & Collins, J. J. (2009), 'A network biology approach to aging in yeast', *Proceedings of the National Academy of Sciences of the United States of America* **106**(4), 1145–50.

Ma'ayan, A., Jenkins, S. L., Neves, S., Hasseldine, A., Grace, E. et al. (2005), 'Formation of regulatory patterns during signal propagation in a mammalian cellular network', *Science* **309**(5737), 1078–83.

Martin, S., Zhang, Z., Martino, A. & Faulon, J.-L. (2007), 'Boolean dynamics of genetic regulatory networks inferred from microarray time series data', *Bioinformatics* **23**(7), 866–74.

Mohr, S., Bakal, C. & Perrimon, N. (2010), 'Genomic screening with RNAi: results and challenges', *Annual Review of Biochemistry* **79**, 37–64.

Niederlein, A., Meyenhofer, F., White, D. & Bickle, M. (2009), 'Image analysis in high-content screening', *Combinatorial Chemistry & High Throughput Screening* **12**(9), 899–907.

Nusslein-Volhard, C. & Wieschaus, E. (1980), 'Mutations affecting segment number and polarity in *Drosophila*', *Nature* **287**(5785), 795–801.

Perlman, Z. E., Slack, M. D., Feng, Y., Mitchison, T. J., Wu, L. F. et al. (2004), 'Multidimensional drug profiling by automated microscopy', *Science* **306**(5699), 1194–8.

Rabiner, L. R. (1989), 'A tutorial on hidden Markov models and selected applications in speech recognition', *Proceedings of the IEEE*, **77**(2), 257–86.

Sardiu, M. E., Cai, Y., Jin, J., Swanson, S. K., Conaway, R. C. et al. (2008), 'Probabilistic assembly of human protein interaction networks from label-free quantitative proteomics', *Proceedings of the National Academy of Sciences of the United States of America* **105**(5), 1454–9.

Schadt, E. E., Friend, S. H. & Shaywitz, D. A. (2009), 'A network view of disease and compound screening', *Nature Reviews Drug Discovery* **8**(4), 286–95.

Seinen, E., Burgerhof, J. G., Jansen, R. C. & Sibon, O. C. (2011), 'RNAi-induced off-target effects in *Drosophila melanogaster*: frequencies and solutions', *Briefings in Functional Genomics* **10**(4), 206–14.

Shumate, C. & Hoffman, A. F. (2009), 'Instrumental considerations in high content screening', *Combinatorial Chemistry & High Throughput Screening* **12**(9), 888–98.

Sims, D., Bursteinas, B., Gao, Q., Zvelebil, M. & Baum, B. (2006), 'FLIGHT: database and tools for the integration and cross-correlation of large-scale RNAi phenotypic datasets', *Nucleic Acids Research* **34**(Database issue), D479–83.

Sowa, M. E., Bennett, E. J., Gygi, S. P. & Harper, J. W. (2009), 'Defining the human deubiquitinating enzyme interaction landscape', *Cell* **138**(2), 389–403.

St Johnston, D. & Nusslein-Volhard, C. (1992), 'The origin of pattern and polarity in the *Drosophila* embryo', *Cell* **68**(2), 201–19.

Stuart, J. M., Segal, E., Koller, D. & Kim, S. K. (2003), 'A gene-coexpression network for global discovery of conserved genetic modules', *Science* **302**(5643), 249–55.

Sun, X. & Hong, P. (2007), 'Computational modeling of *Caenorhabditis elegans* vulval induction', *Bioinformatics* **23**(13), i499–507.

Sun, X. & Hong, P. (2009), 'Automatic inference of multicellular regulatory networks using informative priors', *International Journal of Computational Biology and Drug Design* **2**(2), 115–33.

van Someren, E. P., Wessels, L. F. A., Backer, E. & Reinders, M. J. T. (2002), 'Genetic network modeling', *Pharmacogenomics* **3**(4), 507–25.

Vinayagam, A., Stelzl, U., Foulle, R., Plassmann, S., Zenkner, M. et al. (2011), 'A directed protein interaction network for investigating intracellular signal transduction', *Science Signaling* **4**(189), rs8.

Walhout, A. J. M., Reboul, J., Shtanko, O., Bertin, N., Vaglio, P. et al. (2002), 'Integrating interactome, phenome, and transcriptome mapping data for the *C. elegans* germline', *Current Biology* **12**(22), 1952–8.

Werhli, A. V. & Husmeier, D. (2007), 'Reconstructing gene regulatory networks with Bayesian networks by combining expression data with multiple sources of prior knowledge', *Statistical Applications in Genetics and Molecular Biology* **6**, Article15.

Yeger-Lotem, E., Riva, L., Su, L. J., Gitler, A. D., Cashikar, A. G. et al. (2009), 'Bridging high-throughput genetic and transcriptional data reveals cellular responses to alpha-synuclein toxicity', *Nature Genetics* **41**(3), 316–23.

Zanella, F., Lorens, J. B. & Link, W. (2010), 'High content screening: seeing is believing', *Trends in Biotechnology* **28**(5), 237–45.

Zhong, W. & Sternberg, P. W. (2007), 'Automated data integration for developmental biological research', *Development* **134**(18), 3227–38.

Zhu, X., Gerstein, M. & Snyder, M. (2007), 'Getting connected: analysis and principles of biological networks', *Genes & Development* **21**(9), 1010–24.

9   Bayesian inference for model selection: an application to aberrant signalling pathways in chronic myeloid leukaemia

Lisa E. M. Hopcroft, Ben Calderhead, Paolo Gallipoli, Tessa L. Holyoake, and Mark A. Girolami

In the analysis of any data using statistical modelling, it is imperative that the choice of model is informed by expert knowledge and that its adequacy is determined based on the extent to which it captures and describes the patterns observed in the data. This is especially true in systems where a subset of the constituent components may not be known or cannot be observed. In this chapter, we demonstrate how statistical inference can be used to inform model selection and, by identifying where existing models are unable to sufficiently capture observed behaviour, that statistical inference can help indicate which model refinements may be required.

In this chapter, we use Bayesian statistical methodology – specifically, Riemannian manifold population MCMC – to model interactions between molecular species in the JAK/STAT pathway in chronic myeloid leukaemia (CML) and compare two candidate models. We set out the biological context for this inference in Sections 9.1–9.1.4 and describe the two candidate models in Section 9.3. With the biology established, we describe our statistical methodology (Section 9.4) which we successfully apply in a simulation study to provide a proof of concept (Section 9.5), before we consider a subsequent, more biologically realistic dataset (Section 9.6) to assess which model best describes the behaviour observed *in vitro*. We relate the findings from this second synthetic study back to our model and dataset construction, thereby highlighting what further *in vitro* and *in silico* work is required (Section 9.7).

9.1   The oncology of chronic myeloid leukaemia

The condition that we now recognise as chronic myeloid leukaemia (CML) was first described in 1845, in quick succession, by two pathologists, Dr John Hughes Bennett (Bennett 1845) and Dr Rudolf Virchow (Virchow 1845). In the case reported by Bennett, hypertrophy of the spleen and liver was prominent and death was thought to have

Systems Genetics: Linking Genotypes and Phenotypes, ed. F. Markowetz and M. Boutros. Published by Cambridge University Press. © Cambridge University Press 2015.

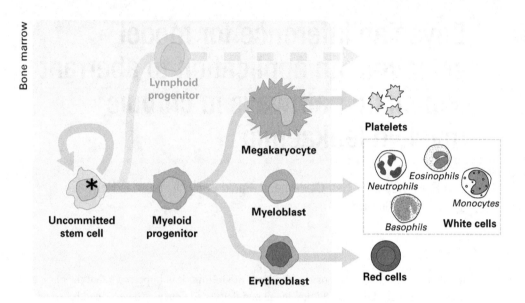

Figure 9.1   Haemopoiesis: the process of blood cell formation. Uncommitted (pluripotent) stem cells
generate daughter cells which successively differentiate into several cell subtypes in the bone
marrow, eventually giving rise to mature, functional cells which leave the bone marrow and enter
the peripheral blood. The first major lineage division is between myeloid and lymphoid
progenitors; CML (chronic *myeloid* leukaemia) predominantly affects the myeloid lineage. The
genomic lesion giving rise to CML (the Philadelphia chromosome, Section 9.1.2) originates
from the uncommitted, pluripotent stem cell (indicated by a star in the diagram). The CML
oncogene is conferred to progeny via the normal process of haemopoiesis. Pluripotent stem cells
can also participate in self-renewal to generate new pluripotent cells or withdraw from the
cell-cycle process altogether, to be described as 'quiescent'.

occurred "from suppuration of the blood". These very early reports, without the aid of
modern laboratory techniques, very accurately described the phenotype of advanced,
untreated CML in which expansion of mature white blood cells takes over the bone
marrow, leading to production of blood in which white cells predominate over red cells
such that the blood resembles puss (a symptom now recognised as leukocytosis).

9.1.1   Blood cell production: haemopoiesis

The process of normal blood cell production (haemopoiesis) is hierarchically organised:
it is initiated in an uncommitted (or pluripotent) stem cell that undergoes a series of divisions and differentiations to produce mature progeny that we recognise in the peripheral
blood as functional white cells, red cells and platelets (Figure 9.1). The numbers of these
mature cells are tightly regulated at the stem and progenitor cell level by a variety of
signalling pathways and interactions with the bone marrow microenvironment that control quiescence versus proliferation, survival versus apoptosis and differentiation versus
self-renewal. This carefully controlled process of haemopoiesis is disrupted in CML.

9.1.2   Clarification of the CML genotype: the *BCR-ABL* oncogene

Chronic myeloid leukaemia was the first example of a single, chromosomal abnormality linked to, and later shown to be causative for, a specific cancer phenotype. In 1960, Nowell and Hungerford, working in Philadelphia, described an acrocentric chromosome in all patients with CML, which became known as the Philadelphia (Ph) chromosome (Nowell & Hungerford 1960) and was later clarified as a shortened chromosome 22 (Figure 9.2). In 1973, it was demonstrated that the Ph chromosome was the product of a balanced translocation between the long arms of chromosomes 9 and 22 (Rowley 1973) (commonly referred to as the t(9;22) translocation). Between 1970 and 1990 it became clear that mutations in normal cellular genes could confer oncogenic potential and de Klein et al. (1982) demonstrated that in CML, c-*ABL*, a tyrosine kinase normally found on chromosome 9, is translocated to the Ph chromosome, adjacent to the breakpoint cluster region *BCR* (Figure 9.3). When investigated by Northern blots it became apparent that, in CML patients, ABL was larger than normal and subsequently shown to represent a chimeric or fused transcript of c-*ABL* and *BCR*, now referred to as *BCR-ABL* (Shtivelman et al. 1985) (Figure 9.4). This transcript is translated into a 210-kDa

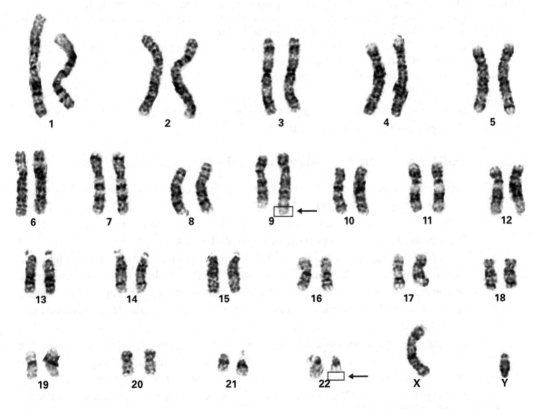

Figure 9.2   A shortened, acrocentric chromosome 22 (referred to as the Philadelphia chromosome) and an elongated chromosome 9 were found to be characteristic of CML patients. They are both highlighted here with arrows.

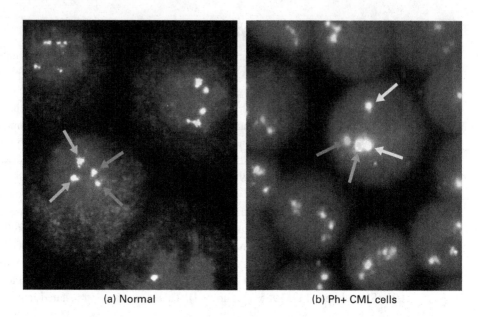

(a) Normal   (b) Ph+ CML cells

Figure 9.3   FISH (fluorescence in situ hybridisation) images of human cells. (a) In normal human cells, the *ABL* (red) and *BCR* (green) genes (two copies of each) are physically separated on chromosomes 9 and 22 respectively; (b) in CML human cells, a yellow signal (due to the combined red and green signal of co-located *ABL* and *BCR*) indicates the presence of the fusion genes *BCR-ABL* and *ABL-BCR*. A black and white version of this figure will appear in some formats. For the colour version, please refer to the plate section.

protein, p^{210}BCR-ABL, a constitutively active tyrosine kinase that is the driver for the CML phenotype (Ben-Neriah et al. 1986).

9.1.3   *BCR-ABL* as an abnormal genotype is sufficient to drive CML phenotype

Further investigation of BCR-ABL kinase activity led to two important discoveries: firstly that BCR-ABL has elevated kinase activity as compared to ABL and secondly that BCR-ABL can transform normal cells into leukaemogenic cells (Konopka et al. 1984, Lugo et al. 1990). The transformation potential of BCR-ABL was confirmed when BCR-ABL was transduced into murine bone marrow cells followed by transplantation into irradiated mice. These mice all developed CML-like disease or acute leukaemia, demonstrating that BCR-ABL as a sole oncogenic event is sufficient to cause disease (Daley et al. 1990, Heisterkamp et al. 1990). More recent studies have concluded that BCR-ABL is capable of transformation and drives leukaemia development only when transduced into haemopoietic stem cells which possess inherent self-renewal capacity; if transduced into more mature blood cells that have differentiated beyond the stage of having self-renewal potential then leukaemia does not occur (Huntly et al. 2004). Extrapolating this to CML in humans, this suggests that CML is initiated when the t(9;22) translocation occurs in a single haemopoietic stem cell – the cell of origin – conferring, via signals derived from BCR-ABL kinase, a growth advantage over normal haemopoietic stem cells in the bone marrow and leading, over time, to myeloid

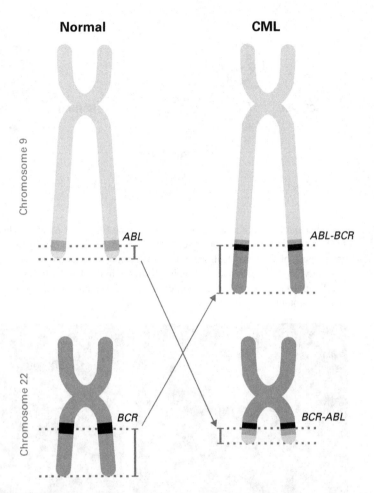

Figure 9.4   The reciprocal t(9;22) translocation between chromosome 9 and chromosome 22 results in the fusion genes *BCR-ABL* and *ABL-BCR* and gives rise to the Philadelphia chromosome shown in Figure 9.2. The constitutively active kinase activity of the BCR-ABL protein causes CML. Although *ABL-BCR* is expressed, it is not currently known to have any role in CML (Melo et al. 1993).

cell expansion and recognisable leukaemia. The progress made between 1960 and 1990, demonstrating the link between oncogenes and kinase biochemistry, identified BCR-ABL as an excellent potential therapeutic target. Imatinib – a drug designed to inhibit the BCR-ABL kinase (Figure 9.5) – emerged in the 1990s and since its introduction to the clinic in 1998 it has become the standard of care for CML patients worldwide (Druker et al. 1996, Druker et al. 2001, Druker et al. 2006).

9.1.4   Understanding the oncology of CML with a view to improving treatment

Chronic myeloid leukaemia is initiated in a normal haemopoietic stem cell upon formation of the Ph chromosome and subsequent expression of the oncogene *BCR-ABL*.

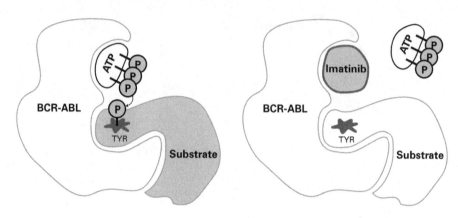

Figure 9.5   The molecular action of imatinib. (a) The oncogene *BCR-ABL* phosphorylates tyrosine residues of substrate proteins. It provides a scaffold to which ATP (adenosine triphosphate) and the substrate can bind, thereby allowing the transfer of a phosphate group and subsequent phosphorylation of the substrate protein. (b) Imatinib inhibits the space otherwise occupied by the ATP molecule, thereby blocking recruitment of ATP and inhibiting phosphorylation of substrate proteins.

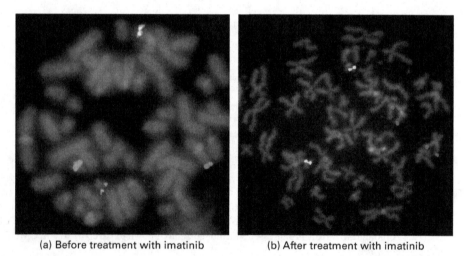

Figure 9.6   A complete cytogenetic response to imatinib treatment in the metaphase cells of a CML patient. The yellow signal in (a) indicates the presence of Philadelphia chromosome (and therefore oncogene *BCR-ABL*); (b) following treatment with imatinib, the Philadelphia chromosome is no longer present. A black and white version of this figure will appear in some formats. For the colour version, please refer to the plate section.

The increased kinase activity of BCR-ABL interferes with numerous signalling pathways culminating in a phenotype with enhanced survival and proliferation, and reduced apoptosis. Patients treated with imatinib rapidly enter a cytogenetic response (the Ph chromosome is no longer detectable in bone marrow cells, Figure 9.6), followed by a major molecular response (*BCR-ABL* transcript levels reduced by 3 logs from 100% at diagnosis to less than 0.1% (Druker et al. 2006)). However, disease levels then plateau, and few, if any, patients are cured. It is clear that this relates to failure to target the rare self-renewing cancer stem cell population which is now the subject of extensive investigation on the pathway to cure (16th Congress of European Haematology Association 2011).

BCR-ABL encodes a large protein with multiple domains. The ABL kinase domain is essential for BCR-ABL-induced leukaemogenesis *in vivo*, thus justifying the use of tyrosine kinase inhibitor (TKI) in the treatment of CML (Zhang & Ren 1998). However, several of the other BCR-ABL domains have been investigated, and their function at least partly unveiled, using mutational analysis. Some of these domains regulate ABL kinase activity, whilst others have been implicated in connecting BCR-ABL to downstream signalling pathways (Gross et al. 1999, Pendergast et al. 1993). These pathways can be broadly described as mitogenic or anti-apoptotic, leading to increased proliferation and reduced killing of leukaemic cells respectively. Most of these pathways are also activated following growth factor stimulation of normal, non-leukaemic haemopoietic stem and progenitor cells (Jin et al. 2006). Their direct activation by BCR-ABL therefore explains the proliferative phenotype of BCR-ABL-driven leukaemias and their growth factor independence *in vitro*.

Given the complexity of signalling in CML cells, novel approaches leading to an enhanced understanding of which pathways or proteins predominantly control leukaemic stem cell survival, and therefore represent potential targets for therapy, are clearly required. Whereas for more differentiated, mature CML cells TKIs induce apoptosis through inhibition of BCR-ABL kinase alone, it is likely that eradication of CML stem cells will only be possible using a combination of different inhibitors targeting multiple pathways. One such pathway is the JAK/STAT pathway.

9.1.5   The JAK/STAT pathway

The JAK/STAT pathway (Figure 9.7) is central to the normal cellular response to growth factors. JAKs are a family of intracellular kinases, normally activated following dimerisation of growth factor receptors upon ligand binding. Once activated, JAK kinases phosphorylate the STAT transcription factors, resulting in STAT dimerisation and nuclear translocation, which in turn leads to activation of a variety of proliferative and anti-apoptotic signals, such as upregulation of the anti-apoptotic protein BCL-XL (Benekli et al. 2003). Amongst the JAK kinases, JAK2 plays a prominent role in the normal response to haemopoietic growth factors by specifically activating the STAT5 protein. Mimicking growth factor signalling, BCR-ABL has been shown to activate both JAK2 and STAT5 in the absence of ligand binding, such that STAT5 is constitutively phosphorylated in CML cells (Chai et al. 1997, Ilaria & Van Etten 1996).

The role of JAK2/STAT5 in BCR-ABL-induced leukaemia has been widely investigated. The possibility that JAK2/STAT5 signalling may represent a potential novel target for leukaemic stem cell eradication is supported by recent evidence suggesting that STAT5 is indispensable for the maintenance of BCR-ABL+ leukaemias (Hoelbl et al. 2010, Walz et al. 2012). High levels of STAT5 have also been proposed as protective for BCR-ABL+ cells upon treatment with TKIs (Warsch et al. 2011). This protection

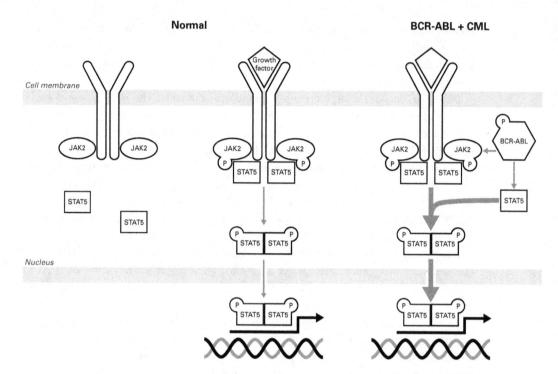

Figure 9.7 The role of JAK2 and STAT5 in CML. Upon activation (phosphorylation) via growth factors, JAK2 phosphorylates the transcription factor STAT5, resulting in STAT5 dimerisation and its subsequent nuclear translocation and upregulation of several anti-apoptotic and proliferative signals. In a Ph+ CML cell, JAK2 is activated by both the constitutively active oncogene BCR-ABL *and* growth factors. Moreover, STAT5 has been shown to be active in the presence of BCR-ABL, *independent* of JAK2 activation. Both mechanisms contribute to pSTAT5 being constitutively active, resulting in upregulation of several downstream, anti-apoptotic and proliferative signalling pathways. This identifies the JAK/STAT pathway as a valid target for CML therapy; to progress in this direction, we must first clarify the precise mechanism(s) by which JAK2 is activated in Ph+ cells.

required STAT5 phosphorylation and effective STAT5 inhibition led to reduced survival of CML cells resistant to TKIs (Nelson et al. 2011). However, before new approaches to inhibit STAT5 phosphorylation can be developed it is critical to gain an improved understanding of exactly how STAT5 is activated in BCR-ABL-expressing leukaemic stem cells.

The relative contributions of JAK2 and BCR-ABL to STAT5 phosphorylation was originally investigated in BCR-ABL+ cell lines. Although BCR-ABL was shown to phosphorylate JAK2, and in turn STAT5, it also appeared to directly activate STAT5 in a JAK2-independent fashion. This observation led to the conclusion that JAK2 was dispensable in the STAT5 activation in BCR-ABL+ cells (Ilaria & Van Etten 1996). However, while this might appear to be the case for BCR-ABL+ cell lines *in vitro*, the situation appears to be more complicated in primary cells and *in vivo*. Primary CML cells have been shown to be capable of autocrine production of haemopoietic growth factors, such as interleukin-3 (IL-3) and granulocyte macrophage colony-stimulating factor (GM-CSF), both of which signal via JAK2 (Jiang et al. 1999, Wang et al. 2007). Moreover, it is conceivable that *in vivo* leukaemic stem and progenitor cells are subjected to cues from the bone marrow stroma which will include the same growth factors. In this respect the use of JAK2 inhibitors with TKI against primary CML cells *in vitro* has been shown to increase CML stem and progenitor cell apoptosis and reduce their proliferative potential, as compared to TKI alone (Hiwase et al. 2010). Renewed interest in JAK2 as a target in CML has also resulted from the discovery that activating JAK2 mutations are frequently detected in BCR-ABL negative myeloproliferative disorders (Baxter et al. 2005, Kralovics et al. 2005), leading to the clinical development of JAK2 inhibitors and making their use in combination with BCR-ABL inhibitors a realistic option for CML patients (Verstovsek et al. 2010).

Alongside its recognised role in transducing growth factor signals, JAK2 also appears to contribute to BCR-ABL leukaemogenesis and particularly to survival of BCR-ABL+ cells, including the maintenance of CML stem and progenitor cells. The exact mechanisms underlying these properties are still not clear. Recent evidence has shown that in BCR-ABL+ cell lines JAK2 forms a complex with BCR-ABL, leading to enhanced stability of the BCR-ABL protein and its activity (Samanta et al. 2006). By disrupting this complex, using either JAK2 knock-down or a JAK2 inhibitor, BCR-ABL kinase activity was reduced. Of further potential therapeutic interest, targeting JAK2 appeared effective even in cases where BCR-ABL kinase was mutated and therefore not responsive to standard TKI, such as imatinib (Samanta et al. 2006, Samanta et al. 2011).

A similar hypothesis is currently being investigated in primary CML cells from chronic phase patients. Preliminary results have shown that BCR-ABL and JAK2 interact in CML stem and progenitor cells via a third protein, AHI-1. *AHI-1* is a novel oncogene which has recently been shown to enhance BCR-ABL transforming activity and activation of downstream signalling pathways, including the JAK/STAT pathway, thus also contributing to resistance of CML stem and progenitor cells to current TKI (Zhou et al. 2008). Disruption of this complex by combined targeting of BCR-ABL and JAK2 kinases resulted in increased apoptosis in primary CML stem and progenitor cells (over and above that observed in following the single-agent TKI treatment) and reduced the number of CML stem cells capable of long-term, multilineage engraftment in immunodeficient mice (DeGeer et al. 2010). These effects appear to be secondary to reduced proliferative and anti-apoptotic signals upon combined inhibition, with the levels of phosphorylated STAT5 significantly reduced.

Other groups have also demonstrated a prominent role for JAK2 in the maintenance of CML stem and progenitor cells. Interestingly, this appears to involve interference with other signalling proteins in addition to STAT5. In particular, JAK2 appears to play a prominent role in inactivating the intracellular phosphatase and tumour suppressor protein phosphatase 2A (PP2A) (Samanta et al. 2009). PP2A activity is reduced in CML stem and progenitor cells and its reactivation by inhibition of JAK2 was demonstrated to lead to the eradication of CML stem and progenitor cells, thus potentially explaining the function of JAK2 inhibitors in CML. Moreover JAK2 appears to play a significant role in the activation of the transcription factor beta-catenin which has also been implicated in the maintenance of leukaemic stem cells. As a result JAK2 inhibitors have been shown to target the most primitive CML stem cells, including the quiescent population, which are spared by standard TKI therapy (Neviani et al. 2010). Taken together, the evidence from several research groups suggests that the interplay between JAK2 and BCR-ABL is more complex than originally thought and highly relevant to CML stem and progenitor cell maintenance and proliferation.

In summary, there is evidence that the JAK/STAT pathway contributes significantly to the oncology of CML, and so presents itself as a candidate pathway for intervention in CML therapy. However, it is also clear that JAK2 and STAT5 are critical to normal haemopoiesis and inhibition of either could have detrimental effects on normal blood cell production (Ward et al. 2000). To progress in the investigation of possible therapies, we must clarify the mechanism by which JAK2 is activated in Ph+ cells and to what extent our current understanding of this pathway accounts for the behaviour of the constituent species. The remainder of this chapter describes our use of Bayesian statistics to address these questions.

9.2   Introduction to model comparison

When modelling with ordinary differential equations (ODEs) we make the assumption that we are modelling an average underlying process and that we observe noisy measurements of this 'true' behaviour. This assumption is valid when we obtain measurements that are averages over a large population of cells, such that the intrinsic noise does not determine the overall dynamics. In contrast, when there are relatively few interacting molecules their intrinsic stochasticity may play a more defining role in determining the overall dynamical behaviour of the system, and in such scenarios stochastic differential equations (SDEs) may be a more appropriate modelling formalism to employ. In the case of this work, we choose to work with ODE models; however, within this class finding the most suitable model to describe a biological system can be very challenging.

Over-parameterisation is a particular problem. We would like to have a model that is complex enough to describe the main features of the dynamical behaviour we observe, but not overly complex such that we might end up describing the noise rather than the underlying process. When using experimentally observed data there is generally no such thing as a 'correct' model, since ultimately we are considering mathematical abstractions and some models may be more or less useful than others in terms of their predictive ability. Ultimately we want to obtain testable predictions that can be used to inform future experimental work in a process of model refinement.

We therefore require a method of systematically comparing and ranking model hypotheses that potentially describe the biological system of interest. A naive approach such as comparing the maximum likelihood of different models is generally not useful, as we will consistently choose the most complex model that may be over-fitting the noisy observations. The Bayesian framework on the other hand provides a natural way of parsimoniously comparing models such that unnecessary complexity is appropriately penalised. By integrating over the posterior probability distribution we obtain the marginal likelihood, from which we can calculate the probability of the model itself given the experimental data. Integration over additional unnecessary parameter dimensions reduces the marginal likelihood so that simpler models with similar predictive performance are preferred; an example of Occam's razor.

Given a set of model hypotheses, in our case encoded using the mathematical formalism of ODEs, we may rank their suitability for describing the data. We note however that this is a relative ranking; by plotting the posterior predictions of our chosen model we may check that this structural hypothesis does describe the observations, or indeed any new observations as they become available. Should our best model not be consistent with additional experimental measurements, we may revise our model hypotheses, perhaps by adding components or changing the structure of interactions, and rerun our Bayesian model ranking procedure. We may inform our selection of model hypotheses by drawing on the expert knowledge and insight of experimental biologists.

9.3   Modelling the JAK/STAT pathway in response to TKI and/or JakI

Uncertainty surrounds many of the mechanisms of molecular interaction that drive the nonlinear dynamical behaviour observed in biological systems. Here, we are interested in describing the interaction of BCR-ABL with JAK2 and STAT5, and predicting the effect of inhibitors TKI and JakI on the overall dynamics. The use of mathematical modelling for this system enables us to probe the possible underlying structures that drive the observed dynamics and make testable predictions. In particular the hypothesised mechanistic structure of a biological system can be naturally represented as a system of differential equations that describes the time evolution of the chemical species involved (Xu et al. 2010). A statistical analysis then allows us to characterise the multiple sources of uncertainty, not only regarding the free parameters of a particular mathematical model, but also the structure of the model itself (Calderhead & Girolami 2011).

Adopting a probabilistic Bayesian approach allows us to infer posterior distributions that describe the probability of parameter values for a particular model given the available data. In addition, we may obtain estimates of the marginal likelihood, which describes the probability of the model itself given the data, averaged over all the possible parameter values. This allows us to rank hypothesised model structures by considering their relative probabilities, and decide which mechanisms of interaction are most supported by the experimental data (Calderhead & Girolami 2009, Vyshemirsky & Girolami 2008).

We demonstrate how this methodological approach may work in practice by considering two hypothetical models to describe a network of interactions involving BCR-ABL, JAK2 and STAT5, and the inhibitors TKI and JakI. Our first hypothesis is based on the assumption that there is no direct interaction between BCR-ABL and JAK2; our second hypothesis captures an additional mechanism by which BCR-ABL and JAK2 may interact with each other (by phosphorylating each other and/or by forming a complex).

The models we employ extend those presented in Swameye et al. (2003) and assume mass action kinetics for each of the reactions taking place. Figures 9.8 and 9.9 show the networks of interactions that are then encoded as systems of coupled differential equations. The first model (Figure 9.8) describes the phosphorylation of STAT5, JAK2 and BCR-ABL and its equations follow as:

$$\frac{d}{dt}(\text{BCR-ABL}) = k2 * \text{pBCR-ABL} - k1 * \text{BCR-ABL} - k7 * \text{BCR-ABL} * \text{TKI}$$
$$+ k8 * \text{TKI-BCR-ABL}$$

$$\frac{d}{dt}(\text{pBCR-ABL}) = k1 * \text{BCR-ABL} - k2 * \text{pBCR-ABL}$$

$$\frac{d}{dt}(\text{JAK2}) = k4 * \text{pJAK2} - k3 * \text{JAK2} - k9 * \text{JAK2} * \text{JAKI}$$
$$+ k10 * \text{JAKI-JAK2}$$

$$\frac{d}{dt}(\text{pJAK2}) = k3 * \text{JAK2} - k4 * \text{pJAK2}$$

$$\frac{d}{dt}(\text{STAT5}) = -k5 * \text{STAT5} * \text{pBCR-ABL} - k6 * \text{STAT5} * \text{pJAK2}$$
$$+ k12 * \text{pSTAT5-NUC}$$

$$\frac{d}{dt}(\text{pSTAT5}) = k5 * \text{STAT5} * \text{pBCR-ABL} + k6 * \text{STAT5} * \text{pJAK2}$$
$$- k11 * \text{pSTAT5}$$

$$\frac{d}{dt}(\text{TKI}) = -k7 * \text{BCR-ABL} * \text{TKI} + k8 * \text{TKI-BCR-ABL}$$

$$\frac{d}{dt}(\text{JAKI}) = -k9 * \text{JAK2} * \text{JAKI} + k10 * \text{JAKI-JAK2}$$

$$\frac{d}{dt}(\text{TKI-BCR-ABL}) = k7 * \text{BCR-ABL} * \text{TKI} - k8 * \text{TKI-BCR-ABL}$$

$$\frac{d}{dt}(\text{JAKI-JAK2}) = k9 * \text{JAK2} * \text{JAKI} - k10 * \text{JAKI-JAK2}$$

$$\frac{d}{dt}(\text{pSTAT5-NUC}) = k11 * \text{pSTAT5} - k12 * \text{pSTAT5-NUC}$$

The second model (Figure 9.9) extends Model 1 by incorporating an additional interaction between BCR-ABL and JAK2, such that they may phosphorylate each other; its equations follow as:

$$\frac{d}{dt}(\text{BCR-ABL}) = k2 * \text{pBCR-ABL} - k1 * \text{BCR-ABL} - k7 * \text{BCR-ABL} * \text{TKI}$$
$$+ k8 * \text{TKI-BCR-ABL} - k13 * \text{BCR-ABL} * \text{pJAK2}$$
$$+ k14 * \text{pBCR-ABL} * \text{pJAK2}$$

$$\frac{d}{dt}(\text{pBCR-ABL}) = k1 * \text{BCR-ABL} - k2 * \text{pBCR-ABL}$$
$$+ k13 * \text{BCR-ABL} * \text{pJAK2} - k14 * \text{pBCR-ABL} * \text{pJAK2}$$

$$\frac{d}{dt}(\text{JAK2}) = k4 * \text{pJAK2} - k3 * \text{JAK2} - k9 * \text{JAK2} * \text{JAKI}$$
$$+ k10 * \text{JAKI-JAK2} - k15 * \text{pBCR-ABL} * \text{JAK2}$$
$$+ k16 * \text{pBCR-ABL} * \text{pJAK2}$$

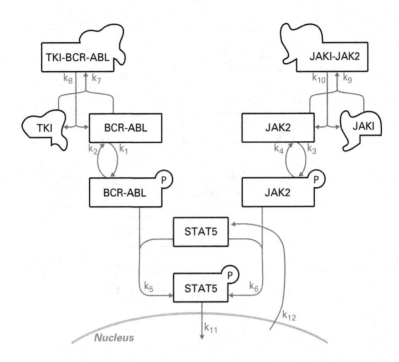

Figure 9.8 Model 1 represents the main mechanisms of phosphorylation of BCR-ABL, JAK2 and STAT5. It also represents the inhibition of BCR-ABL by TKI, the inhibition of JAK2 by JakI and the nuclear transport of phosphorylated STAT5. The grey arrows show how the species are related, and are annotated with k_n values which correspond to coefficients in the mathematical model (p. 172).

$$\frac{d}{dt}(\text{pJAK2}) = k3 * \text{JAK2} - k4 * \text{pJAK2} + k15 * \text{pBCR-ABL} * \text{JAK2}$$
$$- k16 * \text{pBCR-ABL} * \text{pJAK2}$$

$$\frac{d}{dt}(\text{STAT5}) = -k5 * \text{STAT5} * \text{pBCR-ABL} - k6 * \text{STAT5} * \text{pJAK2}$$
$$+ k12 * \text{pSTAT5-NUC}$$

$$\frac{d}{dt}(\text{pSTAT5}) = k5 * \text{STAT5} * \text{pBCR-ABL} + k6 * \text{STAT5} * \text{pJAK2}$$
$$+ - k11 * \text{pSTAT5}$$

$$\frac{d}{dt}(\text{TKI}) = -k7 * \text{BCR-ABL} * \text{TKI} + k8 * \text{TKI-BCR-ABL}$$

$$\frac{d}{dt}(\text{JAKI}) = -k9 * \text{JAK2} * \text{JAKI} + k10 * \text{JAKI-JAK2}$$

$$\frac{d}{dt}(\text{TKI-BCR-ABL}) = k7 * \text{BCR-ABL} * \text{TKI} - k8 * \text{TKI-BCR-ABL}$$

$$\frac{d}{dt}(\text{JAKI-JAK2}) = k9 * \text{JAK2} * \text{JAKI} - k10 * \text{JAKI-JAK2}$$

$$\frac{d}{dt}(\text{pSTAT5-NUC}) = k11 * \text{pSTAT5} - k12 * \text{pSTAT5-NUC}$$

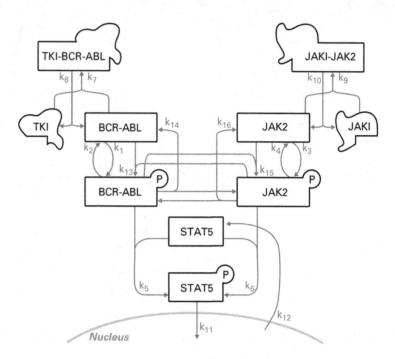

Figure 9.9   Model 2 extends Model 1 (Figure 9.8) by incorporating mechanisms of phosphorylation between BCR-ABL and JAK2, which may change the range of possible system dynamics. This additional component of the model is represented by the variables k_{13}–k_{16} (see p. 172 for a full description of Model 2).

9.4   The statistical methodology: Riemannian manifold population MCMC

Performing Bayesian statistical inference over systems of nonlinear differential equations is a challenging task; the equations are generally analytically intractable, and solving the system for thousands of parameter sets within a Monte Carlo framework is computationally expensive. Markov chain Monte Carlo (MCMC) methods are commonly used to obtain samples from the required probability distribution. In this case, however, the nonlinearity of the model equations can induce strong correlation structure in the resulting probability distribution, which makes drawing samples of the parameters difficult using standard MCMC approaches (Calderhead & Girolami 2011). Recent advances in differential geometric MCMC methods (Girolami & Calderhead 2011) address these problems by making use of the local sensitivity of the system at each point in parameter space, in order to draw samples much more efficiently. In addition, embedding this approach within a population MCMC algorithm allows us to estimate accurately marginal likelihoods via thermodynamic integration (Calderhead & Girolami 2009), which we may use for ranking model hypotheses.

We may begin by writing down Bayes' theorem, which defines the posterior probability distribution as

$$p(\boldsymbol{\theta}|\mathbf{Y}) = \frac{p(\mathbf{Y}|\boldsymbol{\theta})p(\boldsymbol{\theta})}{p(\mathbf{Y})} = \frac{p(\mathbf{Y}|\boldsymbol{\theta})p(\boldsymbol{\theta})}{\int p(\mathbf{Y}|\boldsymbol{\theta})p(\boldsymbol{\theta})d\boldsymbol{\theta}} \tag{9.1}$$

All our subsequent analysis is based on this simple equation. The likelihood of the data given the parameters of our model, $p(\mathbf{Y}|\boldsymbol{\theta})$, is given by the error model we decide to employ and this effectively measures the mismatch between the data $\mathbf{Y}$ and the model prediction using the parameters $\boldsymbol{\theta}$. In practice if the data we observe are measured as the average of a large collection of cells then we may assume normally distributed errors; as a consequence of a large number of small unknown random factors, the Central Limit Theorem states that the overall error may be well approximated by a Gaussian distribution. The likelihood that we employ then follows as

$$p(\mathbf{Y}|\boldsymbol{\theta}) = \prod_n \mathcal{N}(\mathbf{Y}_{\cdot,n}|\mathbf{X}(\boldsymbol{\theta},\mathbf{x}_0)_{\cdot,n}, \mathbf{I}_T\sigma_n^2) \tag{9.2}$$

where $\mathbf{X}(\boldsymbol{\theta},\mathbf{x}_0)_{\cdot,n}$ denotes the solution of the nth equation of the ODE system at T time points with parameters $\boldsymbol{\theta}$ and initial conditions $\mathbf{x}_0$. Likewise, $\mathbf{Y}_{\cdot,n}$ denotes the T observed data points of species n, and σ_n^2 denotes the variance of the independent Gaussian noise at each time point for species n.

The probability distribution $p(\boldsymbol{\theta})$ defines our prior knowledge regarding the likely range of parameters that might be suitable in our model; we may have specific biochemical knowledge of reaction rates, or we may simply choose a positive, long-tailed distribution that reflects our relative ignorance. When analysing multiple models we may employ the same priors for the common parameters.

Bayes' theorem is simply an expression derived using the conditional rule for joint probabilities, and allows inference to take place within the mathematically consistent framework of probability theory. The posterior distribution is usually analytically intractable, as is the case for the models we consider here based on nonlinear differential equations. We can get around this problem by using Monte Carlo estimators; given N samples of $\boldsymbol{\theta}$ drawn from the posterior distribution we can obtain summary statistics, such as the mean or variance of the random variable $\boldsymbol{\theta}$, using the following formula to calculate the expected value of some function $f(\boldsymbol{\theta})$

$$\mu_f = E_{p(\boldsymbol{\theta}|\mathbf{y})}(f(\boldsymbol{\theta})) = \int f(\boldsymbol{\theta})p(\boldsymbol{\theta}|\mathbf{y})d\boldsymbol{\theta} \approx \frac{1}{N}\sum_{n=1}^N \boldsymbol{\theta}_n \tag{9.3}$$

Drawing samples from the posterior distribution provides us with information regarding the global sensitivity of our model given the data. Unlike a local sensitivity analysis whereby each parameter is perturbed with all other parameters fixed at some values, we sample from the posterior distribution by varying all parameters simultaneously. Markov chain Monte Carlo methods are commonly employed to draw correlated samples from the distribution of interest, and these involve simulating a Markov chain which visits different points in the parameter space with a frequency that equals the posterior probability as the number of moves made tends to infinity. A Markov chain will converge to the intended stationary distribution if it satisfies the following condition:

$$\int p(\boldsymbol{\theta})A(\boldsymbol{\theta}^*|\boldsymbol{\theta})d\boldsymbol{\theta} = p(\boldsymbol{\theta}^*) \tag{9.4}$$

Intuitively this states that the average probability of being at a point in the parameter space θ and moving to a specified point θ^* must be equal to the probability of the point θ^* itself. Fortunately if our Markov chain satisfies a *detailed balance* constraint, it will also satisfy the condition given in Equation 9.4, and such a Markov chain is straightforward to construct. Detailed balance is defined as the following condition:

$$p(\theta)A(\theta^*|\theta) = p(\theta^*)A(\theta|\theta^*) \tag{9.5}$$

This simply implies that our Markov chain is reversible; the probability of moving to a particular point θ^* is the same as the the reverse move from that point to the original point θ. The standard Metropolis–Hastings procedure is given by Algorithm 1. The algorithm proceeds by proposing a new point given the current point, and then either moving to that point or staying at the current point dependent on some probability. Usually the transition density is chosen to be an isotropic Gaussian distribution $\mathcal{N}(\theta, \epsilon^2 \mathbf{I})$, where ϵ may be tuned to obtain an acceptance rate of between 30% and 60%. The main difficulty with this approach is that the nonlinearity in the model may induce strong correlation structure in the parameters, thus making it very challenging to propose new points in the parameter space that have high posterior probability and that are thus accepted with high probability. This also becomes increasingly difficult as the dimensionality of the parameter space increases. A poorly tuned Markov chain will tend to remain at the same set of parameters for longer periods of time, increasing the correlation in the samples and reducing the accuracy of the Monte Carlo estimates. However, even when the chain does have a reasonable acceptance rate, the correlation in the samples may still be high if the chain does not move far from the current point at each iteration.

Algorithm 1 Standard Metropolis–Hastings Algorithm

1: Given current state θ, draw proposed state θ^* from transition density $T(\theta^*|\theta)$

2: Calculate the acceptance ratio $R(\theta^*|\theta) = \min\left[1, \frac{p(\theta^*)T(\theta|\theta^*)}{p(\theta)T(\theta^*|\theta)}\right]$

3: Draw $U \sim \text{Uniform}[0, 1]$

4: Let $\theta = \begin{cases} \theta^* & \text{if } U < R(\theta^*|\theta) \\ \theta & \text{otherwise} \end{cases}$

Very recently, ideas from Riemannian geometry have been incorporated into MCMC schemes, and the resulting algorithms dramatically improve the quality of proposed steps of the Markov chain and thus increase the efficiency of performing inference over such nonlinear models (Girolami & Calderhead 2011). The main idea is to represent the parameter space as a Riemannian manifold such that distances between parameter values across the manifold are defined in terms of the changes in the posterior probability, rather than the absolute changes in the parameter values themselves. Different parameters may have to be perturbed by different amounts to obtain the same change in probability, depending on how sensitive the model output is to each of the parameters. Each point on this Riemannian manifold represents a set of parameter values, and a position specific metric tensor defines the local basis under which the direction of moves should be calculated at each point. It was shown in 1945 by C. R. Rao (Rao 1945) that the

Expected Fisher Information provides the link between statistical models and Riemannian geometry, as it satisfies all the required properties of a metric tensor. The Expected Fisher Information can also be considered as providing local sensitivity information at a specific set of parameters, and this is therefore used to guide the Markov chain to its next position. It is defined as

$$G_{i,j}(\boldsymbol{\theta}) = E_{p(\mathbf{x}|\boldsymbol{\theta})}\left(\frac{\partial \mathcal{L}}{\partial \theta^i}^T \frac{\partial \mathcal{L}}{\partial \theta^j}\right) \tag{9.6}$$

$$= \text{Cov}\left(\frac{\partial \mathcal{L}}{\partial \theta^i}, \frac{\partial \mathcal{L}}{\partial \theta^j}\right) \tag{9.7}$$

and we see its equivalence to the covariance of the gradient vectors of the log-likelihood function. This is therefore effectively a rescaling of the basis vectors with respect to changes in probability and this defines a local geometry such that larger steps are made in less sensitive directions and smaller steps are made in directions of high sensitivity. We use a proposal for our MCMC algorithm known as simplified manifold Metropolis-adjusted Langevin algorithm (mMALA),

$$T(\boldsymbol{\theta}^*|\boldsymbol{\theta}) \sim \mathcal{N}\left(\boldsymbol{\theta} + \frac{\epsilon^2}{2}G^{-1}(\boldsymbol{\theta})\nabla_{\boldsymbol{\theta}}\mathcal{L}(\boldsymbol{\theta}), \epsilon^2 G^{-1}(\boldsymbol{\theta})\right) \tag{9.8}$$

The mean of the proposal moves the chain in the direction of steepest gradient with respect to the local geometry, and the random perturbation is also made with respect to the local geometry using a covariance structure given by the inverse of the Expected Fisher Information. We can calculate the Expected Fisher Information analytically for nonlinear ODE models; it follows straightforwardly from the definition of the likelihood function in Equation 9.2,

$$G_{ij}(\boldsymbol{\theta}) = \sum_{n=1}^{N} \mathbf{S}_{\cdot,n}^{i^T} \mathbf{\Sigma}_n^{-1} \mathbf{S}_{\cdot,n}^{j} \tag{9.9}$$

We denote by $\mathbf{S}_{\cdot,n}^i$ the derivative of the ODE solution at each of the T time points for the nth species with respect to the ith parameter, and we now write the covariance matrix from the likelihood function as Σ. We therefore need to solve the ODE system to obtain its first-order sensitivities with respect to the parameters. We follow the approach in Calderhead & Girolami (2011) and this allows us to efficiently obtain samples drawn from the posterior probability distribution which characterises the *global* sensitivity of the model parameters given the data.

As we have multiple structural hypotheses, we also want to perform inference at the model level. This involves integrating out the parameters of our statistical model to obtain the probability of the data for a particular model,

$$p(\mathbf{y}) = \int p(\mathbf{y}|\boldsymbol{\theta})p(\boldsymbol{\theta})d\boldsymbol{\theta} \tag{9.10}$$

We can then compare model hypotheses via Bayes factors, which represent the relative evidence in favour of one model given an alternative. We note that marginal likelihoods are implicitly conditioned on the model M, and so Bayes factors may be calculated as the ratio of posterior probabilities for the two models:

$$\frac{p(M_1|\mathbf{y})}{p(M_2|\mathbf{y})} = \frac{p(\mathbf{y}|M_1)}{p(\mathbf{y}|M_2)} \frac{p(M_1)}{p(M_2)} \tag{9.11}$$

Calculating marginal likelihoods can be very difficult in general, since it involves estimating a high-dimensional integral of an often strongly nonlinear function. However, the combination of thermodynamic integration and differential geometric MCMC embedded within a parallel tempering framework allows us to obtain low variance estimates of the required quantities. For the simulation study we follow the implementation provided in Calderhead & Girolami (2011).

9.5   A proof-of-concept study with synthetic data

As a proof of concept, we obtain measurements by simulating from each mathematical model using randomly generated parameter values. We then use these synthetic data to infer the posterior distribution of parameters and estimate the marginal likelihoods for each model. We examine how well the posterior model predictions describe the data and whether the models can consistently be ranked in the correct order given each dataset.

We generated five datasets from each of the two models (Figures 9.8 and 9.9) using parameter values that were randomly generated from a uniform distribution, $U(0, 5)$. Firstly, the models were run with the initial conditions of BCR-ABL, JAK2 and STAT5 all set to 1, and all other species set to 0, in order to obtain the steady state of the system, which were then taken as the initial conditions for the simulation. The initial conditions of the TKI and JakI were then set to 1 to simulate the introduction of the inhibitors into this steady state system. The dynamics of the system response were recorded for each species over the next 10 units of time, with 20 measurements made for each species. Normally distributed noise, with variance equal to 10% of the variance of the model output for each species, was then added to these measurements, forming the 10 datasets used for the statistical inference. We employed vague gamma priors, $\text{Gam}(1.1, 2)$, for the parameters, and more precise gamma priors for the initial conditions, with means equal to the observed data points at time $t = 0$ and variance equal to that used to create the synthetic data. The chains all converged to the target distribution within 5000 iterations, and 20 000 samples were drawn after this initial 'burn-in' period. Marginal likelihoods were then estimated using the posterior samples from each of the tempered distributions using thermodynamic integration.

We first consider a statistical analysis of these ODE models based on a fully observed system. Of course the assumption that all chemical species are observable is not at all realistic, however it provides us with useful insight into the challenges of performing inference over such models. The estimated marginal likelihoods in Table 9.1 are based on datasets randomly generated from Model 1. We might expect therefore that Model 1 will be favoured more often than Model 2; however, this does of course depend on the parameter values that are randomly chosen and the type of dynamics they induce.

Table 9.1 Log marginal likelihood results using fully observed data from Model 1

Dataset	Model 1	Model 2
1	680.3 ± 1.0	676.7 ± 0.4
2	645.8 ± 0.8	646.1 ± 1.1
3	533.2 ± 0.7	532.9 ± 0.8
4	695.1 ± 1.5	691.6 ± 1.5
5	790.4 ± 1.3	785.6 ± 0.8

Table 9.2 Log marginal likelihood results using fully observed data from Model 2

Dataset	Model 1	Model 2
1	700.4 ± 1.6	702.7 ± 1.6
2	693.8 ± 1.1	703.5 ± 1.7
3	741.8 ± 1.1	744.3 ± 1.4
4	683.9 ± 0.6	687.4 ± 1.3
5	739.1 ± 2.5	807.6 ± 2.2

For datasets 1, 4 and 5, the evidence is clearly in favour of Model 1, while in datasets 2 and 3 both models appear to be able to describe the dynamics equally well. Similarly, the marginal likelihoods in Table 9.2 are based on datasets randomly generated from Model 2. For datasets 2, 3, 4 and 5 there is greater probability that Model 2 better describes the observed dynamics, whereas for dataset 1 there is again slightly more ambiguity as to which model provides the better description. We therefore see that with a fully observed system we may rank model hypotheses successfully given plenty of data, although sometimes there will be ambiguity if both models are able to describe the observed dynamics sufficiently well.

We now consider the same datasets, but with only a subset of the species observed. We use the data for phosphorylated BCR-ABL, JAK2 and STAT5, and assume the other eight species are unobserved. The results for each collection of datasets are given in Tables 9.3 and 9.4. We see that once again the same preference for each model is obtained after performing inference on each dataset, although the weight of evidence in favour of each one is lower than in the fully observed case, as might be expected.

We may compare the predicted model outputs for two cases to obtain more insight into the results of this synthetic study. Let us first consider the second partially observed dataset generated from Model 1. The estimated log marginal likelihoods for each model are very similar, and we can see how both models are able to reproduce the observed dynamics (Figure 9.10). Contrasting this to the fifth partially observed dataset generated from Model 2, the log marginal likelihood for Model 2 is much higher than that of Model 1, and indeed we see that this is because Model 1 cannot reproduce the observed dynamics (Figure 9.11).

The use of Bayesian statistics, and in particular the estimation of marginal likelihoods, allows us to systematically compare model hypotheses encoded as differential

Table 9.3 Log marginal likelihood results using partially observed data from Model 1

Dataset	Model 1	Model 2
1	161.0 ± 0.6	159.0 ± 0.6
2	191.1 ± 0.4	191.8 ± 0.7
3	167.7 ± 0.5	167.5 ± 0.2
4	163.4 ± 0.4	159.1 ± 0.5
5	235.9 ± 0.4	233.9 ± 0.9

Table 9.4 Log marginal likelihood results using partially observed data from Model 2

Dataset	Model 1	Model 2
1	225.9 ± 0.4	226.3 ± 0.8
2	233.3 ± 1.4	239.4 ± 0.6
3	218.0 ± 0.8	220.8 ± 0.4
4	216.7 ± 1.8	219.0 ± 0.6
5	174.4 ± 0.6	229.3 ± 0.4

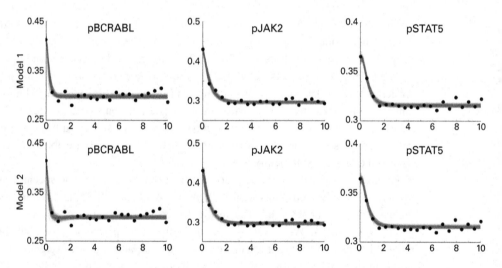

Figure 9.10   The model outputs based on fully observed dataset 2 generated from Model 1. Both Model 1 and Model 2 are able to reproduce the required dynamics to describe the observed data.

equations. Model ranking is often easier with more data; however, inferences may still be drawn even with limited experimental observations. The ability to discriminate models ultimately depends on the observed dynamics and the ability of the model to reproduce that behaviour. Such inferences may then be usefully employed to guide future experimental direction, and will play an important role in driving future research that elucidates the structure of signalling pathways given uncertain and relatively sparse datasets.

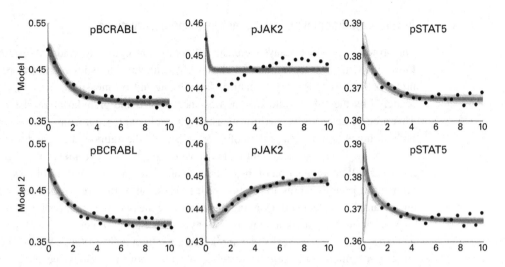

Figure 9.11 The model outputs based on the partially observed dataset 5 generated from Model 2. In this case only Model 2 is able to reproduce dynamics that are complex enough to describe the observed data; the best output from Model 1 is unable to capture the full range of dynamics.

9.6 Beyond a proof of concept: considering a more biologically realistic dataset

In the previous section, we described a simulation study that demonstrated that we are able to correctly identify the provenance of two randomly generated synthetic datasets using our statistical methodology given two plausible models. We can therefore state that the models are adequately descriptive to allow discrimination between them, even when the system is not fully observed. With this established, we now consider to what extent this methodology is applicable beyond entirely synthetic data.

To do this, we have constructed a more biologically realistic dataset. We have considered our experimental observations in combination with data published elsewhere, and assimilated both with our understanding of the biological system as a whole. Moreover, we make the biologically more realistic assumptions that only the three phosphorylated species (pBCR-ABL, pSTAT5 and pJAK2) are observable and even then only at sparse time intervals, reflecting the difficulty of obtaining such measurements experimentally.

The process of generating these data begins with a realistic choice for the time point intervals in the system; in this work, the biological species under investigation will be considered at time point t, where t is measured in hours. In all of our *in vitro* experiments, $0 \leq t \leq 72$, and we are therefore unable to form any hypotheses regarding species behaviour where $t > 72$.

The next section describes our expectations of the consitituent species of this system in response to TKI treatment (dose being equipotent to 5 µM imatinib ®), JakI treatment (dose being equipotent to 1µM INCYTE INCB18424) and a combined TKI/JakI treatment at specific time points. First, however, we describe the *in vitro* methodology.

9.6.1   *In vitro* experimental materials, methods and observations

The protein expression assays were carried out by flow cytometry and/or Western blots. Flow cytometry is a technique that allows physical characteristics of cells to be measured. Cells are exposed to a beam of laser light, and by measuring how this light is scattered by the cell, physical properties such as size and granularity can be inferred. This process can also be used for measuring protein expression using fluorescent antibodies known to bind to a protein of interest. Should the antibody be bound to its target within a cell, the fluorescent light will be detected by the flow cytometer; the intensity of this light will be a reflection of the amount of antibody binding, and, it follows, a measurement of protein expression in that cell. Expression of the protein of interest for the entire sample is calculated by taking the geometric mean of the readings for all the cells.

A Western blot is a complementary assay for measuring protein expression. Proteins are extracted from lysed cells and applied to a medium called a 'gel'. When an electric current is applied to the gel, the protein molecules move at various speeds through the material, as determined by their molecular weight. These size-separated proteins are then transferred onto a membrane, which can be stained with an antibody specific to the protein of interest. Should the protein of interest be present in the sample, a band will appear at the molecular weight corresponding to the protein of interest, reflecting the presence of the bound antibody and thus the presence of the target protein.

Given that antibodies vary with respect to their specificity and affinity for their targets and that different antibodies are used for each protein of interest, it is not possible to quantify protein levels or even comment on relative quantities of different proteins (e.g. levels of protein X are higher than that of protein Y). However, it *is* possible to compare levels of the *same* protein across different time points, if the same antibody has been used.

It is also worth mentioning that reliable antibodies are not currently available for all the species in the model, necessitating the use of surrogate markers. In our experiments, pBCR-ABL is measured by proxy using pCRKL (known to be directly phosphorylated by pBCR-ABL). Similarly, the phosphorylation of JAK2 is inferred from levels of pSTAT5, a direct phosphorylation target of pJAK2. And finally, our data are derived from stem cells isolated from peripheral blood, taken from CML patients.

Values at $t = 0$

When $t = 0$ (before any treatment) it is assumed that BCR-ABL is constitutively active; that is, $\approx 95\%$ of the BCR-ABL protein is phosphorylated (Deininger et al. 2000). It follows that the levels of the non-phosphorylated form are $\approx 5\%$. Similarly, it is assumed that STAT5 is constitutively active at $t = 0$ and that there are negligible levels of the unphosphorylated form of STAT5. The expected levels of pJAK2 before treatment are approximately 20%–50%, therefore levels of JAK2 will be between 50%–80% (Ilaria & Van Etten 1996).

Effect of growth factors

In the presence of growth factors, it is assumed that JAK2 is $\approx 95\%$ activated and that this level of activation is certainly achieved within an hour ($t = 1$), perhaps even within

30 minutes ($t = 0.5$) (Ilaria & Van Etten 1996). We assume that this activity lasts until $t = 72$ (*in vitro* experimental protocol requires treatment with growth factors every 72 hours). We expect that there will be no direct effect of growth factors on BCR-ABL or pBCR-ABL.

Effect of TKI monotherapy

In our experimentation and elsewhere (Jørgensen et al. 2007), we have observed complete inhibition of pBCR-ABL by $t = 4$ which lasts until $t = 24$. It has been shown that 72 hours after treatment with TKI BCR-ABL is again constitutively active (Copland et al. 2006, Hamilton et al. 2006). Similarly, our experimentation with pSTAT5 has demonstrated complete inhibition by $t = 4$ persisting until $t = 24$. By $t = 72$, the levels of pSTAT5 are approaching baseline, with complete reactivation of pSTAT5 occuring when $t > 72$. We do not anticipate that the TKI treatment will have an effect on pJAK2 levels.

Effect of JakI monotherapy

We have observed that the assumed dose of JakI achieves complete inhibition of JAK2 at $16 < t < 24$ (by flow cytometry for the surrogate marker pSTAT5). Once again, levels are approaching baseline by $t = 72$, but complete reactivation is achieved when $t > 72$. A partial inhibition of pSTAT5 is achieved at $t = 24$. It is not possible to determine what this level is with the data that we currently have; however, we can say that the inhibition of pSTAT5 by JakI when $t = 24$ is less than the inhibition achieved by TKI by $t = 4$. The levels of pSTAT5 approach baseline 72 hours after JakI treatment, again recovering at some point where $t > 72$. We have observed by flow cytometry that the JakI treatment does not have an effect on pBCR-ABL levels.

Effect of TKI/JakI combotherapy

We have observed that inhibition of pBCR-ABL following the combination therapy occurs at least as quickly as the TKI monotherapy, but probably earlier (i.e. when $t < 4$), and that this inhibition persists until $t = 72$. Both pSTAT5 and pJAK2 are inactive within an hour of the combination therapy and remain inactive 72 hours later.

9.6.2   Results of modelling the more biologically realistic dataset

The intention here is to assess which model more accurately reflects the behaviour of the constituent species, as described in the previous section (see Table 9.5 for a summary). The likely concentrations described in Section 9.6.1 are most reliable where $t \leq 24$; beyond this time point, the behaviour of the species is less well defined. As such, we will only consider the data where $t \leq 24$ when comparing the performance of the models.

We consider three different datasets. The first represents the likely system response when the JAK inhibitor is introduced into the system, the second represents the introduction of the TKI, and the third is based on the likely response of the system to both JakI and TKI being introduced simultaneously. The estimated marginal likelihoods are

Table 9.5 Summary of likely concentrations of the phosphorylated species in the system, as expressed as % of the total protein level, at different time points and in response to different treatments

Species	Treatment	$t = 0$	$t = 1$	$t = 4$	$t = 16$	$t = 24$	$t = 72$
pBCR-ABL	TKI	95	–	5	–	5	95
pSTAT5	TKI	95	–	5	–	5	95
pJAK2	TKI	20–50	–	–	–	20–50	20–50
pSTAT5	JakI	95	–	–	–	50	≈85
pJAK2	JakI	20–50	–	–	5	5	≈85
pBCR-ABL	TKI+JakI	95	–	5	–	5	5
pSTAT5	TKI+JakI	95	5	–	–	5	5
pJAK2	TKI+JakI	20–50	5	–	–	5	5

Table 9.6 Log marginal likelihood results using biologically realistic dataset with JakI

Model	Mean	Std
Model 1	7.25	0.12
Model 2	2.42	0.20

Table 9.7 Log marginal likelihood results using biologically realistic dataset with TKI Inhibitor

Model	Mean	Std
Model 1	0.74	0.22
Model 2	−0.18	0.25

presented in Tables 9.6, 9.7 and 9.8, and the predictive posterior model outputs and data points are shown in Figures 9.12, 9.13 and 9.14 for each of these datasets respectively.

The simpler Model 1 appears to better describe the dynamics of the system between $t = 0$ and $t = 24$ when the JAK inhibitor is present (Figure 9.12). In both of the other cases, however, the data provide no strong evidence in favour of one model over the other model. Indeed the predicted dynamics for pBCR-ABL and pSTAT5 by both models when TKI is present indicate a rapid increase in concentration within the first hour before quickly decaying to fit the data points (Figure 9.13). This dynamical behaviour is not biologically realistic and suggests that different structural hypotheses need to be considered in order to more accurately describe the underlying process.

Further, we have only considered the dynamics across the time range $0 < t < 24$. It has been observed that certain phosphorylated species return to baseline levels after 72 hours, however the models considered in this analysis both reach a steady state after 24 hours. This suggests that there are additional components missing, most likely a mechanism that allows for decay of the inhibitors over time or accounts for the altered cell population due to cell death after TKI and/or JakI treatment.

Table 9.8 Log marginal likelihood results using
biologically realistic dataset with both TKI and JakI
Inhibitors

Model	Mean	Std
Model 1	3.30	0.17
Model 2	2.81	0.27

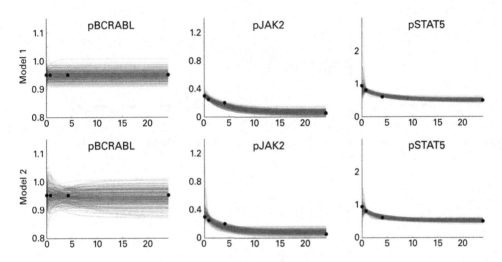

Figure 9.12 The posterior model outputs based on a dataset generated from the literature describing when JakI is present in the system. Both Model 1 and Model 2 are able to reproduce the required dynamics to describe the observed data.

9.6.3 A reconsideration of the model and data: what next?

We have shown that both models are inadequate with respect to capturing the expected dynamics of the system, as described in Section 9.6.1, requiring us to reconsider our current models, methods and/or data. Paying close attention to the results described in Figures 9.12 and 9.13, it is clear that both models struggle to model the behaviour of pBCR-ABL and pJAK2 in response to JakI and TKI respectively. While this could be a failure of the models in representing the interaction of JakI with pBCR-ABL and TKI with pJAK2, it is more likely to be a result of naive assumptions when constructing our biologically realistic data: in both cases, we predicted that the drug would have no effect on the phosphorylated protein species. While it is true that the TKI does not directly target pJAK2 and that the JakI does not directly target pBCR-ABL, it is entirely possible (and, indeed, probable) that the drugs will have indirect effects on these proteins. These indirect effects are not represented in our 'biologically realistic' dataset, and this is likely to be one reason that these models struggle to capture the dynamics of the system.

With this in mind, we should focus our further work on collecting genuine biological data for analysis, ideally incorporating more observations for each species at an

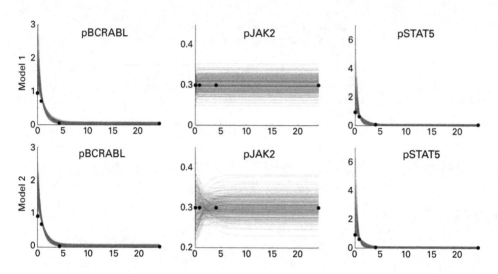

Figure 9.13   The posterior model outputs based on a dataset generated from the literature describing when TKI is present in the system.

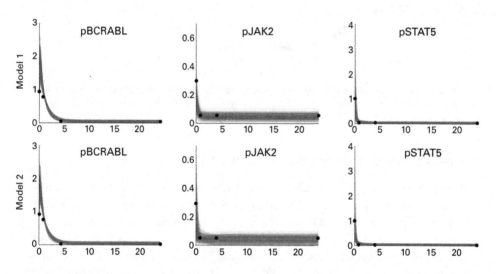

Figure 9.14   The posterior model outputs based on a dataset generated from the literature describing when both TKI and JakI are present in the system.

increased number of time points. It was clear that neither model was able to predict baseline recovery at $t \geq 72$ (irrespective of their performance where $t < 24$) and work towards characterising the appropriate missing component (as discussed briefly in the previous section) is required. All future structural hypotheses will be subjected to further Bayesian analyses of the type described in this chapter to allow model comparison and ranking.

9.7   Discussion

We have described the application of Bayesian inference to the problem of model selection in a biological system, specifically the JAK/STAT pathway in chronic myeloid leukaemia. Using our methodology to estimate marginal likelihoods and Bayes factors, we were able to make some contribution to the investigations into this biological system. Firstly, using synthetic data generated from two candidate models, we were able to demonstrate that we could infer purely from the data which model was more likely to have generated it. We can therefore discriminate between competing model hypotheses, both in fully and partially observed systems. This is an important first step which allows us to progress with the analysis of a more biologically realistic dataset, formed by careful consideration of the literature and our own observations. However, in calculating the marginal likelihoods generated by our methodology, we were unable to identify either model as the more successful in capturing the observed behaviour. Although the simpler model appeared to capture the dynamics of the system in response to JakI treatment, the same model produced biologically unlikely behaviour in response to both the TKI and combination treatment. In considering the results more closely, it is probable that naive assumptions when generating our more biologically realistic synthetic dataset contribute at least in part to the difficulties that the candidate models have in modelling these data. Furthermore, we observed that the models were unable to capture model dynamics for $t > 24$, specifically with respect to baseline recovery of the constituent species.

This work highlights a systematic approach we can employ to tackle this type of problem, and demonstrates the need for an extended model of the JAK/STAT pathway. Further work is necessary to assess how the current model might be extended and/or improved with respect to the observed data. The main challenge is that it is difficult to predict *a priori* how changes in model structure relate to changes in the range of possible dynamics. For each new hypothesised structure we therefore need to rerun the Bayesian analysis to calculate the marginal likelihood for the new model given a dataset. What is clear in the case of our application, however, is that factors influencing baseline recovery (for example, decay of the drug treatments or altered cell population following treatment) are noticeably absent and require to be incorporated. Revisiting the model of this system following these observations will allow us to more accurately capture the behaviour of this potentially therapeutic pathway.

Acknowledgements

We acknowledge the following funding bodies for their support: Cancer Research UK (C11074/A11008); Engineering and Physical Sciences Research Council (EP/I017909/1, EP/E032745/1, EP/E052029/1); Biotechnology and Biological Sciences Research Council (BB/G006997/1) and Medical Research Council (G1000288). Images in Figures 9.2 and 9.6 kindly provided by Avril Morris (Cytogenetics Department, Yorkhill Hospital, Glasgow); images in Figure 9.3 kindly provided by Elaine Allan (Paul O'Gorman Leukaemia Research Laboratory, University of Glasgow).

References

Baxter, E. J., Scott, L. M., Campbell, P. J., East, C., Fourouclas, N. et al. (2005), 'Acquired mutation of the tyrosine kinase JAK2 in human myeloproliferative disorders', *Lancet* **365**, 1054–1061.

Ben-Neriah, Y., Daley, G. Q., Mes-Masson, A. M., Witte, O. N. & Baltimore, D. (1986), 'The chronic myelogenous leukemia-specific P210 protein is the product of the BCR/ABL hybrid gene', *Science* **233**, 212–214.

Benekli, M., Baer, M. R., Baumann, H. & Wetzler, M. (2003), 'Signal transducer and activator of transcription proteins in leukemias', *Blood* **101**, 2940–2954.

Bennett, J. H. (1845), 'Case of hypertrophy of the spleen and liver in which death took place from the suppuration of the blood', *Edinb Med Surg J* **64**, 413–423.

Calderhead, B. and Girolami, M. (2009), 'Estimating Bayes factors via thermodynamic integration and population MCMC', *Computational Statistics and Data Analysis* **53**(12), 4028–4045.

Calderhead, B. and Girolami, M. (2011), 'Statistical analysis of nonlinear dynamical systems using differential geometric sampling methods', *J Roy Soc Interface Focus* **1**(6), 821–835.

Chai, S. K., Nichols, G. L. & Rothman, P. (1997), 'Constitutive activation of JAKs and STATs in BCR-ABL-expressing cell lines and peripheral blood cells derived from leukemic patients', *J Immunol* **159**, 4720–4728.

Copland, M., Hamilton, A., Elrick, L. J., Baird, J. W., Allan, E. K. et al. (2006), 'Dasatinib (BMS-354825) targets an earlier progenitor population than imatinib in primary CML but does not eliminate the quiescent fraction', *Blood* **107**, 4532–4539.

Daley, G. Q., Van Etten, R. A. & Baltimore, D. (1990), 'Induction of chronic myelogenous leukemia in mice by the P210BCR/ABL gene of the Philadelphia chromosome', *Science* **247**, 824–830.

de Klein, A., van Kessel, A. G., Gerard, G., Bartram, C. R., Hagemeijer, A. et al. (1982), 'A cellular oncogene is translocated to the Philadelphia chromosome in chronic myelocytic leukaemia', *Nature* **300**, 765–767.

DeGeer, D., Gallipoli, P., Chen, M., Sloma, I., Jørgensen, H. et al. (2010), 'Combined targeting of BCR-ABL and JAK2 with ABL and JAK2 inhibitors is effective against CML patients' leukemic stem/progenitor cells', *Blood* **116**, 1393–1394.

Deininger, M. W., Vieira, S., Mendiola, R., Schultheis, B., Goldman, J. M. et al. (2000), 'BCR-ABL tyrosine kinase activity regulates the expression of multiple genes implicated in the pathogenesis of chronic myeloid leukemia', *Cancer Res* **60**, 2049–2055.

Druker, B. J., Guilhot, F., O'Brien, S. G., Gathmann, I., Kantarjian, H. et al. (2006), 'Five-year follow-up of patients receiving imatinib for chronic myeloid leukemia', *N Engl J Med* **355**, 2408–2417.

Druker, B. J., Talpaz, M., Resta, D. J., Peng, B., Buchdunger, E. et al. (2001), 'Efficacy and safety of a specific inhibitor of the BCR-ABL tyrosine kinase in chronic myeloid leukemia', *N Engl J Med* **344**, 1031–1037.

Druker, B. J., Tamura, S., Buchdunger, E., Ohno, S., Segal, G. M. et al. (1996), 'Effects of a selective inhibitor of the Abl tyrosine kinase on the growth of BCR-ABL positive cells', *Nat Med* **2**, 561–566.

Girolami, M. & Calderhead, B. (2011), 'Riemann manifold Langevin and Hamiltonian Monte Carlo methods', *J Roy Statist Soc: Series B (Statistical Methodology)* **73**, 123–214.

Gross, A. W., Zhang, X. & Ren, R. (1999), 'Bcr-Abl with an SH3 deletion retains the ability to induce a myeloproliferative disease in mice, yet c-Abl activated by an SH3 deletion induces only lymphoid malignancy', *Mol Cell Biol* **19**, 6918–6928.

Hamilton, A., Elrick, L., Myssina, S., Copland, M., Jørgensen, H. et al. (2006), 'BCR-ABL activity and its response to drugs can be determined in CD34+ CML stem cells by CrkL phosphorylation status using flow cytometry', *Leukemia* **20**, 1035–1039.

Heisterkamp, N., Jenster, G., ten Hoeve, J., Zovich, D., Pattengale, P. K. et al. (1990), 'Acute leukaemia in BCR/ABL transgenic mice', *Nature* **344**, 251–253.

Hiwase, D. K., White, D. L., Powell, J. A., Saunders, V. A., Zrim, S. A. et al. (2010), 'Blocking cytokine signaling along with intense BCR-ABL kinase inhibition induces apoptosis in primary CML progenitors', *Leukemia* **24**, 771–778.

Hoelbl, A., Schuster, C., Kovacic, B., Zhu, B., Wickre, M. et al. (2010), 'STAT5 is indispensable for the maintenance of Bcr/abl-positive leukaemia', *EMBO Mol Med* **2**, 98–110.

Huntly, B. J. P., Shigematsu, H., Deguchi, K., Lee, B. H., Mizuno, S. et al. (2004), 'MOZ-TIF2, but not BCR-ABL, confers properties of leukemic stem cells to committed murine hematopoietic progenitors', *Cancer Cell* **6**, 587–596.

Ilaria, R. L. & Van Etten, R. A. (1996), 'P210 and P190(BCR/ABL) induce the tyrosine phosphorylation and DNA binding activity of multiple specific STAT family members', *J Biol Chem* **271**, 31 704–31 710.

Jiang, X., Lopez, A., Holyoake, T., Eaves, A. & Eaves, C. (1999), 'Autocrine production and action of IL-3 and granulocyte colony-stimulating factor in chronic myeloid leukemia', *Proc Natl Acad Sci USA* **96**, 12 804–12 809.

Jin, A., Kurosu, T., Tsuji, K., Mizuchi, D., Arai, A. et al. (2006), 'BCR/ABL and IL-3 activate Rap1 to stimulate the B-Raf/MEK/Erk and Akt signaling pathways and to regulate proliferation, apoptosis, and adhesion', *Oncogene* **25**, 4332–4340.

Jørgensen, H. G., Allan, E. K., Jordanides, N. E., Mountford, J. C. & Holyoake, T. L. (2007), 'Nilotinib exerts equipotent antiproliferative effects to imatinib and does not induce apoptosis in CD34+ CML cells', *Blood* **109**, 4016–4019.

Konopka, J. B., Watanabe, S. M. & Witte, O. N. (1984), 'An alteration of the human C-ABL protein in K562 leukemia cells unmasks associated tyrosine kinase activity', *Cell* **37**, 1035–1042.

Kralovics, R., Passamonti, F., Buser, A. S., Teo, S.-S., Tiedt, R. et al. (2005), 'A gain-of-function mutation of JAK2 in myeloproliferative disorders', *N Engl J Med* **352**, 1779–1790.

Lugo, T. G., Pendergast, A. M., Muller, A. J. & Witte, O. N. (1990), 'Tyrosine kinase activity and transformation potency of BCR-ABL oncogene products', *Science* **247**, 1079–1082.

Melo, J. V., Gordon, D. E., Cross, N. C. & Goldman, J. M. (1993), 'The ABL-BCR fusion gene is expressed in chronic myeloid leukemia', *Blood* **81**, 158–165.

Nelson, E. A., Walker, S. R., Weisberg, E., Bar-Natan, M., Barrett, R. et al. (2011), 'The STAT5 inhibitor pimozide decreases survival of chronic myelogenous leukemia cells resistant to kinase inhibitors', *Blood* **117**, 3421–3429.

Neviani, P., Harb, J., Oaks, J., Walker, C., Santhanam, R. et al. (2010), 'BCR-ABL1 kinase activity but not its expression is dispensable for Ph plus quiescent stem cell survival which depends on the PP2A-controlled Jak2 activation and is sensitive to FTY720 treatment', *Blood* **116**, 227–228.

Nowell, P. C. & Hungerford, D. A. (1960), 'Minute chromosome in human chronic granulocytic leukemia', *Science* **132**, 1497.

Pendergast, A. M., Quilliam, L. A., Cripe, L. D., Bassing, C. H., Dai, Z. et al. (1993), 'BCR-ABL-induced oncogenesis is mediated by direct interaction with the SH2 domain of the GRB-2 adaptor protein', *Cell* **75**, 175–185.

Rao, C. R. (1945), 'Information and the accuracy attainable in the estimation of several parameters', *Calcutta Math Bull* **37**, 81–91.

Rowley, J. D. (1973), 'New consistent chromosomal abnormality in chronic myelogenous leukemia identified by quinacrine fluorescence and giemsa staining', *Nature* **243**, 290–293.

Samanta, A. K., Chakraborty, S. N., Wang, Y., Kantarjian, H., Sun, X. et al. (2009), 'JAK2 inhibition deactivates Lyn kinase through the SET-PP2A-SHP1 pathway, causing apoptosis in drug-resistant cells from chronic myelogenous leukemia patients', *Oncogene* **28**, 1669–1681.

Samanta, A. K., Lin, H., Sun, T., Kantarjian, H. & Arlinghaus, R. B. (2006), 'Janus kinase 2: a critical target in chronic myelogenous leukemia', *Cancer Res* **66**, 6468–6472.

Samanta, A., Perazzona, B., Chakraborty, S., Sun, X., Modi, H. et al. (2011), 'Janus kinase 2 regulates BCR-ABL signaling in chronic myeloid leukemia', *Leukemia* **25**, 463–472.

Shtivelman, E., Lifshitz, B., Gale, R. P. & Canaani, E. (1985), 'Fused transcript of ABL and BCR genes in chronic myelogenous leukaemia', *Nature* **315**, 550–554.

Swameye, I., Müller, T., Timmer, J., Sandra, O. & Klingmüller, U. (2003), 'Identification of nucleocytoplasmic cycling as a remote sensor in cellular signaling by databased modeling', *Proc Natl Acad Sci USA* **100**, 1028–1033.

Verstovsek, S., Kantarjian, H., Mesa, R. A., Pardanani, A. D., Cortes-Franco, J. et al. (2010), 'Safety and efficacy of INCB018424, a JAK1 and JAK2 inhibitor, in myelofibrosis', *N Engl J Med* **363**, 1117–1127.

Virchow, R. (1845), 'Weisses Blut', *Frorieps Notizen* **36**, 151–156.

Vyshemirsky, V. & Girolami, M. (2008), 'Bayesian ranking of biochemical system models', *Bioinformatics* **24**, 833–839.

Walz, C., Ahmed, W., Lazarides, K., Betancur, M., Patel, N. et al. (2012), 'Essential role for STAT5a/b in myeloproliferative neoplasms induced by BCR-ABL1 and Jak2V617F in mice', *Blood* **119**, 3550–3560.

Wang, Y., Cai, D., Brendel, C., Barett, C., Erben, P. et al. (2007), 'Adaptive secretion of granulocyte-macrophage colony-stimulating factor (GM-CSF) mediates imatinib and nilotinib resistance in BCR/ABL+ progenitors via JAK-2/STAT-5 pathway activation', *Blood* **109**, 2147–2155.

Ward, A. C., Touw, I. & Yoshimura, A. (2000), 'The JAK-STAT pathway in normal and perturbed hematopoiesis', *Blood* **95**, 19–29.

Warsch, W., Kollmann, K., Eckelhart, E., Fajmann, S., Cerny-Reiterer, S. et al. (2011), 'High STAT5 levels mediate imatinib resistance and indicate disease progression in chronic myeloid leukemia', *Blood* **117**, 3409–3420.

Xu, T., Vyshemirsky, V., Gormand, A., Kriegsheim, A. V., Girolami, M. et al. (2010), 'Inferring signaling pathway topologies from multiple perturbation measurements of specific biochemical species', *Science Signaling* **3**, 20.

Zhang, X. & Ren, R. (1998), 'BCR-ABL efficiently induces a myeloproliferative disease and production of excess interleukin-3 and granulocyte-macrophage colony-stimulating factor in mice: a novel model for chronic myelogenous leukemia', *Blood* **92**, 3829–3840.

Zhou, L. L., Zhao, Y., Ringrose, A., DeGeer, D., Kennah, E. et al. (2008), 'AHI-1 interacts with BCR-ABL and modulates BCR-ABL transforming activity and imatinib response of CML stem/progenitor cells', *J Exp Med* **205**, 2657–2671.

10 Dynamic network models of protein complexes

Yongjin Park and Joel S. Bader

Current views of biological networks and pathways are primarily static, comprising databases of curated pathways (Croft et al. 2011, Liberzon et al. 2011) or of pairwise interactions, primarily between proteins (Stark et al. 2006). Many methods have been developed to cluster, partition, or segment an interaction network into putative complexes (Bader & Hogue 2003, Clauset et al. 2004, Rivera et al. 2010). Recent comparisons suggest that hierarchical stochastic block models provide the most accurate reconstruction of real protein complexes from interaction data (Clauset et al. 2008, Park & Bader 2011). These static views, however, fail to capture the rich dynamic network structure of a cell. Accounting for dynamic changes in protein complexes is crucial to building accurate models of cellular state.

Our approach uses stochastic block models, in which interactions are conditionally independent given group membership. These models can be hierarchical, with larger complexes containing sub-complexes with more fine-grained interaction probabilities, and complexes can themselves be contained in larger structures with coarse-grained interaction probabilities. Networks are observed at specific time points, termed 'snapshots,' and the goal is to infer or estimate the time-dependent block model given the snapshots. The model itself is generative. While others have explored the properties of networks generated from a pre-specified model (Leskovec et al. 2005), the focus here is on network model inference.

The observed snapshots are a series of T time-ordered graphs. The goal is to infer a corresponding sequence of time-evolving stochastic block models, $\{M^{(t)} : t = 1, \ldots, T\}$, where each $M^{(t)}$ is a good network-generative model for $G^{(t)}$. Many methods maximize the model for each snapshot independently, obtaining $\hat{M}(t)$ as $\arg\max_M P(M|G^{(t)})$, then attempt to stitch together the results.

10.1 Dynamic network data

In some disciplines temporal networks $G^{(t)}$ are directly observable through timestamps of vertices or edges accumulated over a certain period of time (Leskovec et al. 2005) (e.g., author–author citation networks). However time-evolving networks of biological

Systems Genetics: Linking Genotypes and Phenotypes, ed. F. Markowetz and M. Boutros. Published by Cambridge University Press. © Cambridge University Press 2015.

systems are to be inferred from time-series measurements such as gene expression microarrays. For a large system such measurements are unavoidably noisy and ultra-high-dimensional but with a limited sample size. Therefore direct inference from temporal expression arrays can easily create a hard problem, of which solution is unidentifiable and computationally intractable. Despite recent advances of statistical learning methods (Banerjee et al. 2008, Friedman et al. 2008, Song et al. 2009), still the structural learning problem remains an open question especially for large biological networks. Moreover the algorithms can only declare possible statistical dependency which are not necessarily evidence for direct interactions.

Instead we generated dynamic biological networks by combining experimental gene expression time-series data with static protein interaction networks. Our method assumes that presence of a protein is related to transcriptional abundance of the corresponding transcript at a nearby time, with possible delays due to translation and protein lifetimes. More realistic models are possible and should yield more accurate results. While it would be preferable to use protein abundances directly, current experimental methods cannot generate these data proteome-wide. We also consider that a compendium of pairwise interactions is a superset of dynamic edges. In analogy to communication networks, the Internet contains many possible dynamic information flows of blogs and web pages, but only a portion of blogs are linked at a certain period of time (Rodriguez et al. 2010). Other researches in systems biology have also revealed useful connections between a static network and dynamic gene expressions (Jansen et al. 2002, Han et al. 2004, Luscombe et al. 2004, Komurov & White 2007, Dittrich et al. 2008).

10.1.1   Basic principles

A static network G consists of V vertices and E edges. Vertices correspond to proteins, and edges represent possible protein–protein physical interactions (PPI). For gene expression microarrays we denote by a matrix X. Time-series measurements of the expression levels of N genes across T time points generates a $N \times T$ matrix X (see Figure 10.1A). Each element X_{ut} corresponds to the expression of gene u at snapshot t. A set of generated T dynamic networks is $\{G^{(t)} : t = 1, \ldots, T\}$. Each single network $G^{(t)} = (V^{(t)}, E^{(t)})$ consists of undirected and unweighted binary edges $E^{(t)}$ and vertices $V^{(t)}$. For an arbitrary pair, $t \neq t'$, $G^{(t)}$ and $G^{(t')}$ can have different vertices and edges. Generalizations to additional vertex classes (transcripts, genes, metabolites) and edge types (directed, weighted, epistatic, or regulatory interactions) follow directly but are not considered here.

We project out dynamic networks under two principles: dynamics of vertices and edges. We used the first scheme on *Arabidopsis* root development networks (Birnbaum et al. 2003) and the second on the Yeast Metabolic Cycle (YMC) dataset (Tu et al. 2005). Both methods can produce false positive and negative edges. More realistic mRNA and protein decay model will produce more accurate snapshots of networks. Also, less noisy measurement methods such as short read sequencing of transcripts will enhance the resolution both qualitatively and quantitatively. The problem remain largely open, yet, at least in our studies, our methods recovered most general features of dynamic networks.

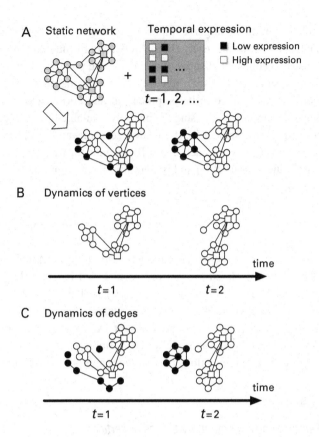

A   Static network   Temporal expression

■ Low expression
□ High expression

$t = 1, 2, \dots$

B   Dynamics of vertices

time

$t = 1$ $t = 2$

C   Dynamics of edges

time

$t = 1$ $t = 2$

Figure 10.1   Dynamic network data generation.

We fixed the inferred networks as datasets and perform dynamic network clustering and subsequent analyses.

Dynamics of vertices

Not all genes are active at all times, thus not all proteins are available to interact with other proteins. A low abundance of mRNA therefore implies absence of its product. For each time point, we generated a temporal subnetwork by removing those absent proteins (see Figure 10.1B). Although high mRNA transcripts do not necessarily mean high concentration of proteins, this scheme gives a good and simple approximation of dynamic networks when the cutoff level is immediately available from experiments and expert knowledge. However, it is possible to adjust the cutoff controlling false positives and statistical powers.

Dynamics of edges

Early systematic analysis (Jansen et al. 2002) revealed that highly correlated mRNA expression profiles were more often seen between proteins in the same complex. While functionally homogeneous complexes (permanent) shared tightly co-regulated mRNA profiles, genes in larger complexes, consisting of smaller subunits, showed discordant patterns within each other. Motivated by this analysis, we basically retain the consistent edges, both active (two interacting proteins are expressed) and inactive (neither protein is expressed); yet we discard inconsistent ones (see Figure 10.1C). The matrix X is assumed to be preprocessed and normalized. For instance, we performed quantile-normalization using the publicly available R-package `gcrma` (Wu et al. 2004). Next it is row-standardized to have zero mean, $\sum_t X_{ut} = 0$, and equal variance, $\sum_t X_{ut}^2 = T - 1$, for each gene. To account for transient complexes and cases where delays due to translation and protein lifetime are important, correlations were averaged over a bandwidth h,

$$\tilde{X}_{uv}(t) = \sum_{s=1}^{T} w(t - s, h) X_{us} X_{vs} \qquad (10.1)$$

with the Gaussian kernel function $w(\Delta t, h) \propto \exp(-|\Delta t|/h)$ and normalized to 1. Each edge is then declared present or absent based on the value of $\tilde{X}_{uv}(t)$: for each snapshot $t = 1, \ldots, T$, a dynamic edge $e_{uv}(t) = 1$ if and only if $\tilde{X}_{uv}(t) > 0$ and $e_{uv} = 1$ in a static network. This procedure retains edges at time t where both proteins are present ($X_{us}, X_{vs} > 0$) or both absent ($X_{us}, X_{vs} < 0$) for times s close to time t. While this method is appropriate for periodic processes, other methods for extracting time-dependent interactions may be more appropriate for more general processes.

10.1.2   Examples

Plant root development: dynamics of vertices

The root is an ideal model for development because temporally staged samples are easily obtained by cutting further back from the root tip, and distinct cell and tissue types are observed radially outward from the root center. A classic study mapped gene expression activity in five spatial regions across three developmental stages (Birnbaum et al. 2003), yielding 15 spatiotemporal snapshots. High-confidence interactions for the corresponding proteins (confidence value ≥ 10) were extracted from TAIR Interactome 2.0 (Geisler-Lee et al. 2007). For this superposition of all genes active anywhere in the root map, we iteratively deleted network vertices with degree less than or equal to 3 until no more vertices could be removed. The resulting network had 332 vertices and 1163 edges. Subnetworks were then generated by extracting the active genes (expression level ≥ 75 as reported by Birnbaum et al. [2003]) and their interactions for each of the 15 snapshots. Each snapshot had approximately 150 to 220 genes and five interactions per gene (Table. 10.1).

Yeast metabolic cycle: dynamics of edges

The YMC networks were generated combining a large-scale protein–protein interaction dataset (Stark et al. 2006) with data from YMC gene expression microarrays over 36 time points (Tu et al. 2005). The BioGrid interactions were required to be physical and supported by two or more publications, a criterion used previously by others to extract about 13 000 interactions (Bandyopadhyay et al. 2010). Moreover, since not all genes are

Table 10.1 The spatiotemporal variation of active subnetworks. The numbers of active genes at each position are shown without parentheses; the numbers of active interactions are shown within the parentheses

	Stele	Endoderm	Endo+Cortex	Epiderm	Lateral root cap
Stage 3	217 (569)	215 (565)	225 (603)	219 (586)	211 (543)
Stage 2	182 (415)	185 (432)	193 (462)	188 (440)	172 (391)
Stage 1	150 (328)	151 (331)	156 (354)	144 (324)	135 (285)

periodically expressed, we restricted our focus on significantly periodic genes defined in Tu et al. (2005).

For each time point, edges (u, v) of periodic genes are present if the edge score (Eq. 10.1) is above threshold (> 0). Yet results were stable for less stringent thresholds, $\tilde{X}_{uv}(t) > -0.5$. While the bandwidth h also can be tuned but rather set to 1.5, results were quantitatively similar for h from 1.2 to 2. Smaller values of h result in stricter co-expression requirements and result in a sparser network. Generated networks typically contained approximately 3000 vertices with average degree ≈ 1.8. Here we show that introducing explicit coupling between time points improves dynamic network clustering.

10.2 Block models of a network

10.2.1 Flat stochastic block model

A stochastic block model M is a generative model for a network G. The number of vertices is V and the number of possible blocks, groups, or clusters is K. Typically, indices $u, v, w \in 1 \ldots V$ denote vertices, and indices $i, j, k \in 1 \ldots K$ denote clusters. The notation $u \in i$ indicates that vertex u is in cluster i, and n_i is the number of vertices assigned to cluster i. For instance, a simple network can be generated from a flat stochastic block model with $K = 3$ (Figure 10.2A and C).

The probability that a vertex is in cluster k is π_k, the parameter for the kth cluster in a multinomial distribution, with $\sum_k \pi_k = 1$. The parameter $\theta_{ij} = \theta_{ji} \in [0, 1]$ gives the probability of an undirected, unweighted edge between any pair of vertices $u \in i$ and $v \in j$, modeled as independent Bernoulli trials for each pair. This model M generates a network G by first sampling the membership of each vertex u with probability π_k for cluster k, then sampling each edge $e_{uv} = 0$ or 1 as a Bernoulli trial with success probability θ_{ij} for $u \in i$ and $v \in j$. Note this Bernoulli model has a biomodal nature, as seen in Figure 10.2B, and captures both assortative (edge-enriching) and disassortative (edge-depleting) patterns without modification. This basic building block for a block model gives quite a flexible framework for an arbitrary heterogeneous network, distinctively modeling between-group relations and within-group edge probabilities.

Edge counts are summarized at the cluster level as $e_{ij} = \sum_{u \in i, v \in j} e_{uv}$ for $i \neq j$, or $\sum_{u \leq v \in i} e_{uv}$ for $i = j$. It is convenient to keep track of the corresponding number of

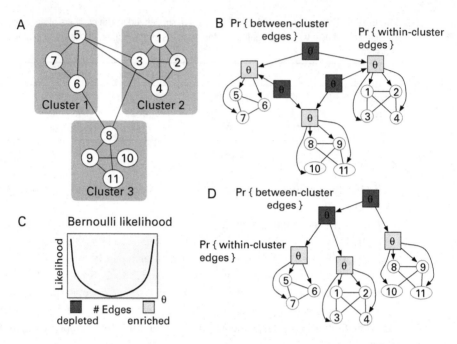

Figure 10.2   Block models of a network. (A) A toy example. (B) Bernoulli likelihood increases if either edges are depleted or enriched. Bright colors represent edge enrichment ($\theta \approx 1$); dark colors represent depleted edges ($\theta \approx 0$). (C) A flat stochastic block model. (D) A hierarchical stochastic block model.

non-edges, or 'holes,' h_{ij}, with $e_{ij} + h_{ij} = t_{ij}$, the total possible number of edges. For $i \neq j$, $t_{ij} = n_i n_j$. For $i = j$, $t_{ii} = n_i(n_i \pm 1)/2$, with the '+' term for graphs with self-edges and the '−' for graphs without self-edges. Using these sufficient statistics, the probability of a network G given the structural model and parameters is

$$P(G|\{\theta, \pi\}, M) = \prod_{k=1}^{K} \pi_k^{n_k} \prod_{i \leq j} \theta_{ij}^{e_{ij}} (1 - \theta_{ij})^{h_{ij}}. \tag{10.2}$$

10.2.2   Hierarchical stochastic block model

We can extend the notion of a model M to a hierarchical random graph (HRG) based on a model that successively merges pairs of groups (Clauset et al. 2008). This original model generates a binary dendrogram T. Each node r in this dendrogram represents the joining of graph vertices $L(r)$ underneath the left sub-tree and vertices $R(r)$ underneath the right sub-tree. With the same Bernoulli probability model, $P(e, h) = \theta^e (1 - \theta)^h$, as a building block, e_r and h_r are defined as the total number of edges and holes crossing between the left and right sub-trees. We also generalize for tree structures that are not completely branching, yielding tree nodes that collect multiple graph vertices into a single group. While the flat model (Eq.10.2) defines $\binom{K}{2}$ group–group relations distinctively, the hierarchical model defines a binary tree that recursively bisects (or merges) until terminal K groups at the leaves. One can think of this as a divide-and-conquer scheme (Figure 10.2D). The hierarchical model reduces model complexity of the flat model ($O(K^2) \rightarrow O(K)$ of parameters) while considering heterogeneity of a complex network.

Letting $M \equiv T$, the likelihood $P(G|M)$ of a hierarchical model T is

$$P(G|\{\theta,\pi\},M) = \prod_{k=1}^{K} \pi_k^{n_k} \prod_{k \in \text{leaves}}^{K} \theta_k^{e_k}(1-\theta_k)^{h_k} \tag{10.3}$$

$$\prod_{r \in \text{non-leaves of tree}} \theta_r^{e_r}(1-\theta_r)^{h_r}$$

where $e_k = e_{kk}$ denotes a number of edges and $h_k = h_{kk}$ holes under group k.

10.3 Learning algorithms

Sampling trees with Markov chain Monte Carlo (MCMC) (Hastings 1970) provides excellent results for predicting missing links by accumulating $\hat{\theta}_r$ values for link probabilities between left and right sub-trees (Clauset et al. 2008). We have found that extending the MCMC approach to genome-scale networks is computationally burdensome. Approximation methods, such as a variational Bayes approach (Park et al. 2010), can reduce computational costs, but still require a good initial estimate of tree structure. We consider agglomerative approaches for finding a model M that optimize the objective function $P(G|\{\theta,\pi\},M)$. We first derive a scalable method for a single snapshot network, which we term hierarchical agglomerative clustering (HAC) (Park & Bader 2011). We then extend the single-snapshot method to a dynamic algorithm that can find dynamic models of time-ordered multiple networks (Park Bader). We term this dynamic hierarchical agglomerative clustering (DHAC).

10.3.1 Hierarchical agglomerative clustering (HAC)

Initially all vertices are individual distinct groups with each one parameter of within- and between-cluster probabilities. We then reduce the number of clusters by merging the best pair and build a tree until no further pair can be found. Suppose currently there are K top-level clusters numbered $1 \ldots K$ within the R total tree nodes. This model, M, has $K(K-1)/2 + R$ total parameters. Merging two of the top-level nodes (and retaining the structure underneath each) gives a model with $(K-1)(K-2)/2 + (R+1)$ parameters, a reduction of $K-2$ parameters. Later we revisit the tree and find the best collapsing point at which we determine an optimal number of bottom clusters (Figure 10.3A).

There is a subtle but crucial difference between this agglomerative algorithm, which assumes a full block model for the top-level nodes, and the more standard approach with a star-like structure at the top with a single parameter governing the interactions

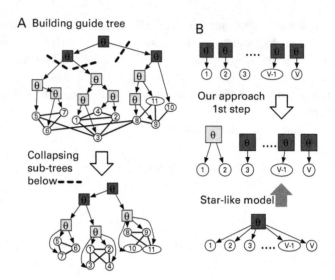

Figure 10.3   Hierarchical agglomerative clustering. (A) Collapsing of a maximum-likelihood guide tree. (B) HAC algorithm starts from fully factored model reducing model complexity at each round of merging, while an algorithm starting from a star-like model gradually increases model complexity.

between all pairs of top-level nodes (Figure 10.3B). A star-like model with K top-level and R total nodes has $R + 1$ parameters, and merging two groups increases the number of parameters by 1. The increase in parameters at each step, coupled with a maximum-likelihood model, is liable to over-fit the group structure.

Maximum-likelihood guide tree

Vertices are merged into increasingly large clusters based on the model likelihood with maximum-likelihood parameters $\hat{\pi}_k = n_k/V$ and $\hat{\theta}_{ij} = e_{ij}/t_{ij}$. Without loss of generality suppose we merge clusters 1 and 2 into a new cluster $1'$, defining a new model M'. The change in log-likelihood is

$$\lambda_{12}^S = \ln \left[\frac{n_{1'}^{n_{1'}}}{n_1^{n_1} n_2^{n_2}} \right] + \ln \left[\prod_{j \neq 1,2} \frac{P_{1'j}^{ML}}{P_{1j}^{ML} P_{2j}^{ML}} \right] \tag{10.4}$$

where $P_{ij}^{ML} \equiv e_{ij}^{e_{ij}} h_{ij}^{h_{ij}} / t_{ij}^{t_{ij}}$ and the superscript S indicates a single snapshot. The first term, arising from the multinomial cluster membership model and favoring balanced merges, was not included in HAC. This probability model maintains the fine-scale group structure underneath the new cluster: the terms P_{11}^{ML}, P_{12}^{ML}, and P_{22}^{ML} are present before and after the merge.

Bayesian collapsing criterion

A Bayes factor selects the model complexity (Kass & Raftery 1995). Integration of parameter θ on a single Bernoulli likelihood with a uniform prior, or Beta(1, 1), results in $\int_0^1 \theta^e (1-\theta)^h d\theta = \text{Beta}(e+1, h+1)$, or $e!h!/(e+h+1)!$ for integer values. Therefore the marginal likelihood is

$$P(G|M) = \prod_{k \leq k'} \int_0^1 d\theta P(e_{kk'}, h_{kk'} | \theta) P(\theta | 1, 1)$$

$$= \prod_{k \leq k'} \text{Beta}(e_{kk'} + 1, h_{kk'} + 1).$$

A similar procedure integrating out the nuisance parameters π_k with an uninformative prior would yield the additional contribution $\prod_k \Gamma(n_k + 1)/\Gamma(V + K)$. Alternatively, integrating out the nuisance parameters using a strong prior, $P(\{\pi_k\}) \propto \prod_k \pi_k^\nu$ with pseudocount $\nu \gg V$, yields a contribution that is independent of $\{n_k\}$. The Bayesian likelihood for the edge terms provided sufficient collapsing; we did not include the vertex assignment term in the Bayesian likelihood, equivalent to a strong prior. The Bayesian log-likelihood ratio for collapsing groups 1 and 2 into $1'$ is

$$\phi_{12}^S = \ln \left[\frac{P_{1'1'}^B}{P_{11}^B P_{12}^B P_{22}^B} \prod_{j \neq 1,2} \frac{P_{1'j}^B}{P_{1j}^B P_{2j}^B} \right] \tag{10.5}$$

where the superscript B indicates Bayesian, S indicates a single snapshot, and $P_{ij}^B \equiv \text{Beta}(e_{ij} + 1, h_{ij} + 1)$. This score is additive, and summing over all ϕ scores from the bottom clusters (individual vertices) upwards is equivalent to the log-likelihood ratio for the model with collapsed versus uncollapsed fine structure, with the collapsed vertices being the top-level groups in a stochastic block model. The guide tree is collapsed from the bottom up, in the order that groups were merged, to identify a local optimum of the cumulative ϕ score.

10.3.2   Dynamic hierarchical agglomerative clustering (DHAC)

We first attempted to couple adjacent snapshots by a $K \times K$ transition kernel matrix where K is a number of clusters. A typical example is the Hidden Markov Model (HMM) (Rabiner 1989). However we found that models for different snapshots are variant in complexity (K), the size of transition matrix can easily explode $O(K^2)$, and it remain nearly diagonal for smooth transition networks. Unless we imposed a strong constraint on the matrix the model was prone to over-fit data without showing proper temporal dependency. Instead we kernelized scores for merging and collapsing with an adaptive bandwidth to couple network clusters at nearby time points, which is similar to the method for regulatory network inference (Song et al. 2009).

Kernel-reweighted scores
Kernalization of the scores λ (Eq.10.4) and ϕ (Eq.10.5) couples nearby snapshots, also providing noise reduction and smoothing. Merging and collapsing scores were

kernelized using Gaussian radial basis functions with width parameter h, $w(\Delta t, h) \propto \exp\{-|\Delta t|/h\}$, where for simplicity Δt is the difference in snapshot indices. The kernelized merging score $\lambda^K(t)$ and collapsing score $\phi^K(t)$ for the tth snapshot (K denotes kernelized) are

$$\lambda_{12}^K(t; h) = \sum_{s=1}^{T} w(t - s, h) \, \lambda_{12}^S(s) \tag{10.6}$$

$$\phi_{12}^K(t; h) = \sum_{s=1}^{T} w(t - s, h) \, \phi_{12}^S(s). \tag{10.7}$$

Although the same clustering is used across all T time points, the scores will differ when proteins (or interactions) are present at one time point and absent at another. Kernels are normalized as $\sum_{s=1}^{T} w(t - s, h) = 1$. As $h \to 0$, $\lambda^K \to \lambda^S$ and $\phi^K \to \phi^S$. Since λ^S is statistically consistent (Park & Bader 2012), λ^K is statistically consistent as $n_k \to \infty$ and $h \to 0$. Collapsing is then performed as for single snapshots, stopping at the maximum of the bottom-up sum, termed $\phi^K(t; h) = \sum_{(i,j) \in \text{collapsed}} \phi_{ij}^K(t; h)$. The overall algorithm is summarized in Algorithm 1.

In the DHAC-local method, the bandwidth parameter h for snapshot t was selected from a grid search over h values $0.5, 1.0, 1.5, \ldots, 3.5$ to maximize $\phi^K(t; h)$, with smaller h favored when the network changes quickly. For the network considered here, $h \approx 1$ to 2 depending on t. Alternatively, a constant value of h may be used for all values of t, which we termed DHAC-constant. We set $h = 1$ for DHAC-constant, although in principle h could be optimized by maximizing $\sum_t \phi^K(t; h)$. In practice, results were very robust to the value of h, and the performance of DHAC-local was nearly identical to DHAC-constant with $h = 1$ (results not shown).

Algorithm 1 DHAC

---

    for $t \leftarrow 1 \ldots T$ **do**

        Set each vertex to be a single cluster

        Let $\phi_{\text{cum}} \leftarrow 0$ be cumulative model comparison score (Eq. 10.7)

        Compute merging scores (Eq. 10.6) of pairs having an edge or one or more shared neighbors

        repeat

            Pick a pair i, j of maximum $\lambda_{ij}^K(t; h)$

            Update scores of affected pairs after merging i, j

            Merge i, j to i'

            Compute merging scores i', j for all j with $e_{i'j} > 0$ or with $\sum_k e_{i'k} e_{kj} > 0$

            Update $\phi_{\text{cum}}(t; h) \leftarrow \phi_{\text{cum}} + \phi_{ij}^K(t; h)$

        until no more pairs to merge

        Output group structure $M(t; h)$ at which $\phi_{\text{cum}}(t; h)$ was maximum

    end for

10.3.3   Post-processing and visualization

Cluster matching

DHAC outputs T models, $\{M_1, \ldots, M_T\}$, and many groups will change slowly between time points. The total number of groups may differ between time points, however, and even if the number of groups and the group membership are nearly identical, group order may be permuted across time points. Matching similar groups across time points remains a general problem for dynamic networks.

For $T = 2$ groups, reasonable yet *ad hoc* procedures are to match groups based on shared members, Jaccard correlation of shared neighbors, or maximum weighted matching of shared neighbors or other pairwise scores (Bayati et al. 2008). Here we extend these ideas to multipartite matching based on a novel probabilistic model that introduces some rigor to the time-course matching problem.

The goal is to find most probable mapping of cluster i at time t to a globally consistent index k. Let $z_{ik}^{(t)} = 1$ if cluster i of snapshot t is assigned to k, and 0 otherwise, with normalization $\sum_k z_{ik}^{(t)} = 1$. Conversely, the sum over local clusters, $\sum_i z_{ik}^{(t)}$, is not fixed because the global cluster may be absent at time t (sum = 0) or it may be broken into multiple smaller clusters (sum > 1).

Each cluster i contains original network vertices $\{u\} \subseteq V$, and $n_{ij}^{(t)}$ counts the number of shared members between group i at time t and group j at time $t + 1$. The probability that a vertex makes a transition from global state k to state k' between two snapshots is $\psi_{kk'}$, with normalization $\sum_{k'} \psi_{kk'} = 1$. For simplicity, $\psi_{kk'}$ is independent of t. When groups do not change over time, $\psi_{kk'} = \delta_{kk'}$, 1 if $k = k'$ else 0. Similarly, the time-independent parameter v_{uk} is the probability that vertex u is in global group k, with $\sum_k v_{uk} = 1$.

The matching probability under consistent indexing is

$$P(\{M_t\}, \{z_{ij}^{(t)}\}|v, \psi) = \prod_{k=1}^{K} \prod_{t=1}^{T} \prod_{i \in S_t} \prod_{u \in C_i} v_{uk}^{z_{ik}^{(t)}} \times \tag{10.8}$$

$$\prod_{k=1}^{K} \prod_{k'=1}^{K} \prod_{t=1}^{T-1} \prod_{i \in S_t} \prod_{j \in S_{t+1}} \psi_{kk'}^{n_{ij}^{(t)} z_{ik}^{(t)} z_{jk'}^{(t+1)}}$$

where S_t denotes the set of clusters at snapshot t and C_i the set of vertices in one of these clusters.

We solved the *maximum a posteriori* (MAP) inference problem using expectation-maximization (EM). The M-step updates are

$$v_{uk} \propto \sum_{t=1}^{T} \sum_{i \in S_t} z_{ik}^{(t)} I\{u \in C_i\}, \tag{10.9}$$

$$\psi_{kk'} \propto \sum_{t=1}^{T-1} \sum_{i \in S_t} \sum_{j \in S_{t+1}} n_{ij} z_{ik}^{(t)} z_{jk'}^{(t+1)}. \tag{10.10}$$

The E-step for $z_{ik}(t)$ is more complicated. If the state at time t is represented as the assignment matrix $\{z_{ik}(t)\}$, then the probability structure is a HMM. This state space is large, however, on the order of $K^K \sim K!$, because each of the approximately K clusters at time t may be assigned to one of K global clusters, and the transition matrix is of order K^{2K}. Instead, we simplify the state space by considering each $z_{ik}(t)$ independently and introducing additional couplings that create loops in the corresponding graphical model, no longer permitting a dynamic programming solution. When groups are stable over time, however, the topology is close to a tree structure and belief propagation (BP) works well (Yedidia et al. 2005).

For max-product BP algorithm we reformulate the above Markov Random Field, or joint probability (Eq. 10.8), constructing a factor graph consisting of factors (hyper-edges) and variables (latent variables). Latent variables $z_i^{(t)}$ take on values from $1, \ldots, K$, or succinctly $[K]$. In other words, $z_i^{(t)}$ provides the index k of the global cluster for which $z_{ik}^{(t)} = 1$. Parameters $\{v\}$ are used to represent singleton factors and $\{\psi\}$ pairwise factors. A certain latent variable $z_i^{(t)}$ depends on neighboring pairwise factors $N(i, t-1)$ from the previous snapshot and $N(i, t+1)$ from the subsequent snapshot. MAP inference is carried out by sending messages from i to j via pairwise factor e. The update equations of the message $m_{i \to e}$ from variable i at time t to factor e and then the message $m_{e \to j}$ from e to variable j at time $t+1$ is

$$m_{i \to e}(k) \propto \prod_{u \in C_i} v_{uk} \prod_{f \in N(i,t-1) \cup N(i,t+1) \setminus \{e\}} m_{f \to i}(k) \qquad (10.11)$$

$$m_{e \to j}(k) \propto \max \left\{ l \in [K] : \psi_{lk}^{n_{ij}^{(t)}} m_{i \to e}(l) \right\}. \qquad (10.12)$$

For variable j at time $t-1$, the message $m_{e \to j}$ is

$$m_{e \to j}(k) \propto \max \left\{ l \in [K] : \psi_{kl}^{n_{ji}^{(t-1)}} m_{i \to e}(l) \right\}. \qquad (10.13)$$

The belief b_i of a certain variable i at snapshot t is the product of incoming messages,

$$b_i(k) \propto \prod_{e \in N(i,t-1) \cup N(i,t+1)} m_{e \to i}(k), \qquad (10.14)$$

normalized as $\sum_k b_i(k) = 1$. To prevent the MLEs and BP steps from overshooting, parameters and messages were updated as 1/10 of the full change, with updates to messages performed on a logarithmic scale (Koller & Friedman 2009). We call this EM method MATCH-EM (Algorithm 2).

Algorithm 2 MATCH-EM

    Initial greedy matching
    Initialize ν and ψ
    repeat
      repeat
        while forward and backward visit of factors **do**
          Calibrate messages i to j (Eqs. 10.11, 10.12, 10.13)
        end while
        for each variable i **do**
          Update belief b_i (Eq. 10.14)
        end for
      until convergence of BP
      Update latent variables $z_{ik} = 1$ with $k = \underset{l}{\mathsf{argmax}}\, b_i(l)$ and $z_{ik'} = 0$ for other $k' \neq k$.

      Update $\hat{\nu}, \hat{\psi}$ by MLE (Eqs. 10.9, 10.10)
    until convergence of EM

10.4   Results

10.4.1   *Arabidopsis* root development

Hierarchical clustering and spatiotemporal mapping

A dynamic hierarchical model (DYHM) (Park et al. 2010) produces hierarchical cluster assignments for each of the 15 spatiotemporal samples. A reduced view of the results, averaging the inferred memberships over the 15 samples, is provided (Figure 10.4A). The node shading represents the averaged interaction enrichment. Leaf nodes, shaped as squares, are groups of clustered genes. These leaves are indexed from 1 (leftmost) to 64 (rightmost) for later reference. Zoomed-in views below illustrate how selected clusters evolve over space and time in increasing resolution (Figure 10.4B and C).

This tree view shows that most of the groups are assortative (enriched for self-interactions), which is typical of protein complexes. Some leaf nodes assemble hierarchically into larger assortative modules, and these components often share similar biological functions. For instance, four of the small nuclear RNA/RNP complexes (snRNA/P) are located adjacently and form a clade (terminal leaves #39–40). Cladistic assignments are also observed for EIF (eukaryotic translation initiation factor) complexes (leaves #1–4) and Splicing/Ribosome complexes (leaves #41–48).

An overview of terminal groups shows how each of the 64 clusters varies over the 15 spatiotemporal snapshots in terms of occupancy and within-cluster interactions. Several of the clusters correspond to protein complexes that appear constitutively active, whose transcripts would typically be filtered out as unchanging. Examples are #7 (membrane fusion), #10 (RNA Pol II), #14 (syntaxin and SNARE proteins), and #26 and #33 (proteasome). A more dynamic pattern is observed for clusters that are conditionally

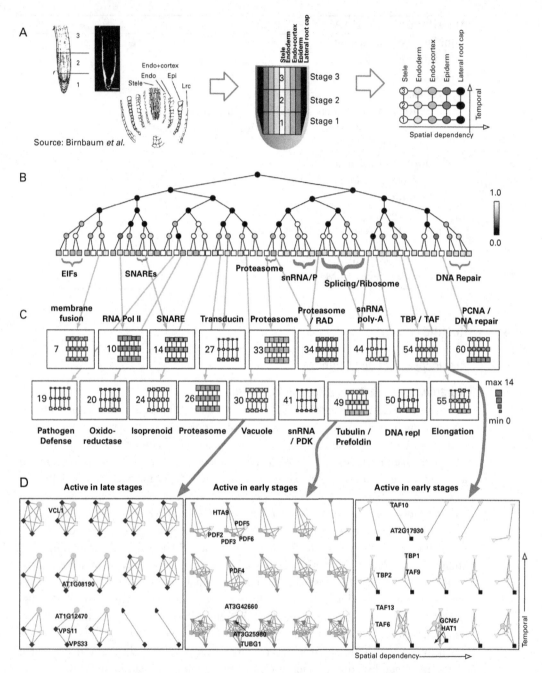

Figure 10.4   Dynamic clusters in *Arabidopsis* root development. (A) Lateral root sections correspond to distinct tissues, and vertical sections correspond to to distinct developmental stages. (B) Average hierarchical decomposition of 15 networks. (C) The evolution of each cluster is displayed over the five tissues and three stages. Size indicates the number of proteins within the cluster. (D) Selected micro-views on network dynamics. The leftmost example shows delayed activity of two genes in the developmental process. The other two examples include complexes that are more active at early stages. Subnetworks in each panel were drawn in identical topology. Gene names are labeled once. See text for details of selected clusters.

activated, most often with complex members present at early times and then absent at later times to yield a smaller core complex. Examples are #44 (mRNA polyadenylation), #49 (a core of prefoldin and the H2A.Z histone variant HTA9 has additional tubulin-related complex members during stage 1), and #60 (a PCNA DNA repair complex is present in stage 1 but vanishes in stages 2 and 3). These observations are consistent with the inference from mRNA data of rapid mitotic activity during stage 1 (Birnbaum et al. 2003).

TATA box-binding protein complex.

A detailed view of cluster #54, involved in transcription from TATA box promoters, highlights this pattern of dynamic complex membership (rightmost of Figure 10.4D). TATA box-binding protein associated factors (TAFs) have time-specific and tissue-specific activity (Tamada et al. 2007). One member of the TAF family, TAF10 (aka AT4G31720, TFIID15), has preferential and transient expression during the middle developmental stages of plant organs. Disrupting this tight regulation causes pleiotropic phenotypic changes and abnormal morphologies (Tamada et al. 2007).

The majority of the genes in cluster #54 are TAFs, including TAFII15/TAF10, TAFII21/TAF9, and TAFII59/TAF6. In the root expression map, TAF10 is a core member of this complex, while other members are transient. Along the temporal axis, the TAF10–TAF9–TFIID-1 complex is present during early root development, persists partially through stage 2, and in the mature root only TAFII15, TBP2, and the uncharacterized PIK-related kinase AT2G17930 remain. TAFs provide DNA-binding specificity for TFIIDs, which bind to the basal transcriptional machinery (Lago et al. 2004). The TAF6 (TAFII59) protein appears to be present primarily in stage 1, although absent from the stele. This factor has a core interaction motif required for H3/H4 heterodimerization (Lago et al. 2004), which suggests regional epigenetic modification in early development. At the early stage, this complex also has HAT1 as a member, a histoneacetyltransferase that is a positive regulator of transcription in root morphogenesis.

10.4.2   Yeast Metabolic Cycle (YMC) dynamics

YMC transcriptional profiling reveals three dominant metabolic states: reductive building (RB, 977 genes); reductive charging (RC, 1510 genes); and oxidative (OX, 1023 genes) (Tu et al. 2005). Almost a half of all genes oscillate along this cycle, indicating that a broad class of processes are involved but making it difficult to extract specific dynamical modules from expression data alone.

The expression data was combined with physical interactions to yield 36 time-varying network snapshots, corresponding to three complete cycles of 12 snapshots each. The network was made less sparse by applying an iterative degree cutoff (≥ 3), reducing the size of the network from ≈ 3000 to ≈ 1000 and increasing the mean vertex degree from 1.8 to >3. Networks were clustered by DHAC-local, and clusters were matched across time points using MATCH-EM.

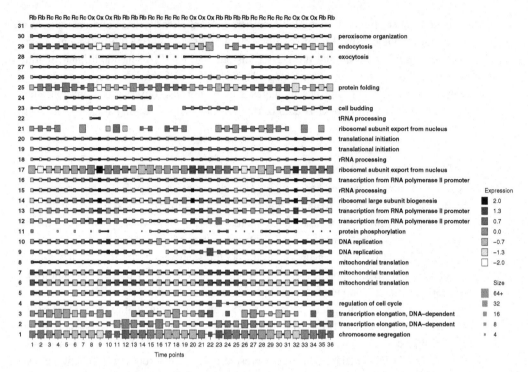

Figure 10.5   Dynamic network clustering reveals a detailed global view of periodic protein complexes during the yeast metabolic cycle. Squared nodes represent clusters matched across time points, showing only clusters having at least three genes/proteins. Cluster order: Clusters are organized by peak activity in RB phase (#1 to #10), OX phase (#11 to #20), and RC phase (#23 to #31). Node size: Number of genes/proteins contained in this cluster. Node color: Average standardized gene expression level at time t. Edge width: Jaccard coefficient (or coherence) between clusters of adjacent snapshots. Gene Ontology: Cluster-specific GO keywords were identified by hypergeometric tests. The right panel shows the top three enriched GO categories with p-value ≤ 0.05.

We checked robustness using a bootstrapping procedure in which a fraction α of edges are randomly rewired according to the degree-consistent configurational model (Karrer et al. 2008). We used $\alpha = 0.01$ and performed >500 bootstraps, with about 80% co-membership conserved across bootstraps at each snapshot.

Macroview of YMC complexes
We recovered 31 dynamic complexes with at least three proteins (Figure 10.5). Many of the complexes have cluster-specific gene ontology (GO) keywords (with p-value $\leq$ 0.05). Organizing clusters by average gene expression at each time point separates those that are active in each phase. RB clusters, #1 to #10, are related to cell cycle checkpoints and mitochondrial translation. OX clusters, #11 to #20, include ribosome metabolism, DNA replication/repair, and translation. RC clusters, #23 to #31, include stress response and transport.

Most of the complexes can be matched across the entire time-course, but some disappear then reappear. An example is complex #4, annotated for DNA repair, that is most active at the end of each 12-point cycle. This behavior required the MATCH-EM algorithm for globally consistent clusters, and would have been impossible to resolve given matching to nearest neighbors alone.

We ascertained whether the complexes predicted by our methods correspond to known complexes obtained from manual curation, CYC2008, or from high-throughput experiments, YHTP2008 (Pu et al. 2009). The 408 manual and 400 high-throughput complexes were filtered to retain the periodic proteins from YMC data, and then the catalog complex with the best Jaccard correlation was identified for each predicted complex. Of the 31 predicted complexes, 14 are poorly represented in the catalogs (Jaccard correlation <20%), 11 are only moderately similar (correlation $\geq$20% and <80%), and 6 have a good match (correlation $\geq$80%). The predicted complexes with poor overlap often recombine subunits from multiple catalog complexes (see #16 below).

To test the effects of the filtering, we also performed clustering using all 63 410 BioGrid interactions and including all genes with YMC data, periodic or non-periodic, yielding a network of 54 758 interactions among 4987 proteins. Clustering this network and retaining complexes with at least three proteins and edge density > 0.1 yields 20 to 40 clusters at each snapshot with 900 $\pm$ 100 proteins included. Most clusters in the unfiltered network contain a high-degree core from the filtered network. Occasionally multiple cores are combined by low-degree connections, making the cluster count smaller than in the filtered network. The overlap with protein complex catalogs is similar to the unfiltered network.

Microviews of YMC dynamics

The protein complex dynamics provide a rich view of YMC providing new biological insight, as demonstrated by in-depth analysis of clusters #7, the mitochondrial ribosome, and cluster #16, the nuclear pore.

Mitochondrial ribosome complex (#7) The mitochondrial ribosome is generally assumed to be RB-specific, with transcription switched on briefly at the transition from OX to RB (Figure 10.6A). This complex contains primarily RSMs (ribosomal small subunit of mitochondria) and MRPs (mitochondrial ribosomal protein), known components of the mitochondrial ribosome (Saveanu et al. 2001).

Underneath this general pattern, however, RSM22 shows systematic expression ahead of other components. At time points $t = 9$, $t = 20$, and $t = 32$, RSM22 is active while other proteins are not transcribed. RSM22 is a nuclear-encoded putative S-adenosylmethionine (SAM) methyltransferase (Petrossian & Clarke 2009), and methylation of the $3'$ end of the rRNA of the small mitochondrial subunit is required for the assembly and stability of the mitochondrial ribosome (Metodiev et al. 2009). Deleting RSM22 yields a viable cell with non-functional mitochondria. Together, these results suggest the hypothesis that early expression of RSM22 may provide the methylation activity necessary for assembly of the mitochondrial ribosome.

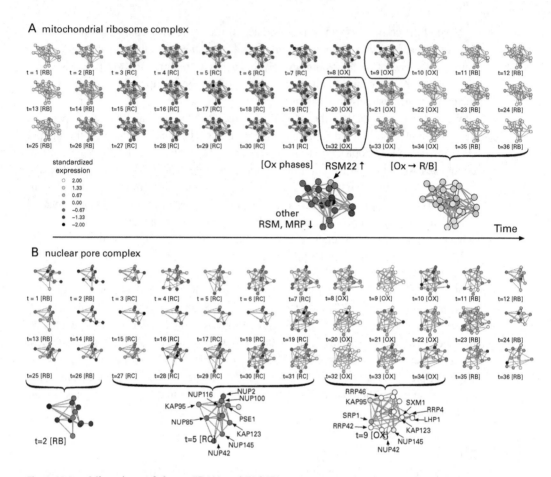

Figure 10.6 Microviews of cluster #7 (A) and #16 (B).

Nuclear pore complex (#16) Most genes in the nuclear pore complex are OX-responsive and the complex is most active at $t = 9, 20, 32$ (Figure 10.6B). Unlike the mitochondrial ribosome, where the entire complex is generally transcribed in synchrony, this complex shows a smaller co-expressed core that is complemented with transient members during the OX phase.

The co-expressed core includes nuclear pore complex (NPC) and Karyopherin (KAP) proteins (Pemberton et al. 1998, Strambio-De-Castillia et al. 2010). The physical structure of the NPC comprises mostly NUP proteins. Among the proteins included in cluster #16, NUP2, NUP100, and NUP116 shape the Phe-Gly passage of the NPC (Strambio-De-Castillia et al. 2010). In contrast, KAP proteins are not considered structural but rather mediate export and import of RNA and proteins (Strambio-De-Castillia et al. 2010, Grünwald et al. 2011). KAP123 and PSE1 specifically transport ribosomal proteins (Schlenstedt et al. 1997). During the OX phases, SRP1 and SXM1 are additionally recruited. These KAP proteins recognize either nuclear localization sequences (NLS) or nuclear export sequences (NES) and direct transport into or out of the nucleus (Pemberton et al. 1998).

Other transient memberships add another interesting story. RRP4 and RRP42 are a part of the exosome that edits RNA molecules $3' \rightarrow 5'$ (Mitchell et al. 1997). Our clustering predicts that these proteins transition between the nuclear pore and other complexes during the cycle. CSL4 was recently reported to interact with RNA and is a possible exosome component (Liu et al. 2006). LHP1 is a La protein that binds to RNA polymerase III transcripts and small ribonuclear proteins (snRNPs), working as a molecular chaperone to protect and terminate the $3'$ end of transcripts (Yoo & Wolin 1994). These results are consistent with the hypothesis that RNA processing is tightly coupled to transport through the nuclear pore to the cytoplasm (Strambio-De-Castillia et al. 2010), but also suggest that dynamic reorganization of the nuclear pore occurs during the metabolic cycle. Additional evidence is the appearance of a second expression peak involving a subset of nuclear pore components at the start of the RB phase, which has not been previously described.

10.5 Discussion

Network clustering and dynamic network clustering problems have been increasingly necessary as large-scale network datasets are collected in diverse disciplines. Our algorithm optimizes the likelihood modularity, which is asymptotically consistent (Bickel & Chen 2009). Recent machine learning and physics approaches are based on probabilistic graphical models such as latent Dirichlet allocation (Blei et al. 2003, Airoldi et al. 2008, Ball et al. 2011). Dynamic extensions have been proposed (Fu et al. 2009), but prior to our work have been impractical except for very small networks with around 100 vertices and under 10 latent classes. Even efficient variational methods such as Variational Bayes Modularity (VBM) (Hofman & Wiggins 2008) have scaling that is far worse than a near linear or at least quadratic run time in number of nodes and edges.

Our DHAC algorithm scales as $O(EJ \ln V)$, the same as HAC (Park & Bader 2011), with a constant prefactor for the number of time points. This provides an excellent trade-off for genome-scale problems. The yeast network considered here with 2000 vertices required about 5 minutes on a single 2 GHz processor. A full-genome network with 10 000 to 100 000 vertices could be analyzed in a day to a week on a single processor, but in practice would be much faster because each time point could be run in parallel.

The cluster matching algorithm MATCH-EM is a second contribution that provides a solution to the general problem of tracing the evolution of a set of groups or clusters over time. It generalizes a previous belief propagation method for bipartite matching (Bayati et al. 2006). The bipartite max-product algorithm is exact with a worst-case run-time of $O(K^3)$ for K classes. Our generalization has an additional linear factor of the number of time points. While it is no longer guaranteed to converge to the exact solution, for biological networks here it converges rapidly with good results.

Our methods applied to real biological data provide new insight. Many transcription time-course experiments reveal waves of correlated gene expression, with no standard methods to parse a large set of correlated genes into well-defined protein complexes. The DHAC method is a general solution to this problem and provides a multi-resolution view of dynamic expression and organization of the proteome. Focusing on specific predicted complexes reveals possible mechanisms of regulation and control. Our analysis of the yeast metabolic cycle identifies protein complexes with asynchronous gene expression, which suggests RSM22 as an RNA methyltransferase whose early expression may be required to assemble and stabilize the mitochondrial ribosome.

Our methods permit proteins to switch between complexes over time, which we see in the dynamics of the nuclear pore. Hierarchical methods like DHAC also provide a natural multi-scale description of complexes, sub-complexes, and proteins. A separate challenge is introducing mixed membership, with the same protein serving as a subunit in two distinct protein complexes (Palla et al. 2005). Cluster membership can be represented as a non-negative vector whose elements sum to 1, and variational inference then can generate mixed-membership clusters. Unfortunately, variational mixed-membership methods are slow and, for most biological networks, have produced worse results than the methods presented here.

Several improvements on DHAC are possible. One challenge is protein abundance data availability. For periodic data, the methods presented here that intersect a protein interaction database with transcription correlation appear to work well. A more general solution may be to use a standard transcription–translation model, $\dot{P}(t) = \beta R(t) - \alpha P(t)$, where $R(t)$ is the measured transcriptional abundance, $P(t)$ is the abundance of the corresponding protein, and β and α are production and degradation rates. In this approach, protein interactions would be retained if each of the interacting pair was above a baseline expression level. This model generates exponentially weighted smoothing of protein abundance, similar to the exponential kernel we used for smoothing. Since the exponential smoothing kernel already works well, we anticipate that results should be robust to choices of β and α, with the possibility of using consensus values for most proteins.

References

Airoldi, E. M., Blei, D. M., Fienberg, S. E. & Xing, E. P. (2008), 'Mixed membership stochastic blockmodels', *Journal of Machine Learning Research* **9**, 1981–2014.

Bader, G. D. & Hogue, C. W. V. (2003), 'An automated method for finding molecular complexes in large protein interaction networks', *BMC Bioinformatics* **4**, 2.

Ball, B., Karrer, B. & Newman, M. E. J. (2011), 'Efficient and principled method for detecting communities in networks', *Physical Review E* **88**, 1.

Bandyopadhyay, S., Mehta, M., Kuo, D., Sung, M. K., Chuang, R. et al. (2010), 'Rewiring of genetic networks in response to DNA damage', *Science* **330**, 1385–1389.

Banerjee, O., El Ghaoui, L. & d'Aspremont, A. (2008), 'Model selection through sparse maximum likelihood estimation for multivariate Gaussian or binary data', *Journal of Machine Learning Research* **9**, 485–516.

Bayati, M., Shah, D. & Sharma, M. (2006), 'A simpler max-product maximum weight matching algorithm and the auction algorithm', *IEEE International Symposium on Information Theory* **2005**, 1763–1767.

Bayati, M., Shah, D. & Sharma, M. (2008), 'Max-product for maximum weight matching: convergence, correctness, and LP duality', *IEEE Transactions on Information Theory* **54**, 1241–1251.

Bickel, P. J. & Chen, A. (2009), 'A nonparametric view of network models and Newman–Girvan and other modularities', *Proceedings of the National Academy of Sciences of the USA* **106**, 21 068–21 073.

Birnbaum, K., Shasha, D., Wang, J., Jung, J., Lambert, G. et al. (2003), 'A gene expression map of the *Arabidopsis* root', *Science* **302**, 1956–1960.

Blei, D. M., Ng, A. Y. & Jordan, M. I. (2003), 'Latent Dirichlet allocation', *Journal of Machine Learning Research* **3**, 993–1022.

Clauset, A., Moore, C. & Newman, M. E. J. (2008), 'Hierarchical structure and the prediction of missing links in networks', *Nature* **453**, 98–101.

Clauset, A., Newman, M. E. J. & Moore, C. (2004), 'Finding community structure in very large networks', *Physical Review E* **70**, 066111.

Croft, D., O'Kelly, G., Wu, G., Haw, R., Gillespie, M. et al. (2011), 'Reactome: a database of reactions, pathways and biological processes', *Nucleic Acids Research* **39** (Database issue), D691–D697.

Dittrich, M. T., Klau, G. W., Rosenwald, A., Dandekar, T. & Müller, T. (2008), 'Identifying functional modules in protein–protein interaction networks: an integrated exact approach', *Bioinformatics* **24**, i223–i231.

Friedman, J., Hastie, T. & Tibshirani, R. (2008), 'Sparse inverse covariance estimation with the graphical lasso', *Biostatistics* **9**, 432–441.

Fu, W., Song, L. & Xing, E. P. (2009), 'Dynamic mixed membership blockmodel for evolving networks', *Proceedings of the 26th Annual International Conference on Machine Learning*, pp. 329–336.

Geisler-Lee, J., O'Toole, N., Ammar, R., Provart, N., Millar, A. et al. (2007), 'A predicted interactome for *Arabidopsis*', *Plant Physiology* **145**, 317–329.

Grünwald, D., Singer, R. H. & Rout, M. (2011), 'Nuclear export dynamics of RNA–protein complexes.', *Nature* **475**, 333–341.

Han, J.-D. J., Bertin, N., Hao, T., Goldberg, D. S., Berriz, G. F. et al. (2004), 'Evidence for dynamically organized modularity in the yeast protein–protein interaction network', *Nature* **430**, 88–93.

Hastings, W. K. (1970), 'Monte Carlo sampling methods using Markov chains and their applications', *Biometrika* **57**, 97–109.

Hofman, J. M. & Wiggins, C. H. (2008), 'Bayesian approach to network modularity', *Physical Review Letters* **100**, 258701.

Jansen, R., Greenbaum, D. & Gerstein, M. (2002), 'Relating whole-genome expression data with protein–protein interactions', *Genome Research* **12**, 37–46.

Karrer, B., Levina, E. & Newman, M. E. J. (2008), 'Robustness of community structure in networks', *Physical Review E* **77**, 046119.

Kass, R. E. & Raftery, A. E. (1995), 'Bayes factors', *Journal of the American Statistical Association* **90**, 773–795.

Koller, D. & Friedman, N. (2009), *Probabilistic Graphical Models: Principles and Techniques*, Cambridge, MA: MIT Press.

Komurov, K. & White, M. (2007), 'Revealing static and dynamic modular architecture of the eukaryotic protein interaction network', *Molecular Systems Biology* **3**, 110.

Lago, C., Clerici, E., Mizzi, L., Colombo, L. & Kater, M. M. (2004), 'TBP-associated factors in *Arabidopsis*', *Gene* **342**, 231–241.

Leskovec, J., Kleinberg, J. & Faloutsos, C. (2005), 'Graphs over time', *Proceedings of the 11th ACM SIGKDD International Conference on Knowledge Discovery in Data Mining*, pp. 177–187.

Liberzon, A., Subramanian, A., Pinchback, R., Thorvaldsdóttir, H., Tamayo, P. et al. (2011), 'Molecular signatures database (MSigDB) 3.0', *Bioinformatics* **27**, 1739–1740.

Liu, Q., Greimann, J. C. & Lima, C. D. (2006), 'Reconstitution, activities, and structure of the eukaryotic RNA exosome', *Cell* **127**, 1223–1237.

Luscombe, N. M., Babu, M. M., Yu, H., Snyder, M., Teichmann, S. A. et al. (2004), 'Genomic analysis of regulatory network dynamics reveals large topological changes', *Nature* **431**, 308–321.

Metodiev, M. D., Lesko, N., Park, C. B., Cámara, Y., Shi, Y. et al. (2009), 'Methylation of 12S rRNA is necessary for in vivo stability of the small subunit of the mammalian mitochondrial ribosome', *Cell Metabolism* **9**, 386–397.

Mitchell, P., Petfalski, E., Shevchenko, A., Mann, M. & Tollervey, D. (1997), 'The exosome: a conserved eukaryotic RNA processing complex containing multiple $3'-5'$ exoribonucleases', *Cell* **91**, 457–466.

Palla, G., Derényi, I., Farkas, I. & Vicsek, T. (2005), 'Uncovering the overlapping community structure of complex networks in nature and society', *Nature* **435**, 814–818.

Park, Y. & Bader, J. S. (2011), 'Resolving the structure of interactomes with hierarchical agglomerative clustering', *BMC Bioinformatics* **12**(Suppl. 1), S44.

Park, Y. & Bader, J. S. (2012), 'How networks change with time', *Bioinformatics* **28**, i40–i48.

Park, Y., Moore, C. & Bader, J. S. (2010), 'Dynamic networks from hierarchical Bayesian graph clustering', *PloS One* **5**, e8118.

Pemberton, L. F., Blobel, G. & Rosenblum, J. S. (1998), 'Transport routes through the nuclear pore complex', *Current Opinion in Cell Biology* **10**, 392–399.

Petrossian, T. C. & Clarke, S. G. (2009), 'Multiple motif scanning to identify methyltransferases from the yeast proteome', *Molecular & Cellular Proteomics* **8**, 1516–1526.

Pu, S., Wong, J., Turner, B., Cho, E. & Wodak, S. J. (2009), 'Up-to-date catalogues of yeast protein complexes', *Nucleic Acids Research* **37**, 825–831.

Rabiner, L. R. (1989), A tutorial on hidden Markov models and selected applications in speech recognition', *Proceedings of the IEEE*, **77** 257–286.

Rivera, C. G., Vakil, R. & Bader, J. S. (2010), 'NeMo: network module identification in cytoscape', *BMC Bioinformatics* **11**(Suppl. 1), S61.

Rodriguez, M. G., Leskovec, J. & Krause, A. (2010), Inferring networks of diffusion and influence', *Proceedings of the 16th ACM SIGKDD international conference on Knowledge Discovery and Data Mining*, pp. 1019–1028.

Saveanu, C., Fromont-Racine, M., Harington, A., Ricard, F., Namane, A. et al. (2001), 'Identification of 12 new yeast mitochondrial ribosomal proteins including 6 that have no prokaryotic homologues', *Journal of Biological Chemistry* **276**, 15 861–15 867.

Schlenstedt, G., Smirnova, E., Deane, R., Solsbacher, J., Kutay, U. et al. (1997), 'Yrb4p, a yeast ran-GTP-binding protein involved in import of ribosomal protein L25 into the nucleus', *EMBO Journal* **16**, 6237–6249.

Song, L., Kolar, M. & Xing, E. P. (2009), 'KELLER: estimating time-varying interactions between genes', *Bioinformatics* **25**, i128–i136.

Stark, C., Breitkreutz, B.-J., Reguly, T., Boucher, L., Breitkreutz, A. et al. (2006), 'BioGRID: a general repository for interaction datasets', *Nucleic Acids Research* **34**(Database issue), D535–D539.

Strambio-De-Castillia, C., Niepel, M. & Rout, M. P. (2010), 'The nuclear pore complex: bridging nuclear transport and gene regulation', *Nature Reviews Molecular Cell Biology* **11**, 490–501.

Tamada, Y., Nakamori, K., Nakatani, H., Matsuda, K., Hata, S. et al. (2007), 'Temporary expression of the TAF10 gene and its requirement for normal development of *Arabidopsis thaliana*', *Plant Cell Physiology* **48**, 134–146.

Tu, B. P., Kudlicki, A., Rowicka, M. & McKnight, S. L. (2005), 'Logic of the yeast metabolic cycle: temporal compartmentalization of cellular processes', *Science* **310**, 1152–1158.

Wu, Z., Irizarry, R. A., Gentleman, R., Martinez-Murillo, F. & Spencer, F. (2004), 'A model-based background adjustment for oligonucleotide expression arrays', *Journal of the American Statistical Association* **99**, 909–917.

Yedidia, J., Freeman, W. & Weiss, Y. (2005), 'Constructing free-energy approximations and generalized belief propagation algorithms', *IEEE Transactions on Information Theory* **51**, 2282–2313.

Yoo, C. J. & Wolin, S. L. (1994), 'LA proteins from *Drosophila melanogaster* and *Saccharomyces cerevisiae*: a yeast homolog of the LA autoantigen is dispensable for growth', *Molecular and Cellular Biology* **14**, 5412–5424.

11 Phenotype state spaces and strategies for exploring them

Andreas Hadjiprocopis and Rune Linding

11.1 . Introduction

Proteins and their interactions determine how cells behave. Genes are the blueprints for protein synthesis; their activation or suppression determines the absence or presence of a protein which in turn can give rise to further activation or suppression of other genes and proteins. This chemical chain reaction usually involves positive and negative feedback loops and is subjected to stochastic noise and the influence of environmental factors. Moreover, epistasis – the cancellation or modification of a gene's contribution to the phenotype by other genes – is generally the rule rather than the exception in genetics.

Despite all these factors, amazingly robust cell behavior is apparent in many biological systems although it is difficult to be modeled and/or predicted. Building the topology and quantifying the direct and indirect cause–effect (stimulus, expression, activation, behavior) relationships of the reactions leading to the phenotypes – in general, genetic regulatory networks (GRN) – is challenging in at least three ways.

Firstly, how are these relationships described? Traditionally, mathematical models are expressed in terms of transfer functions relating inputs to outputs expressed as a composition of differential equations with a time dimension. We argue though that the cellular signaling networks are probabilistic in nature and that diffusion-based models remain challenging due to lack of knowledge of essential system parameters, such as rate constants. Most importantly, treating intracellular protein and gene interactions as in-vitro chemical reactions might not be safe because the usual assumption of diffusion dynamics namely that of the free movement of a sufficiently large number of molecules is usually the exception rather than the rule due to the very small number of reactant molecules in highly confined and crowded space. Moreover, the concentration of a protein is highly dependent on subcellular localization and thus the picture of the cell as a homogeneous mix container is simply wrong. Of even more profound impact, many signaling systems are centered on or around scaffolding proteins mimicking solid-state chemical environments and have little or no resemblance to diffusion limited systems. For these reasons, significant statistical fluctuations in the behavior of the reactions are common and the use of diffusion limited modeling approaches not appropriate.

Systems Genetics: Linking Genotypes and Phenotypes, ed. F. Markowetz and M. Boutros. Published by Cambridge University Press. © Cambridge University Press 2015.

More recently, computational models in the form of Boolean networks, Petri nets, and Interacting (Probabilistic) State Machines have been proposed (Fisher & Henzinger 2007) as an alternative way of modeling biological processes with the added benefits of abstraction, high-level reasoning, and mature process calculus tools to aid analysis.

Secondly, our ability to gain understanding and abstract from derived molecular mechanisms and regulatory networks topology is really limited because of the sheer complexity and abundance of low-level information inherent in these networks. In this respect, a way to reduce complexity is by treating the parts of these networks where the interactions within are much larger compared to the interactions between, as semi-autonomous modules – essentially, as black boxes. Modular response analysis (Brüggeman et al. 2002) offers such a framework. Building on this, Zamir & Bastiaens (2008) developed a methodology for reverse-engineering network structure in order to analyze how perturbations propagate in a network. Modularization at a higher level is also key for reusing parts of the derived regulatory networks.

Finally, our ability to understand GRN networks is impacted by the fact that kinase activation can initiate different cellular decisions depending on the pre-activation state of the network. In Janes et al. (2005), it was shown that JNK activation can be anti- or pro-apoptotic depending on network state when cells received growth factor cues. Therefore, to describe and predict a cellular response to a perturbation, studies must be carried out in the context of the cell's multivariate network state (Linding 2010).

11.2   Phenotype: a constructive generality

Genotype can be seen as a set of instructions carried within an organism's genetic code, the DNA. This is straightforward. Phenotype's definition however is an example of constructive generality. According to the Free Dictionary, phenotype is *"the set of observable characteristics of an individual (e.g. cell) resulting from the interaction of its genotype with the environment."*

The question is at what temporal and spatial scales is such a characteristic observable, hence, measurable? Furthermore, should the measurements be integrated over time and cell population size and at what differential amount (resolution)?

Many cell processes are oscillatory. The frequency and amplitude of these oscillations as well as other qualitative characteristics may also be considered as phenotype. When cells are studied individually or in sub-populations, these oscillations are observed. However, when a population of cells is studied as a whole, via, say a Western blot, it is quite possible that the observed result of these processes averages (in contrast to the median) to some behavior that does not exist at all anywhere in the population (Batchelor et al. 2009). It is also possible that for some of these oscillatory processes and depending on cell (a)synchronicity, the observed population average is constant over time whereas the signal from individual cells or sub-populations is variable (Spiller et al. 2010). On the other hand, cells rarely do exist in isolation or alone and thus understanding interactions of cells at individual and population level is critical, particularly in areas such as cancer and tumor biology.

In Section 11.7, it is proposed that what it is perhaps a more suitable assessment of cellular phenotypes is pertaining changes to phenotype rather than absolute phenotypes; defining cell behavior as a trajectory through phenotype states or ensembles of states may be more useful for biological systems.

11.3   Cellular noise

The question why cells with the same genes/genome while exposed to the same environment do not always exhibit the same phenotype is worth considering. The answer to this is most likely not found in further end-point genetic sequencing, though potentially more insight into genome dynamics would be useful. Rather it is the combination of noise and the diversity in behavior of otherwise deterministic dynamical systems; not an anomaly but a fact of life.

Unfortunately the prevailing deterministic thinking in cell biology, as Quaranta & Garbett (2010) put it, has obscured the importance of variability and noise in the cell's behavior and micro-environment by averaging out the phenotypes of large populations of cells. In fact, for many tissues the variation in gene and protein expression among clonal progenitor cells has biological consequences as it was shown that it leads to different functional states (Chang et al. 2008).

The fluctuations in the expression of a gene arise mainly from two sources of stochasticity. Extrinsic sources are external to the cell and due to existing cellular heterogeneity, e.g. cell size, cycle stage, number of ribosomes, and growth rate.

On the other hand, intrinsic sources are internal to the cell. They have to do with the stochasticity of chemical reactions at the molecular level particularly when low copy numbers of reactants are involved. Therefore, chemical reaction events leading to protein expression, such as translation and transcription across different but genomically identical cells, will fluctuate even if the cells are otherwise in (cell cycle) synchronicity (Swain et al. 2002).

Elowitz et al. (2002), studied the variation in expression of two genes (Cyan Fluorescent Protein, CFP, and Yellow Fluorescent Protein, YFP) within single cells of *Escherichia coli* as the amount of intrinsic noise increased. It is important to note how they controlled the levels of intrinsic noise using internal reporter gene construct.

They observed that when intrinsic noise is low, the levels of CFP and YFP vary over time but in complete step within cells causing cells to exhibit a homogeneous color (the synthesis of CFP and YFP frequencies).

Increasing intrinsic noise caused the protein levels to vary independently within cells which resulted in a different color depending on the expression of whether CFP or YFP was at that particular time higher.

Cell shape characteristics such as nucleus size and perimeter shape do affect the levels of intrinsic noise in the cell but, more importantly, it can be argued that they can localize or isolate it to certain cell areas. Cell shape determines the physical characteristics of the volume within which the chemical reactions take place and thus controls the level of intrinsic noise because some areas will be more restricted and crowded than others. This also relates to the solid-state chemical nature of many subcellular molecular compartments.

11.3.1 Maximum information flow and noise

In (Tkačik et al. 2009), the problem of information flow in genetic control circuits is explored. A simplified version of these circuits consists of a transcription factor binding to DNA, regulating the transcription of mRNA and the further synthesis of one or more proteins, which are assumed to not interact with each other. Information is represented by the various molecular events flowing from the input (the transcription factor concentration) to the outputs (the synthesis of the protein molecules). In this model, there are two noise components; the input noise is due to fluctuations in the arrival and concentration of the transcription factor and the output noise which is proportional to the mean levels of expressed protein.

The authors assume (among other things) that the cell optimizes these processes for maximum information transmission at a given maximum input concentration by the adaptability of the process to its input (Tkačik et al. 2008). The suggested framework is to maximize the mutual information between the input and output quantities constrained by the maximum number of molecules at both ends. The expression for the mutual information involves two probability distributions; the (input) transcription factor concentration, and the conditional probability of protein expression levels (outputs) given an input concentration. The overall distribution of protein expression levels is the product of the aforementioned distributions integrated over all input concentrations.

11.4 Genome evolution, protein families, and phenotype

It has been observed that different proteins, even across species, share "similar" amino-acid sections. Such similarities indicate that the species the *homologue* proteins belong to, although now genetically isolated, had in the past shared a common ancestor species or through a past gene duplication identical proteins in the same species had became distinct but "similar" through accumulated mutations.

Although proteins can be described on at least three levels; 3D structure, sequence structure, and function, the main tool for family identification is structure-based sequence alignment as this is, from a computational point of view, the most tractable.

Exactly because of this multiple character of proteins, the notions of similarity/distance and relationships between protein domains or between protein linear motifs become difficult to define. The field of Information Theory has provided a variety of tools and methods for quantifying protein similarity, detecting evolutionary conservation, and mapping protein sequence structure to protein function, for example, the concept of mutual information used in statistical coupling analysis (SCA). In Socolich et al. (2005) it was shown that it is not safe to assume that two residues occupying their respective positions in the structure of a protein are statistically independent events.

SCA was used to identify pairwise dependencies between sites in proteins of the same family by analyzing their statistical profile. The probability that a pair of sites has mutually co-evolved increases as its mutual information, essentially a correlation metric, increases. This implies that protein structural stability (folding) is undermined if a mutation in one of the sites is not compatible with its counterpart – even if they are not near in linear sequence space. Evidence of this coupling has been known for some time (Yanofsky et al. 1964). However, the importance of this research lies in the suggested computational frameworks for detecting and quantifying protein site dependencies as well as sampling artificial proteins from a probability distribution using the Monte Carlo and the Metropolis algorithm. The benefit of using the Monte Carlo method is to be able to *sample* the space of eligible proteins taking into account co-evolution dependencies without constructing the protein model as an actual rule-based system.

Yet, this is just the tip of the iceberg; the heart of the problem is to be able to determine the relationship between form (sequence) and function (impact on phenotype, which in turn most likely depends on the interaction network of the protein at the time of activation/regulation) of proteins emerging from the vast number of combinations of amino-acid interactions and in view of the multitude of different sequences associated with similar function. We argue that the form and function dyad, epitomized by genotype and phenotype, is a recurring theme in biology manifested at many levels as a dynamical system (see Section 11.4.1) whose evolution (function) can potentially span a huge space but in reality, it is restricted to a relatively smaller number of attractor states, as its free parameters (form) change (see also Section 11.5.2). In our opinion, the role of Information Theory and Statistical Mechanics is and will be significant in breakthroughs in cellular biology, as outlined below.

The aim in Bialek & Ranganathan (2007) is to "*have a description of this ensemble of sequences, ideally being able to write down the probability distribution out of which functional sequences are drawn.*" To this end, a link between Monte Carlo sampling of artificial proteins and maximum entropy (ME) models is made. A ME distribution is one which is as random as possible while, at the same time, consistent with prior information, i.e. experimental observations. In Mora & Bialek (2011) an attempt is made to lay the problem within a statistical mechanics (SM) framework because "*maximum entropy models naturally map onto known statistical physics models, which will ease the study of their critical properties.*"

This direction has good potential because it allows a whole class of *emergent* systems as diverse as networks of neurons, ensembles of (amino-acid) sequences, and flocks of birds to be discussed within a single, unified framework with powerful mathematical analysis tools which combine the key concepts of information, entropy, and energy. Moreover, this framework allows for the investigation of scale-free properties (see Section 11.5.1) and self-organized criticality (see Section 11.6.4) in these systems as demonstrated by Mora & Bialek (2011).

11.4.1   Phenotype state spaces

Using the example of the routes a cell may take during differentiation, Waddington (Goldberg et al. 2007, Wang et al. 2011) devised the analogy of a landscape on which a ball under the force caused by some gravitational field rolls down from high altitude, following various (forking) paths with occasionally a brief upward direction given enough kinetic energy. Eventually the ball settles (or is trapped) in a low-energy valley and stays there until there is some change in the landscape which will get it rolling again.

Traditionally, Waddington's epigenetic landscape is employed as a demonstration of the variability but overall finiteness, discreteness, and stability of phenotypes. Similarly, more complex phenotypes such as cell morphology or cell fate (proliferation/apoptosis) processes and development and differentiation may also be considered as operating in a space of phenotypic attractor states. A big challenge in modern biology is to link these to molecular implementations to predict biological systems behavior.

Balázsi et al. (2011) proposes some revisions to Waddington's picture in order to properly describe cellular dynamics and reflect recent research:

- The concentrations of all molecules in the cell and all relevant environmental factors must add one dimension each to the landscape, rendering it super high-dimensional.
- The landscape is not rigid; each point is subjected to molecular noise of various magnitude and properties. In addition, cells or mutations to the genome may reshape the landscape due to cell-to-cell interactions and growth rate dependence of protein concentrations.

In addition to Balázsi's extensions we would argue that a critical parameter is the activation state of a molecule, e.g. a kinase which is crucial to relate its activity to networks and phenotype.

The idea of a landscape, a surface which represents a potential/fitness/energy/probability/etc. spanning the space encompassing all the possible factors that affect it, has been around for some time. Sewall Wright introduced such landscapes to biology (Wright 1932).

In mathematical optimization one seeks to minimize or maximize a fitness function subject to the constraints and inter-dependence of its various input parameters. The fitness function is represented as a $(N + 1)$-dimensional surface over its N inputs. Its minimization is the process by which the deepest valley – the global minimum – is sought. Naturally, one assumes the existence of a fitness function in the first place. This means that the relationship of all inputs is known and can be expressed analytically as a mathematical equation or at least be consistently computable.

Given that the possible search space is potentially extremely large – depending on the number of input dimensions and complexity of the fitness function – one can not rely on *global* algorithms which require enumerating and assessing each possible combination of inputs. In fact, the phrase "each possible combination" has little meaning when we are talking about continuous-valued inputs. Discretization of input parameters may be of some help but the success is problem-dependent as the surface can be extremely multimodal in which case solutions will be missed.

Various *local* search algorithms have been proposed for finding the deepest valley in such landscapes. A widely used one is *gradient descent* or *hill climbing* depending on whether we seek the deepest valley (e.g. minimum energy) or the highest peak (e.g. maximum fitness). They are based on following the steepest way (as in steepest slope, gradient) up or down – in a sense it is the route water would follow, under the influence of gravity, when released from the top of a mountain.

A common characteristic of local algorithms is that they suffer from the problem of *local minima*. Following the steepest path downhill guarantees that the next step will be at the same or lower altitude (fitness) than the current position. It does not guarantee however that the final step will be the lowest over the landscape. Simply because these algorithms are local, they are blind to the wider landscape topography because in order to see it, they will have to calculate it and it is too expensive to compute and store the landscape in more locations other than those on a simple trajectory.

Heuristics are workarounds to the inherent limitations of these (local) search algorithms. They are what we call *rules of thumb* and they may work most of the time but not all the time. So there are numerous heuristics to cope with the problem of local minima mostly based on the observation that optimization is not necessarily a monotonic process, i.e. one has to go down first in order to go further up (e.g. snakes and ladders).

But first consider an analogous problem from metallurgy (there are similar examples in spin-glass theory). Crystal formations in a metal at room temperature are fixed. Many of the properties of the material depend on bond stability which depends on the type of the crystal formations and their potential energy. The lower the energy the more stable the crystal structure. When the material is at high temperature, basically melting, the molecules are free to move around, breaking old and creating new crystal formations of various types. As the temperature is reduced, molecular mobility is less and crystal structures become more permanent bar some local movement – the material slowly solidifies

Molecules are trying to solve a problem; the objective is to minimize the sum of the potential energy of the crystal bonds in the material given a set of N parameters: molecule positions, etc. These are the N dimensions over which the energy landscape lies. The collection of all molecules, the system, seeks the deepest valley of this landscape.

Annealing is a heuristic which improves this process by carefully controlling the temperature which is related to the mobility of molecules which is related to the probability of accepting a relatively bad solution now as a means of leading to a better solution later. This is equivalent to shaking the energy landscape and causing the, by now familiar, ball to jump out from the valley it has currently settled in, with a good chance that when it goes up, it will find its way to an even deeper valley. As the temperature is reduced, the landscape shaking is reduced. It can not give the ball sufficient energy to make these dramatic bounces but it can be enough to move the ball to the left or right a bit in order to stabilize its position locally.

Simulated annealing in mathematical optimization is a similar process by which the landscape is explored in an "*educated*" yet stochastic way. Given a current state, S_0, a candidate state, S_c, and a temperature value, T_t, the probability of moving to the S_c depends on the energies of the two states, E_0 and E_c, and the temperature. At higher temperatures, the likelihood of jumping to the candidate state is very high even if its energy/fitness is not as good as the current state. This allows for sampling the landscape uniformly for good solutions. As the temperature is reduced, however, candidate states with worse fitness than the current one are not likely to be accepted. Because we are using a computer, at each step, the best solution found so far is saved. Although this is not possible in physical systems (e.g. metallurgy), it may be possible in biological systems, in fact there seems to be evidence pointing in this direction in epigenetic studies of glucose metabolism in yeast.

In mathematical optimization, raising the temperature is equivalent to injecting noise into the system. It is like shaking a pinball machine; making the behavior of sampling more random, more explorative, forcing the ball to jump out of local minima traps and via one or more bounces through possibly worse positions, to land on, hopefully, a better solution.

The geometry of the landscape depends on the properties of the system under study – dimensionality, input relationships, etc. – and the success of local search algorithms crucially depends on geometry. Take for example the size of the basin of a valley. If deep valleys (good solutions) have a very narrow basin, then the probability that a ball bouncing randomly around its neighborhood falls in it is low; the solution will be missed. Another landscape feature is its *ruggedness* – at one extreme there is the Fuji-mountain type of landscape where all routes will eventually lead the ball to the best solution. At the other extreme there are the rugged landscapes, characteristic of discrete-input or combinatorial problems, e.g. protein sequence space. Rugged landscapes are characterized by a large number of neighboring peaks and valleys, where a small step has an enormous effect on fitness, thus making it difficult for local search algorithms to succeed. A measure of the ruggedness of a landscape is its *correlation length value* (Stadler & Schnabel 1992). This is a simple metric of the number and distribution of its peaks and valleys in relation to their neighbors. A series of fitness values, i.e. the height of the landscape, is collected by doing a random walk on it; sampling is necessary because it is not practically feasible to exhaustively enumerate the whole landscape. The correlation length value is the average correlation between two such values a number of steps, t, apart. As observed in Hordijk & Kauffman (2005), *"smooth landscapes have long correlation lengths, random landscapes have zero correlation length, and rugged landscapes have correlation lengths that decrease as ruggedness increases"* (see also Section 11.6.6).

11.4.2   Evolution as optimization

Evolution is a way for living organisms to cope with environmental changes and continue living. The key elements are two.

Firstly, a source of stochasticity, noise, which will perturb the current state in order to explore the landscape of solutions as efficiently as possible.

Secondly, some modes of exploration rely on the existence of memory (i.e. genetic blueprint, DNA or epigenetic information, or even biochemical attractors) in order to accumulate fitness and/or recombine fitness parameters with mates thus making this exploration wider. The fitness parameters in genetic memory may be passed on to the following generations or be erased when the organism or cell dies.

How the exploration of the space happens and whether memory is used or not all depends on the lifespan and scale of an organism and the energy needed to be expended on finding a fit solution.

The need for survival in a changing environment is the same for animals as it is for single cells. Each in its own environment will seek to adapt. Cellular intrinsic noise aids robustness and stability within the cell in the face of micro- and macro-environmental changes in the same way as noise in the analogy of the ball rolling on a landscape aids the search for optimality in the vast and complex space of candidate solutions. Phenotypic variation is key to the survival of cells, as is the fact that these phenotypes are short-lived. Noise levels influence the exploration of phenotypes in three interconnecting ways: (i) energy; how much of it is expended in unfit candidates, (ii) time to react/adapt to environmental changes, (iii) variability of phenotypes; how "adventurous" the exploration is.

11.5   Complex networks

One approach to modeling interactions between several players be they species in an evolutionary context or molecules in a biological processes context are the so-called complex networks. Using Graph Theory terminology, Paul Erdös and Alfréd Rényi (Bollobas 2001) derived a framework for analysis of networks with random connectivity – what they called Random Graphs. The emphasis here is on qualitative aspects of the connectivity of interactions between players, e.g. degree, clustering coefficient, etc., rather than how one player interacts with the other (e.g. the Cellular Automata rules, above, or random Boolean networks, below). One such aspect is the probability that a node has a certain number of links – the degree. The degree distribution is a key feature in categorizing networks. For example, in a plain *random regular* network each node has a constant degree, whereas Erdös–Rényi have normal (Gaussian) degree distribution.

11.5.1   Power law forms, small-world and scale-free networks

Scale-free networks (Barabási & Albert 1999) are those whose degree follows a *power law* distribution; the probability of degree k, $P(k)$, is exponentially decreasing with degree, $P(k) = Ak^{-\gamma}$, $\gamma > 0$ is a constant that determines the shape of the distribution; A is a normalizing constant. This means that, depending on γ, a large fraction of nodes have a small number of links and only a few – the so-called hubs – have significantly higher, by orders of magnitude, connectivity. Popularization of the term has certainly benefited from the fact that many random, complex networks occurring in our everyday life, e.g. social interaction networks or the World Wide Web itself, fall in this category. Nonetheless, networks whose fundamental construction properties, e.g. node degree, are drawn from power law distributions are significant because their structure remains (statistically) unaltered when these construction properties change, even by orders of magnitude. This stems from the unique property of power law distributions that $P(nk) = A(nk)^{-\gamma} = n^{-\gamma}Ak^{-\gamma} = n^{-\gamma}P(k)$ which is a scaled version, by $n^{-\gamma}$, of the original distribution, $P(k)$. Hence the term *scale-free* networks.

On the other hand, networks are also characterized by the average shortest path between any two nodes; the least number of intermediate nodes connecting them. This property increases modularity and affects the speed of signal propagation as well as signal-to-noise ratios. A small-world network is one with small average shortest path values (typically proportional to the logarithm of the total number of nodes) and is usually characterized by abundance of sub-networks with high intra-connectivity and low inter-connectivity, mainly via the hubs.

11.5.2   Controllability of complex networks

One interesting question about complex networks is controllability. Given a current state can the network be steered towards a desired state, by which nodes and how? To this end, Barabasi and colleagues (Liu et al. 2011) have used elements of control theory and statistical physics to analyze the controllability of several types of complex networks. A practical finding of their work is that *driver nodes* (those nodes which can control the whole system) in the networks investigated tend to avoid the hubs and are mostly found among the ranks of nodes with low degree. In general, this is an interesting and somewhat counter-intuitive finding laying a general framework for such analysis. However, its applicability to biological networks, e.g. transcriptional networks controlling cellular phenotypes, has been disputed recently by Muller & Schuppert (2011) who argued that these particular types of networks show a high degree of co-regulation, i.e. when regulators partner with each other, as a linear combination, to control common targets (see also Bhardwaj et al. 2010). The result is that the actual gene expression state space is reduced to a combinatorial expression space with relatively low dimensionality; hence controllability can be achieved by only a few nodes.

11.6   Random Boolean networks

11.6.1   Cellular automata

Boolean variables are those which are permitted only two states, usually denoted as 1 / 0 or, more intuitively, ON / OFF. Boolean functions are functions in terms of Boolean variables and the three Boolean operations (conjunction, disjunction, and negation a.k.a. AND, OR, NOT). A Boolean logic element is a computational unit of one or more inputs, one output and a Boolean transfer function which maps the inputs to the output. In simple terms, a Boolean element may be ON or OFF depending on what the state of its inputs are. Once its output state is determined, it will affect all those elements which are connected to turning them ON or OFF respectively.

Early computational models of evolution were based on cellular automata (CA). These consist of a set of Boolean logic elements arranged on a regular grid. The behavior (output state) of each element is determined by the behavior of its immediate neighbors via simple rules, for example *"turn ON if exactly 2 or 3 of your neighbors are ON; otherwise turn OFF."*

Random Boolean networks (RBN) are an extension to CA where the behavior of each node in the network is no longer determined only by its immediate neighbors but also by potentially any other node irrespective of grid distance. In a sense, in RBN there is no grid. CA and RBN are useful in studying bottom-up, emergent behavior stemming from simple elements operating under simple rules. However, emulating emergence in nature using these techniques may be too simplistic and fundamentally constrained by implementation in digital computers.

11.6.2   *NK* automata

Cellular chemical chain reactions are often explained using nomenclature more appropriate for static relationships between chemical reactants and their product under normal laboratory conditions; for example, often they are wrongly called cascades or pathways (Jørgensen & Linding 2010).

In reality, chemical reactions within the cell are highly dynamical and non-linear, they influence each other via cross-talk, and occur in conditions far from ideal (laboratory); any attempt to describe them with static, deterministic terminology will lead to oversimplification.

Cell states and the transition between them have also received the same treatment of being characterized by oversimplistic, top-down models where in fact they are attractors of bottom-up, stochastic processes with complex emergent properties.

One of the first attempts to put them in a more "appropriate" framework was done by Kauffman (Kauffman 1969, Kauffman 1993) who used random Boolean networks to model cell state and regulatory interactions in terms of epigenetic factors, genes, and proteins.

Kauffman's *NK* automaton consists of N Boolean logic elements with K inbound connections. The state of an *NK* automaton at a given time is basically the state of each of its elements. Such a system is dynamic in the sense that it contains feedback loops. When the output of an element changes in response to its inputs changing, it will spawn a number of changes in other elements to which it connects. These changes in turn will cause the inputs to the first element to change and so on. Determining the stability of such a system is tricky.

By analogy, cell state or phenotype is determined by the state of the individual elements (epigenetic factors, genes, or proteins). However, in the case of the cell, a state, say proliferation or cell-cycle check points, may correspond to a number of actual molecular states because the available means of observation and poor resolution limit our ability to differentiate between cell states.

The *NK* automaton serves less as a cell modeling paradigm and more as a conceptual framework where the behavior and, in particular, criticality of a dynamical distributed system are investigated. An *NK* automaton has 2^N possible states – the enumeration of all possible combinations of its elements being ON or OFF. This number grows exponentially with the number of elements; a relatively small N yields a prohibitively large number of possible states. However, a lot of possible states may never be reached.

The free parameters of the NK automaton are N, K, and the Boolean functions of each element (i.e. the "rules"). We tend to study the system through aggregating a fairly large number of simulations and not on a one-off basis. For example, we try to draw conclusions about the long-term average behavior of a system when the complexity of the topology, or the complexity of the rules is of a certain degree and drawn from a certain distribution. The initial conditions (i.e. starting state) are important in determining which cycle the system will settle in, only in cases when there is more than one cycle. Because these systems are completely deterministic, once inside a cycle, the starting state is completely unimportant. For example, if state A leads to B which leads to C, then starting either from A or B, the end state will still be C. This also means that it does not matter via which state the system enters a cycle should there be more than one entry state.

The most important parameter is K – this is the number of other elements an element is influenced by. Effectively, it controls the complexity of the system by influencing the number of cycles and cycle lengths. Whereas the parameter N lays down the players by controlling the total number of possible states, K effectively controls the complexity of the system by arranging these 2^N possible states in loops and attractors. However, as N becomes large the way to find these attractors is usually by random sampling which gives a statistical indication of their existence. For example, for a long time it was wrongly thought that for the NK automata of $K = 2$, the number of attractors was proportional to $\sqrt{N}$ (Socolar & Kauffman 2003). Just because a specific attractor is difficult to discover by random sampling (i.e. it is small) it does not mean that it can not be manifested. This last point is important for coupled biological dynamical systems (as in cells in the same tissue). Through evolution they have adapted to a very probable attractor behavior, but the system splinters when a rare yet perfectly possible attractor appears.

At this point it is useful to introduce another aspect of complexity beyond the number of cycles and cycle lengths. This is the similarity between states, more importantly between states of the same cycle. How dramatic are the state transitions within a cycle? There are many distance metrics which output the distance between two points in a space, for example the Euclidean metric. A metric which is appropriate for Boolean (or discrete-valued in general) variables and spaces is the Hamming distance. It is defined as the number of bits by which the two states differ. Other metrics do exist and can be used instead.

A value of $K = 1$ yields an uninteresting unit length cycle, meaning that the system settles to a single state. In biological systems this is the equivalent of death – nothing changes. On the other hand, large values of K yield extremely long cycles through a very large number of states. The system is very complex, almost pseudo-random although it is totally deterministic – this is a common feature of chaotic systems. K values of 2, 3, 4 yield a small number of cycles and cycle lengths.

There are no magic numbers, this is just what makes sense to the human brain and its own pattern recognition capacity, given the temporal and spatial resolution of observation and measurement. Extracting general rules from the behavior of these networks may be more useful. It is also time to rethink how a cellular phenotype corresponds to the behavior of such automata. Is a phenotype mapped to a certain state occurring at the ith step of the cycle? Or is phenotype a subset of cycle states or indeed the whole cycle itself? After all a phenotype has an effect inside and outside the cell. It is useful to study these systems interacting. If there was ever a resemblance between a random Boolean network, an *NK* automaton, a cellular automaton, etc. with the mechanism inside a cell, then it is worth investigating what happens when a lot of these systems are coupled as in interacting cells within the same tissue. The objective is for each unit alone and the ensemble as a whole to be stable yet to exhibit rich and diverse behavior which will allow for it to adapt to environmental changes, to share resources, and in general to sustain existence not only in single units but as an ensemble. The question is what are the requirements in terms of their complexity – i.e. the number of cycles and cycle lengths? What is an optimum level of interaction between units?

11.6.3   Computation at the edge of chaos

The relationship between complexity and synthesis (as in self-organization) was studied by John von Neumann and Stanislaw Ulam in the 1940s and 50s in the context of their proposed self-reproducing automata (before the discovery of the structure of the DNA!). They observed that there exists a critical threshold of complexity below which the process of synthesis is degenerative but above which, synthesis may become "explosive." But there is another issue here which is as important as complexity and this is stability and criticality. A small perturbation to a dynamical system can have a temporary effect which soon disappears as the system returns back to stability or can have a substantial effect and cause the system to become unstable. There are one or more parameters of a dynamical system which control its behavior to small perturbations. For some critical value of these properties, the system's response to a small perturbation switches from stable to chaotic.

Chris Langton in his paper "Computation at the edge of chaos" (Langton 1990) proposed that von Neumann's threshold is more likely to be a region; too little or too much complexity can kill synthesis. Langton made a distinction between two kinds of cellular automata; those that exhibit a very ordered, predictable behavior and soon die or cycle between a few states (stable); and those that exhibit a totally unpredictable behavior (chaotic). He introduced a parameter (λ) as the fraction of rules which produce a "live" automaton over the total number of rules and argued that this parameter controlled the criticality of the generated automata. Some range of λ values produced really interesting automata with rich and diverse behavior – clearly distinct from that at the ordered regime but without becoming chaotic either. This region was called *"edge of chaos"*; what M. Waldrop calls *"the zone between stagnation and anarchy"* (Waldrop 1992).

On the point of system stability, it may seem counter-intuitive but strong stability will cause a system to be less sensitive to perturbations which effectively causes it to not react to environmental change. For this reason, behavior at the edge of chaos is even more interesting.

One way to quantify complex behavior (as in output repertoire, e.g. in a morphological space) is by using Kolmogorov's notion of complexity. It is defined as the shortest description (program) which can replicate an output pattern exactly. Consider a system consisting of a uniform random generator which produced 1 million ones and zeros. According to Kolmogorov, its behavior is as complex as it gets because there is no program which can be used to produce the sequence exactly without resorting to just printing the sequence itself hard-coded in the program. Now consider a sorted version of this random sequence. A very simple program (print 500 000 zeros; print 500 000 ones) can be used to reproduce it, hence its complexity is minimal. Kolmogorov's notion of complexity is one of many metrics which can rank the complexity of the behavior of cellular automata and classify them as "boring," "interesting," "chaotic," or anything between.

11.6.4 Self-organized criticality and power law forms

The term *self-organized criticality* is associated with systems of locally interacting agents where occasionally a critical state is reached, there is a phase transition, and global dynamics emerge. The example usually cited is that of creating a sandpile by dropping sand, one grain at a time, onto a surface, at random locations. At the beginning, each new grain settles in the vicinity of where it was dropped through local interactions. At this point, small, local avalanches can be observed. This continues until the slope of the pile reaches a critical value. This is when the drop of a single grain of sand will cause all sorts of avalanches including global ones. The system changed by itself (self-organization) from local to global dynamics (criticality). Furthermore, computer simulations of simplified sandpile models have revealed the distribution of avalanche magnitudes to generally follow power law forms (see Section 11.5.1). Subsequently, power laws were (re-)discovered in other toy or real world systems, e.g. in the frequency and magnitude of earthquakes (Gutenberg–Richter), the Bak–Sneppen model of evolution (Bak 1996), etc. In Mora & Bialek (2011) a link is made between statistical mechanics and biological systems whose events exhibit power law distributions (see Section 11.4).

11.6.5 *NK* automata variations

These automata systems are programmed for and run on digital, discrete time computers which contain a small number of processors. This means that the system is not reacting instantaneously to any input changes. Instead the processor is updating each sequentially, one after the other in a queue, and synchronously in step with its clock. This may have significant effects to the behavior of the system. For example, it was observed that the synchronous update scheme introduces a large number of unstable attractors (Greil & Drossel 2005). Simulation with parallel computers may or may not be able to overcome this limitation depending on their architecture, but there is definitely a limitation due to the small number of processors. A more natural solution would be an analogue, parallel computer consisting of many thousands of very simple analogue computational units (basically analogue electronics circuits) communicating via wireless broadcast. Insight from biological neural circuits may be helpful.

Networks which contain a mixture of real number and Boolean variables are probably a more realistic implementation as they may be used to introduce protein expression levels. Accordingly the rules must be such so as to accommodate both types of variables. Alternatively, a type of variable which is a cross between Boolean and real is called *fuzzy*. In the framework called fuzzy logic, qualitative terms like *low*, *medium*, and *high* are mapped to real value ranges. Fuzzy calculus allows fuzzy transfer functions. Fuzzy logic has been used to model various signaling systems (Aldridge et al. 2009, Morris et al. 2011).

Other additions include:

1. Stochasticity: noise as described earlier – signals may be distorted by this noise or never arrive, rules may also be affected by noise, connectivity may be altered by noise.
2. Probabilistic rules: the rules are probability distributions rather than deterministic functions.
3. Dynamic/context-based topology: connectivity between elements changes over time, old links are broken, new links are created deterministically, over time or depending on the context (current state of the network, external influences).
4. (Random) time delays in the propagation of signals: signals may take a while to arrive at their destination depending on current state or just randomly. They may never arrive.
5. The incorporation of some kind of memory: the current state may depend not only on current but also past states.

Kadanoff et al. (2002) contains a review of research in some of the above points.

11.6.6   Boolean networks and fitness landscapes

In Hordijk & Kauffman (2005) a link is made between a fitness landscape, the players that affect it and also each other, and the *NK* automaton. The free parameters of the landscape is the set of N genes that form a space into which the fitness landscape spans. Each gene combination – a genotype – is assigned a fitness value. This is represented by the height of the landscape at that particular point, which is the sum of individual gene contributions. Epistasis is modeled by the factor K which is the number of genes affecting a given gene.

The *ruggedness* of the landscape increases with K. When each gene is affected by all other ($K = N - 1$) genes, then the landscape is random with zero correlation length which means that moving from one genotype to another very nearby, the change in fitness is likely to be huge. Contrast this with small values of K which yield landscapes of long correlation lengths meaning that a small change in genotype will most likely have a proportionally small change, positive or negative, in fitness. Undoubtedly, the methods to explore the landscape and maximize fitness will differ according to ruggedness.

As a means of modeling co-evolution among different species, Kauffman (1995) suggests that genes from one species affect the fitness of another species via some kind of inter-species, external epistasis. This is, for example, the case where a prey species develops a particular skin color as a defense mechanism for escaping its predator. In effect, the prey has deformed the fitness landscape of the predator. The predator is no longer as fit as it used to be because, say, it can not spot that color easily, from a distance.

The *NKC* model consists of a number of *NK* automata, one for each species. Each automaton has *N* genes (Boolean elements) and each gene is affected by *K* other genes of the same species as before. However, each gene is also affected by *C* external genes, from the other species.

The *edge of chaos* is a recurring theme in dynamical systems and appears also in coupled fitness landscapes. Depending on the parameters *K* and *C*, co-evolution may behave in the ordered or chaotic regimes or in between, at the edge of chaos. In the ordered regime all species reach some acceptable fitness level and happily co-exist without further increasing their fitness. This happens when *K* is high or *C* is low. If *K* is high then a species landscape is extremely rugged so the effect of losing fitness because of the effect of other species (via *C*) can quickly be compensated as a new fitness peak is bound to be in the neighborhood. Likewise, if *C* is low, then the influence of one species on another's fitness landscape is minimal; therefore if a species finds a fitness peak, it is difficult to lose it because of the effect of another species.

In the chaotic regime, the species are changing constantly and never settle down because they affect each other too much. The effect of a low *K* value is that the fitness landscape will be quite smooth with very few fitness peaks very far apart. Evolution for this species will be a long and slow process. If another species continuously deforms this landscape (via *C*) the long and slow process of evolution will never bear any results as it will continuously be rebooted, and the players expend more energy in "sabotaging" each other than in improving their own fitness. It is claimed that maximal fitness for all species involved lies in the region between order and chaos – the *edge of chaos*. This region can be reached by adjusting *K* and *C* values.

It can't be stressed enough that automata serve as a tool of investigating dynamical systems via simulation, rather than as faithful models of nature. Let us not forget that gene regulation networks are the result of years of evolution and the manifestation of the laws of physics. This is in contrast with the random connectivity of RBN, although there is research in using experimental data to form connectivity (Harris et al. 2002).

11.7   Genomic state spaces

The term state space is often associated with dynamical systems and ergodic theory. This is the relationship between all the parameters (degrees of freedom) that determine the behavior of a system or process. In the classical example of a pendulum, the two parameters that determine fully its behavior are the angle from the vertical and angular velocity (which is the rate of change of the angle with time). These two parameters make up the state space of the system which can be visualized as a (two-dimensional) plot of velocity versus angle. At the intersection of each possible combination of velocity and angle value, a mark is made. This will reveal the trajectories of the system or where the system goes next given a current state. This is much more useful than a time plot because it gives a view of the system completely independent of initial conditions.

Huang (2010) constructs a "genomic" state space for a cell where the free parameters of the system are all the genes in the cell's genome. A gene expression pattern is a point in this vast space. Given no underlying gene regulatory interactions this space would be occupied totally by expression patterns – any expression pattern would be possible.

In reality there are several constraints in the interaction of genes, be they physical or chemical, what is called "gene regulation." The result of this is that only certain gene expression patterns are possible; only some states in the state space can ever be occupied.

A trajectory in this state space is a movement from one gene expression pattern to a neighboring one and may constitute a phenotype change. This change is independent of time; there is no time variable in the state space construction. All it is, is that it is possible to move from state A to state B if states A and B are connected via a trajectory. Whether this is statistically significant or not is another issue, related to the assumption of ergodicity.

An "attractor" in a state space is a closed-loop trajectory which traps the state of the system in a cyclical behavior. For obvious reasons, a finite state space must have at least one reachable attractor. The set of states which lead to the attractor, but not necessarily part of it, form the so-called "basin of attraction." In the case of a ball inside a bowl, the attractor is the single point at the bottom of the bowl. The basin of attraction is each point on the inside of the bowl.

It must be noted that a state space is not the same as Waddington's epigenetic landscape, the main difference being that the latter has one extra dimension and this is the fitness or some other potential. That's why a state space has trajectories, whereas a fitness landscape has hills, valleys, canals, and ravines.

The problem with building a genomic state space is that it requires gene expression data be obtained accurately for a single cell over a long time and at high temporal resolution. If the system is ergodic (see Section 11.7.1), then the genomic profile of a lot of single cells at a specific time is equivalent to that of a single cell over a long time. This may be easier but practical issues still remain.

11.7.1   Ergodic theory

The question whether observations made on a population at a particular time instance as a snapshot are equivalent to observations made on a small subset of the population averaged over a long time interval always arises or must arise in experimental setups. In effect, one asks whether what has been observed at a snapshot is not significantly different to the long-term (over time) behavior of each member of the population.

The answer is, of course, yes and no. Under some conditions, the observations are equivalent. Such a system is called ergodic. Ergodic theory's main question is to find those conditions, if any, under which ergodicity holds.

Those points (states) in the state space which are occupied by a unique trajectory are ergodic sets as they satisfy the main prerequisite which is that once in, the probability of leaving it is zero. An alternative, equivalent prerequisite is that each point (state) in an ergodic set must happen with equal frequency with any other.

In many biological experiments where some aspects of a phenotype corresponding to a given genotype are measured, it is necessary to question the relationship between time and space averages and thus ask whether observing a snapshot of a lot of cells, as in measuring the phenotype of thousands of cells in a static microscope picture, is equivalent to following a few of these cells around and analyzing their phenotypes over time.

It is thus possible to gain insight into the behavior of individual cells as estimated from the population average while systematically down-sampling the fraction of cells analyzed of the population with an ensemble technique (e.g. mass spectrometry).

To this end, stochastic profiling is a method to quantify single-cell heterogeneities without the need to measure expression in individual cells (Janes et al. 2010). Instead, measurements are made over a number of smaller samples of cells collected randomly from the large population. Heterogeneously expressed genes will have a larger variance when compared with homogeneously expressed genes acting as a reference (and identified by smaller variance). This is a better alternative either to observing the whole population, which invariably averages out any single-cell heterogeneity, or to observing individual cells which entails large measurement errors.

Acknowledgements

AH wishes to thank Dr. Herbert Wiklicky for the useful comments he made on the manuscript and during their many discussions.

References

Aldridge, B. B., Saez-Rodriguez, J., Muhlich, J. L., Sorger, P. K. & Lauffenburger, D. A. (2009), 'Fuzzy logic analysis of kinase pathway crosstalk in TNF/EGF/insulin-induced signaling', *PLoS Computational Biology* **5**(4), e1000340+.

Bak, P. (1996), *How Nature Works: The Science of Self-organized Criticality*, Springer, New York.

Balázsi, G., van Oudenaarden, A. & Collins, J. J. (2011), 'Cellular decision making and biological noise: from microbes to mammals', *Cell* **144**(6), 910–925.

Barabási, A. L. & Albert, R. (1999), 'Emergence of scaling in random networks', *Science* **286**(5439), 509–512.

Batchelor, E., Loewer, A. & Lahav, G. (2009), 'The ups and downs of p53: Understanding protein dynamics in single cells', *Nature Reviews Cancer* **9**(5), 371–377.

Bhardwaj, N., Carson, M. B., Abyzov, A., Yan, K.-K., Lu, H. et al. (2010), 'Analysis of combinatorial regulation: Scaling of partnerships between regulators with the number of governed targets', *PLoS Computational Biology* **6**(5), e1000755+.

Bialek, W. & Ranganathan, R. (2007), 'Rediscovering the power of pairwise interactions'.

Bollobas, B. (2001), *Random Graphs (Cambridge Studies in Advanced Mathematics)*, 2nd edn, Cambridge University Press.

Brüggeman, F. J., Westerhoff, H. V., Hoek, J. B. & Kholodenko, B. N. (2002), 'Modular response analysis of cellular regulatory networks', *Journal of Theoretical Biology* **218**(4), 507–520.

Chang, H. H., Hemberg, M., Barahona, M., Ingber, D. E. & Huang, S. (2008), 'Transcriptome-wide noise controls lineage choice in mammalian progenitor cells', *Nature* **453**(7194), 544–547.

Elowitz, M. B., Levine, A. J., Siggia, E. D. & Swain, P. S. (2002), 'Stochastic gene expression in a single cell', *Science* **297**(5584), 1183–1186.

Fisher, J. & Henzinger, T. A. (2007), 'Executable cell biology', *Nature Biotechnology* **25**(11), 1239–1249.

Goldberg, A., Allis, C. & Bernstein, E. (2007), 'Epigenetics: A landscape takes shape', *Cell* **128**(4), 635–638.

Greil, F. & Drossel, B. (2005), 'Dynamics of critical Kauffman networks under asynchronous stochastic update', *Physical Review Letters* **95**(4), 048701+.

Harris, S. E., Sawhill, B. K., Wuensche, A. & Kauffman, S. (2002), 'A model of transcriptional regulatory networks based on biases in the observed regulation rules', *Complexity* **7**(4), 23–40.

Hordijk, W. & Kauffman, S. A. (2005), 'Correlation analysis of coupled fitness landscapes', *Complexity* **10**, 41–49.

Huang, S. (2010), 'Cell lineage determination in state space: A systems view brings flexibility to dogmatic canonical rules', *PLoS Biology* **8**(5), e1000380+.

Janes, K. A., Albeck, J. G., Gaudet, S., Sorger, P. K., Lauffenburger, D. A. et al. (2005), 'A systems model of signaling identifies a molecular basis set for cytokine-induced apoptosis', *Science* **310**(5754), 1646–1653.

Janes, K. A., Wang, C.-C. C., Holmberg, K. J., Cabral, K. & Brugge, J. S. (2010), 'Identifying single-cell molecular programs by stochastic profiling.', *Nature Methods* **7**(4), 311–317.

Jørgensen, C. & Linding, R. (2010), 'Simplistic pathways or complex networks?', *Current Opinion in Genetics & Development* **20**(1), 15–22.

Kadanoff, L., Coppersmith, S. & Aldana, M. (2002), 'Boolean dynamics with random couplings', *ArXiv Nonlinear Sciences e-prints* 0204062.

Kauffman, S. (1995), *At Home in the Universe: The Search for the Laws of Self-Organization and Complexity*, Oxford University Press, New York.

Kauffman, S. A. (1969), 'Metabolic stability and epigenesis in randomly constructed genetic nets', *Journal of Theoretical Biology* **22**(3), 437–467.

Kauffman, S. A. (1993), *The Origins of Order: Self-Organization and Selection in Evolution*, Oxford University Press, New York.

Langton, C. (1990), 'Computation at the edge of chaos: phase transitions and emergent computation', *Physica D: Nonlinear Phenomena* **42**(1–3), 12–37.

Linding, R. (2010), 'Multivariate signal integration', *Nature Reviews Molecular Cell Biology* **11**(6), 391.

Liu, Y. Y., Slotine, J. J. & Barabasi, A. L. (2011), 'Controllability of complex networks', *Nature* **473**(7346), 167–173.

Mora, T. & Bialek, W. (2011), 'Are biological systems poised at criticality?', *Journal of Statistical Physics* **144**, 268–302.

Morris, M. K., Saez-Rodriguez, J., Clarke, D. C., Sorger, P. K. & Lauffenburger, D. A. (2011), 'Training signaling pathway maps to biochemical data with constrained fuzzy logic: quantitative analysis of liver cell responses to inflammatory stimuli', *PLoS Computational Biology* **7**(3), e1001099+.

Muller, F.-J. & Schuppert, A. (2011), 'Few inputs can reprogram biological networks', *Nature* **478**(7369), E4.

Quaranta, V. & Garbett, S. P. (2010), 'Not all noise is waste', *Nature Methods* **7**(4), 269–272.

Socolar, J. E. S. & Kauffman, S. A. (2003), 'Scaling in ordered and critical random Boolean networks', *Physical Review Letters* **90**(9), 098701.

Socolich, M., Lockless, S. W., Russ, W. P., Lee, H., Gardner, K. H. et al. (2005), 'Evolutionary information for specifying a protein fold', *Nature* **437**(7058), 512–518.

Spiller, D. G., Wood, C. D., Rand, D. A. & White, M. R. H. (2010), 'Measurement of single-cell dynamics', *Nature* **465**(7299), 736–745.

Stadler, P. F. & Schnabel, W. (1992), 'The landscape of the traveling salesman problem', *Physics Letters A* **161**, 337–344.

Swain, P. S., Elowitz, M. B. & Siggia, E. D. (2002), 'Intrinsic and extrinsic contributions to stochasticity in gene expression', *Proceedings of the National Academy of Sciences of the USA* **99**(20), 12 795–12 800.

Tkačik, G., Callan, C. G. & Bialek, W. (2008), 'Information flow and optimization in transcriptional regulation', *Proceedings of the National Academy of Sciences of the USA* **105**(34), 12 265–12 270.

Tkačik, G., Walczak, A. M. & Bialek, W. (2009), 'Optimizing information flow in small genetic networks', *Physical Review E* **80**, 031920.

Waldrop, M. M. (1992), *Complexity: The Emerging Science at the Edge of Order and Chaos*, Simon & Schuster, New York.

Wang, J., Zhang, K., Xu, L. & Wang, E. (2011), 'Quantifying the Waddington landscape and biological paths for development and differentiation', *Proceedings of the National Academy of Sciences of the USA* **108**(20), 8257–8262.

Wright, S. (1932), 'The roles of mutation, inbreeding, crossbreeding and selection in evolution', *Proceedings of the 6th International Congress of Genetics* **1**, 356–366.

Yanofsky, C., Horn, V. & Thorpe, D. (1964), 'Protein structure relationships revealed by mutational analysis', *Science* **146**(3651), 1593–1594.

Zamir, E. & Bastiaens, P. I. (2008), 'Reverse engineering intracellular biochemical networks', *Nature Chemical Biology* **4**(11), 643–647.

12 Automated behavioural fingerprinting of *Caenorhabditis elegans* mutants

André E. X. Brown and William R. Schafer

Rapid advances in genetics, genomics and imaging have given insight into the molecular and cellular basis of behaviour in a variety of model organisms with unprecedented detail and scope. It is increasingly becoming routine to isolate behavioural mutants, clone and characterise mutant genes and discern the molecular and neural basis for a behavioural phenotype. Conversely, reverse genetic approaches have made it possible to straightforwardly identify genes of interest in whole-genome sequences and generate mutants that can be subjected to phenotypic analysis. In this latter approach, it is the phenotyping that presents the major bottleneck; when it comes to connecting phenotype to genotype in freely behaving animals, analysis of behaviour itself remains superficial and time-consuming. However, many proof-of-principle studies of automated behavioural analysis over the last decade have poised the field on the verge of exciting developments that promise to begin closing this gap.

In the broadest sense, our goal in this chapter is to explore what we can learn about the genes involved in neural function by carefully observing behaviour. This approach is rooted in model organism genetics but shares ideas with ethology and neuroscience, as well as computer vision and bioinformatics. After introducing *Caenorhabditis elegans* as a model, we will survey the research that has led to the current state of the art in worm behavioural phenotyping and present current research that is transforming our approach to behavioural genetics.

12.1 The worm as a model organism

Caenorhabditis elegans is a nematode worm that lives in bacteria-rich environments such as rotting fruit and has also been isolated from insects and snails which it is thought to use for longer-range transportation (Barriere & Felix 2005, Lee et al. 2011). In the laboratory, it is commonly cultured on the surface of agar plates seeded with a lawn of the bacterium *Escherichia coli* as a food source. On plates, worms lie on either their left or right side and crawl by propagating a sinuous dorso-ventral wave from head to tail. When immersed in a liquid, they can also swim (Fig. 12.1).

As a model organism for the genetics of behaviour, development and neuroscience *C. elegans* offers several advantages. It progresses through its four larval stages in three

Systems Genetics: Linking Genotypes and Phenotypes, ed. F. Markowetz and M. Boutros. Published by Cambridge University Press. © Cambridge University Press 2015.

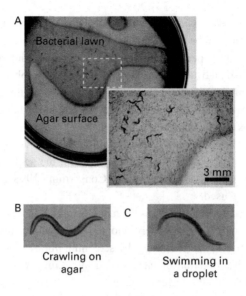

Figure 12.1 (A) *C. elegans* is typically cultured on agar plates seeded with bacteria as a food source. Because of their small size, hundreds of worms can be grown on a standard 5 cm petri dish. (B) Close-up of a worm crawling on an agar plate. (C) When placed in a drop of liquid, *C. elegans* starts to swim. Swimming is characterised by more rapid body bending with a longer wavelength. A black and white version of this figure will appear in some formats. For the colour version, please refer to the plate section.

and half days with each animal yielding about 300 progeny. Hermaphrodites can reproduce either by cross- or self-fertilisation; thus, mating behaviour is not essential for the viability of mutant strains. Indeed, paralysed animals with almost no nervous system function can be propagated in the laboratory, allowing the analysis of mutants defective in fundamental neuronal molecules such as synaptic vesicle proteins and voltage-gated channels. Both molecularly and anatomically, *C. elegans* is exceptionally well characterised. It was also the first multicellular organism to have its genome sequenced (*C. elegans* Sequencing Consortium 1998) and its well-annotated genome as well as information on mutants is available on the community website WormBase.org (Yook et al. 2012). It remains the only organism whose connectome (the wiring diagram of all its 302 neurons) has been completely mapped at electron microscopic resolution (White et al. 1986). The connectome and many other details of the worm's anatomy and physiology are available on WormAtlas.org. Finally, the developmental lineage of each of its 959 adult cells is known and is highly repeatable from one individual to the next (Sulston 1977). In other words, given a cell in the adult, one can look up the complete list of divisions that led to that cell from the single-celled embryo.

12.1.1 The genetics of *C. elegans*

The genetic study of *C. elegans* began in earnest with Brenner's 1974 paper in which he screened chemically mutagenised populations of worms for morphological and locomotory defects (Brenner 1974). Males occur spontaneously at low rates ($\sim$0.1%) and this rate can be increased using heat shock. These males can then be used in genetic crosses and to perform complementation tests and mapping. Using this approach, Brenner isolated numerous mutants with visible phenotypes, including mutants with movement defects comprising 77 complementation groups. Since then, many more mutants have been identified, cloned and often further characterised molecularly using variations on this approach.

Forward genetic approaches continue to play an important role in uncovering the function of nervous system genes, but with the sequencing of the *C. elegans* genome in 1998 (*C. elegans* Sequencing Consortium 1998), reverse genetics is also being extensively used.

Targeted gene deletion has been demonstrated in worms (Frøkjaer-Jensen et al. 2010), but has not yet been widely adopted. In practice, reverse genetics in *C. elegans* is done principally in two ways: by screening libraries of mutagenised genomes for those containing a deletion of interest and by feeding worms with bacteria that express an appropriate small interfering RNA.

In the first approach, a population of worms is mutagenised using psoralen and ultraviolet irradiation which produces more small and medium deletions than the more standard mutagen ethyl methanesulfonate (EMS). The population is screened using PCR primers designed around the gene to be deleted until a match is found. Including library construction, a strain carrying a desired deletion allele can be isolated in about two months (Ahringer 2006). The procedure has been standardised and is now performed by the *C. elegans* Gene Knockout Consortium as a service to the worm research community. Knockouts of new genes, or new alleles of previously knocked out genes, can be ordered online. Knockout and other strains can be ordered for a nominal fee from the *C. elegans* Genetics Center at the University of Minnesota.

Gene knockdown based on RNA interference (RNAi) can be achieved in *C. elegans* simply by feeding worms with bacteria that have been modified to express double stranded RNA (dsRNA) complementary to the gene to be knocked down (Fraser et al. 2000). This technique works because most nematode cells express the SID-1 transporter, which enables systemic uptake of dsRNA triggers for RNAi. Libraries expressing dsRNA fragments of nearly all *C. elegans* open reading frames have been used to conduct near genome-wide knockdown screens for developmental and morphological phenotypes (Kamath et al. 2003). Unfortunately, neurons do not express SID-1, which has limited the applicability of the original method to studying behaviour and nervous system function. Several approaches have been attempted to increase the efficiency of feeding RNAi with varying degrees of success (Simmer et al. 2002, Kennedy et al. 2004, Lehner et al. 2006). Recently, Calixto et al. demonstrated robust feeding RNAi using worms engineered to express SID-1 transgenically in specific groups of neurons (Calixto et al. 2010). This exciting development may prove to be a key enabling technology for a genome-wide investigation of genes affecting behaviour.

With these techniques, as well as the promise of population genomics, the tools for perturbation studies in *C. elegans* currently significantly outstrip behavioural phenotyping methods in both throughput and sophistication. However, this is changing, as we will describe after a brief introduction to some of what is already known about *C. elegans* behaviour.

12.1.2   Classic approaches to behaviour

Nematode behaviour has been studied by researchers for well over 50 years. Early work examined diverse species and their entire range of behaviour from escape from the egg, through moulting, feeding and mating. Several modes of locomotion were observed from crawling on solid surfaces to swimming and even jumping. Croll thoroughly reviewed this early work in Croll (1975a). For a more current survey of work on nematode behaviour, see Gaugler & Bilgrami (2004).

Furthermore, much of this work was not simply descriptive. Quantitative analysis and modelling was done using computer simulations of aspects of nematode exploration (Croll & Blair 1973), mechanical models of swimming motion (Wallace 1968) and quantification of many worm locomotion features including wavelength, frequency and duration of reversals (Croll 1975b). These are all still areas of active interest. Perhaps most relevant to our discussion here is the work on track analysis in *C. elegans* in which single or multiple worms are allowed to crawl on a fresh agar surface and inscribe a small crevice. At the end of the observation time, the agar is transferred to a photographic enlarger and projected onto film. This provided a permanent record of the data and made subsequent analysis easier (Fig. 12.2).

Track analysis gave quantitative insight into possible dispersal strategies and the behavioural mechanisms of chemotaxis. Interestingly, this method is amenable to high-throughput experiments with very low equipment cost, although as far as we know this has never been attempted. Many worms could be tracked simultaneously, limited only by time to transfer single worms to plates, and the analysis could be done on a single desktop computer. Nonetheless, track analysis does have its disadvantages. For example, worms can only be followed until they reach the edge of their arena and there is no information about the dynamics of the motion other than the total time taken for the path.

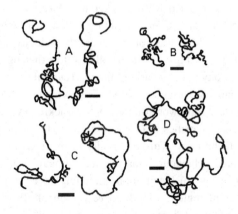

Figure 12.2   Worms leave tracks in an agar plate and these were used for track analysis. Adapted from Croll (1975b). Pairs of tracks labelled A–D are taken from the same individual at different times to illustrate the observed variation.

12.2   High-throughput data collection and information extraction

There are two classes of problems that must be solved to bring behavioural phenotyping to a level that is truly commensurate with that of current molecular tools. The first is collecting rich high-throughput data and the second is extracting meaning from this wealth of data in an unbiased and quantitative way. There are several important advantages to both of these aspects of automated behavioural fingerprinting. The first is simply that it is essential to collect data from a large number of strains to get a sense for the phenotypic landscape and to cover as many interesting genes as possible. The second is that a carefully engineered data collection pipeline will be naturally standardised and should be more reproducible over time, between operators, and between laboratories. Perhaps most importantly, quantitative feature extraction and data analysis lead to abstraction. Once behavioural data have been abstracted, the precise source of the data matters less than their form, meaning that the extensive tools of statistics and especially bioinformatics can be brought to bear, potentially revealing subtle relationships between phenotypes that may reflect underlying genetic connections.

As we will see in the next two sections many of these problems have been solved at the demonstration level, but the real power of this approach to behavioural genetics has yet to be realised.

12.2.1   Worm tracking: throughput and resolution

Worm tracking simply refers to following a moving worm over time, typically recorded in video. The approaches taken so far in the field can be divided according to resolution: low-resolution tracking allows many worms to be captured in a single field of view (multi-worm tracking) but results in a less detailed picture of each worm's behaviour (Fig. 12.3A), while high-resolution single-worm tracking provides more detail (Fig. 12.3B) but requires the system to follow a single worm in the camera's field of view for long enough to collect a reasonable amount of behaviour.

The first automated worm tracker was described in 1985 in a remarkable paper by Dusenbery (1985a). Using computer and video equipment that are rudimentary by today's standards he was able to record the velocity and reversal rate of 25 worms at 1 Hz, updated in real time. He and collaborators then used the system to study *C. elegans'* response to oxygen and carbon dioxide (Dusenbery 1985b) and to a variety of chemicals (Williams & Dusenbery 1990). Following this pioneering effort, there was a period with relatively little work on automated tracking even though there was rapid progress in behavioural genetics in *C. elegans* during the same period that could have benefited from the technology. This eventually changed as more groups discovered the utility of automated behavioural analysis and numerous groups used some kind of tracking to investigate worm locomotion, including speed to detect variability in wild isolates from different regions (de Bono & Bargmann 1998), reversals and turns to understand chemotaxis (Pierce-Shimomura et al. 1999) or simply using tracking to record long periods of behaviour for subsequent manual annotation (Waggoner et al. 1998). Since then there has been significant interest in developing more user-friendly and/or

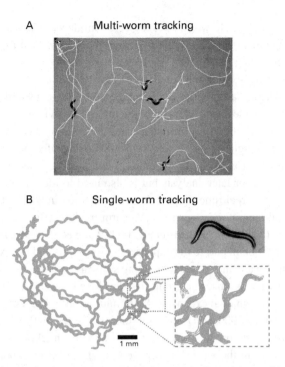

A   Multi-worm tracking

B   Single-worm tracking

1 mm

Figure 12.3   (A) Multiple worms are simultaneously tracked at low resolution using a multi-worm tracker. The results shown here were generated using the Parallel Worm Tracker from the Goodman laboratory (Ramot et al. 2008). (B) A single worm can be tracked using a motorised stage to keep the worm in the camera's field of view. The worm's outline and skeleton are determined in each frame. By recording the distance travelled by the stage, a detailed track can be reconstructed including high-resolution posture information (outlines and skeletons from each frame are overlaid here). A black and white version of this figure will appear in some formats. For the colour version, please refer to the plate section.

standardised approaches to tracking and analysis both for multi-worm and single-worm trackers.

We are aware of three general-purpose multi-worm trackers designed for broad use. NeMo tracks worm locomotion and includes a graphical user interface that allows users to adjust tracking parameters and correct errors (Tsibidis & Tavernarakis 2007). It also includes functionality for the subsequent analysis of behavioural data. The Parallel Worm Tracker uses a simple threshold to distinguish worms from background and works with bright field and dark field videos (Ramot et al. 2008). This system also includes a graphical user interface for rejecting or splitting bad tracks and outputs a variety of movement parameters for subsequent analysis. Because both groups make their source code (written in Matlab) freely available, users can add new locomotion metrics relatively straightforwardly. The most sophisticated multi-worm tracker to date, specifically designed for high throughput, uses Labview to capture data from a 4-megapixel camera as well as to do the first stage of data processing in real time (Swierczek et al. 2011).

The first single-worm trackers were introduced to allow higher-resolution tracking of an individual for long periods of time; these were used for subsequent manual analysis of egg-laying behaviour (Waggoner et al. 1998) or for automated analysis of location data (Pierce-Shimomura et al. 1999, Hardaker et al. 2001). Because only a single worm is tracked at a time, this approach is more time-consuming than multi-worm tracking, but it provides a more nuanced picture of behaviour and makes it possible to extract detailed features of worm shape and locomotion, which we will discuss in the next section. Single-worm trackers typically consist of a motorised stage mounted on a dissecting microscope with a camera. A computer records video data from the camera that are stored for later analysis but is also used to identify the worm and update the stage position in real time to keep the worm centred in the field of view. Although there are several laboratories using single-worm trackers (Baek et al. 2002, Cronin et al. 2006, Feng et al. 2004, Stephens et al. 2008, Wang et al. 2009), they have not yet been widely adopted due in part to the expense of the systems and the expertise required to configure the hardware and write the tracking software.

To address this, our group has developed a user-friendly single-worm tracker built around a small inexpensive USB microscope. The entire system can be purchased for under $5000 USD and will soon be paired with a feature-rich analysis package. A parts list, instructions for set-up and the tracking software itself can be downloaded from the worm tracker website of the MRC Laboratory of Molecular Biology (http://www.mrc-lmb.cam.ac.uk/wormtracker/). Because the system is relatively inexpensive, we have been using it to run eight single-worm trackers in parallel and thus increase the throughput of single-worm tracking.

This highlights the basic trade-off between single- and multi-worm tracking – throughput versus resolution – and how these limitations are being addressed from both sides (Fig. 12.4). In the case of multi-worm tracking, higher-resolution cameras are making it possible to extract skeleton-based features of worms even in a field of view with many individuals (Swierczek et al. 2011), while less expensive hardware coupled with free software is making it possible to increase the throughput of the inherently high-resolution single-worm approach. Still, with current technology, multi-worm trackers can collect data from a larger number of worms more quickly and single-worm trackers still have a resolution advantage so the two methods will probably coexist for some time.

12.2.2   Data reduction and abstraction: segment, skeletonise, featurise

Some early applications of worm tracking were simply for data collection with the analysis still done manually from recorded video. To really take advantage of the wealth of data provided by trackers, it is necessary to develop algorithms to extract worms from background and to ultimately convert the data into a form that is interpretable by humans to give insight and guide future experiments.

Using appropriate lighting conditions (Yemini et al. 2011a) it is usually possible to achieve sufficient contrast, even with worms on a bacterial food lawn, that a simple global threshold does a good job of identifying worms. For multi-worm trackers, this is sometimes coupled with a size range (Ramot et al. 2008) to eliminate large or small

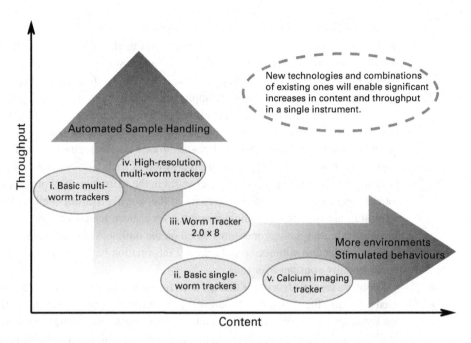

Figure 12.4   Schematic illustration of the current trade-off between content and throughput in behavioural phenotyping. Integrating automated sample handling will significantly improve throughput for imaging systems that can run sufficiently autonomously because picking worms to plates is time-consuming and labour-intensive. Content can be significantly increased using better imaging methods, especially functional neural imaging. In the near future it should be possible to extend both content and throughput by adapting and integrating existing technology. A black and white version of this figure will appear in some formats. For the colour version, please refer to the plate section.

background features that are included by the threshold. For single-worm tracking, taking the largest connected component in the thresholded image is often sufficient. Still, there are circumstances in which more robust approaches are helpful. For example, worms in more naturalistic environments, in microfluidic devices, in thick food or interacting with each other present challenges for the simplest thresholding approaches.

In an effort to make a more universal system for worm identification, Sznitman et al. have taken a multi-environment model estimation approach to worm segmentation (Sznitman et al. 2010c). In the first frame of the video only, users must identify the worm to train a Gaussian mixture model used to classify background and foreground (worm) pixels. This model is then used in subsequent frames to identify the worm which is extracted for further analysis. The same system was able to extract swimming and crawling worms and worms in microfluidic devices (Sznitman et al. 2010c).

In addition to the pixel intensities within a frame, there is also significant information in the correlation between frames. This has been exploited in worm segmentation using a Kalman filter (Fontaine et al. 2006) or recursive Bayesian filter (Roussel et al. 2007) coupled with a worm model. Both approaches allow for the separation of overlapping worms which could improve multi-worm trackers which often simply drop worms that are touching each other from the analysis (Ramot et al. 2008). It could also help track worms in, for example, a thick bacterial lawn that can sometimes obscure portions of the worm. Fontaine et al. (2006) illustrated their method by studying mating, a behaviour that naturally requires partially overlapping individuals!

A series of outlines over time is still not particularly useful without further processing. For lower-resolution multi-worm trackers, the next step in analysis is typically to quantify aspects of locomotion and posture based on the segmentation. For example, velocity is simply the change in centroid position of each blob over time. Reversals and turns can be identified as sharp angle changes in the worm's path. Worm posture can be roughly approximated by the eccentricity of the worm's equivalent ellipse. For higher-resolution multi-worm and single-worm trackers, more features are readily calculated. In addition to morphology measures based on the outline, skeletonisation leads to a further data reduction without loss of relevant information because worms' width profiles are essentially constant over time.

There are several algorithms that are used for skeletonisation. The first and still widely used computes the skeleton by thinning the thresholded worm to a single pixel and pruning branches based, for example, on taking the longest path through the structure. This works reasonably well, but because of the possible ambiguity of choosing the correct branch at the ends, it is not ideal for picking up more subtle head-foraging motions (Huang et al. 2008). It is also possible to use a chamfer distance transform coupled with curvature to estimate a skeleton (Sznitman et al. 2010c). An attractive approach given its speed and simplicity has recently been reported based on finding the points of highest curvature on the outline and tracing the midline connecting these points (Leifer et al. 2011). Because the tip of the head is centred on a local curvature maximum, it does a good job of picking up subtle head motions. At the resolution of a typical single-worm tracker, worm skeletons are on the order of 100 pixels long. This is almost certainly an over-sampling given that C. elegans has 22 rows of body wall muscles and therefore significantly fewer degrees of freedom than 100, even including the head which is capable of more complicated motions.

Skeleton curvature can be used to identify even subtle differences in body posture and changes in skeleton curvature over time should in principle be able to distinguish different kinds of uncoordination. Furthermore, looking at specific sequences of postures has allowed the detection of known behaviours including reversals and reorientations called omega turns (Huang et al. 2006).

A complementary approach to C. elegans behavioural representation has been described by Stephens et al. (2008). They segment and skeletonise the worm in each video frame, but instead of looking at the skeleton positions they analyse skeleton angles, rotated by the mean angle, yielding a position and orientation independent representation of body postures over time. They found that the covariance matrix of these angles has a relatively smooth structure and that just four eigenvectors can capture 95% of the shape variance of worms crawling off food. They call these four principal components eigenworms, which are essentially four basis shapes that can be added together to reconstruct worm postures. This compact representation led to interesting results on dynamical models of *C. elegans* and even to the emergence of stereotyped behaviour without the requirement of a central pattern generator (Stephens et al. 2011a). More recently, we have shown that the eigenworms derived from wild-type animals can also be used to capture the postures of mutant worms, even those that are highly uncoordinated (Brown et al. 2012). This extends the applicability of the eigenworm representation and provides a common and compact basis for capturing worm postures.

12.2.3   Towards an open worm analysis platform

Given the variety of systems that have been developed for the automated analysis of *C. elegans* behaviour from videos, we argue that now is a reasonable time to consider coordinating the efforts of individual groups and creating an open platform for worm behaviour. In part this could consist simply of shared knowledge of protocols and hardware design. More importantly though, a set of standard functions available in an easy-to-use and extendable package would help lower the barrier to entry to new groups and focus researchers with interest and skills on developing useful new analysis tools rather than re-inventing the wheel at the start of each new project. In the ideal case, the project would look something like ImageJ, the open source image analysis package developed at the National Institutes of Health (Abramoff et al. 2004). It contains many core functions in an easy-to-use interface but most importantly, it allows the incorporation of plugins and has been adopted by an active group of users and developers. Of course, the possible user base for such a package will be much less than for ImageJ, but its focus will be correspondingly narrower and hopefully still manageable.

12.3   Linking behaviours and genes

12.3.1   Insights from quantification

Sometimes knocking out a gene results in little or no observable behavioural consequence. In these cases, careful quantification can confirm a phenotype suggested by human observation or even reveal completely new phenotypes. For example, worms lacking an ion channel called *trpa-1* move well and seem healthy when observed under a dissecting scope (Kindt et al. 2007). However, on closer inspection you might notice an abnormally large head swing (sometimes called foraging motion). By quantifying the rate of head swings a subtle but reproducible phenotype emerged and this defect in foraging helped direct studies that revealed new aspects of *trpa-1* function (Kindt et al. 2007).

A related situation arose in the study of proprioception, or sense of body position, in *C. elegans*. Mutant worms lacking an ion channel called *unc-8* were previously reported to have no visible phenotype (Park & Horvitz 1986), but upon closer inspection were found to have a shallower body bend during locomotion than wild-type worms. This visual impression was confirmed using track analysis (Tavernarakis et al. 1997). Because of the neural circuits where *unc-8* is expressed and its homology to MEC-4 channels involved in mechanosensation, it was hypothesised to have a role in proprioception and the propagation of the travelling wave worms use for locomotion. Similarly, mutants lacking an ion channel called *trp-4* show a posture defect but still move well. However, in contrast to *unc-8* animals, *trp-4* worms have deeper body bends than the wild-type (Li et al. 2006). Several experiments now strongly suggest that *trp-4* encodes a mechanosensitive ion channel with a role in proprioception (Kang et al. 2010).

There are two aspects of these studies that are particularly interesting. First, phenotypic quantification was used to confirm subtle phenotypes which then guided functional experiments. Second, although the phenotypes were subtle, they were picked up by human observers so it was clear what needed to be quantified. To see if there are other channels in the trp and deg channel families (*unc-8* is a DEG/ENaC channel and *trp-4* a TRP channel) that affect curvature, we tracked worms (Yemini et al. 2011b) with mutations in one or sometimes several TRP or DEG/ENaC channels. Several of these mutants were significantly more or less curved than wild-type, possibly suggesting more channels with roles in proprioception (Fig. 12.5). Note that *unc-8* and *trp-4* are at the extremes of the curvature ranges, perhaps explaining why these channels were the ones initially picked up by human observation. Although the other curvature phenotypes have not been previously reported, quantitative analysis can reliably detect them and they may prove just as useful in guiding functional studies.

12.3.2   Unbiased reverse genetic analysis of *C. elegans* behaviour

As the curvature example suggests, even looking at a single feature can reveal previously unobserved phenotypes. What might we learn if we instead took an unbiased look at many mutants and many features? There have been several papers that demonstrate the potential of this approach at a relatively small scale.

The first application of automated tracking and machine vision-based analysis to *C. elegans* was reported by Baek et al. in 2002 using single-worm tracking and algorithms to extract morphological, locomotion and posture data (Baek et al. 2002). Using video data recorded from five mutant strains, they built a classification tree that could reliably distinguish the different types. Shortly thereafter, Geng et al. used the same system with eight mutant strains to show that the data could not only be used to classify worms based on their genotypes, but also cluster them based on their phenotypic similarity (Geng et al. 2004). Although related, clustering phenotypically similar strains is more relevant than accurately dividing a mixed population into genotype classes because a worm's genotype can be easily and accurately determined directly. Furthermore, although it may seem reasonable to assume that features with high classification accuracy will also be informative for phenotypic clustering, this is not necessarily the case and even for the small number of strains reported in these papers, somewhat different features were found to be important for the clustering compared to the classification tasks.

Since then, there have been several papers that report automated algorithms to quantify the behaviour of *C. elegans* mutants. Without being exhaustive, we will highlight some notable examples. A similar approach, but with different, complementary features

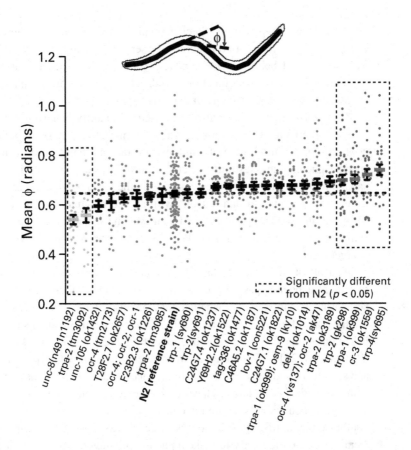

Figure 12.5   Some mutant strains are more or less curved than the N2 reference strain (wild type). Each point in the plot represents an individual worm tracked for 15 minutes. In each frame the change in tangent angle ϕ is averaged over the skeleton. These approximately 25 000 measurements are averaged to give each point shown in the figure. Black bars are means and error bars show standard errors. The black dashed line shows the N2 mean. Strains that are significantly more or less curved than N2 are indicated with the dashed boxes. A black and white version of this figure will appear in some formats. For the colour version, please refer to the plate section.

was described by Cronin et al. in 2006 (Cronin et al. 2006). They applied their system to a small number of mutants as a proof of concept and showed that they could detect differences in mutant sensitivities to chemicals including arsenite and aldicarb. More recently, Sznitman et al. (2010a) and Krajacic et al. (2012) have taken a novel approach to feature extraction based on a detailed mechanical model of swimming *C. elegans* that takes into account both the worm body and its interaction with the surrounding fluid. What makes their approach unique is that the features they extract – worm stiffness and tissue viscosity – have a direct physical interpretation that cannot be directly seen in the videos. They show that some mutants with defective muscle can be distinguished from wild-type and from each other based on these estimated parameters.

Stephens et al. have used their eigenworm representation combined with stochastic modelling to derive unique metrics that summarise worm behaviour and can distinguish mutant from wild-type worms (Stephens et al. 2011a). As with the approach of Sznitman et al., these biophysics-based features may serve as useful complements to more empirical measures of behaviour for genetics. They will be especially useful to the extent that they guide subsequent functional studies by relating a mutant phenotype to specific aspects of a worm's physiology or nervous system.

Multi-worm trackers have also been used to quantify behaviours of mutants and to look at the effects of, for example, drugs (Ramot et al. 2008) and temperature (Biron et al. 2006). As in the case of single-worm trackers, these have been used to answer specific phenotypic questions or serve as proof-of-concept experiments on relatively small numbers of mutants. Swierczek et al. studied the response of populations of worms to repeated plate tap to find mutants with altered habituation (Swierczek et al. 2011). They also showed the applicability of their system for studies of chemotaxis. They found some interesting hits in their sample of 33 mutant strains and this will hopefully lead to a larger-scale study.

What all of these studies have in common is their use of human-defined features to represent phenotypes. However, we do not know if these features are optimal for comparing phenotypes nor whether they are useful for discovering new behaviours. We have therefore also taken a complementary route, using an unsupervised search for behavioural motifs – short subsequences of repeated behaviour – to define phenotypes (Brown et al. 2012). We projected worm skeletons onto wild-type-derived eigenworms and searched for closely repeated subsequences of different lengths and repeated this for many individuals in a large behavioural database. Some of the motifs represent subtle or irregular behaviours that are nonetheless closely repeated at two different times. We combined the motifs into a dictionary and each video was then compared to each of the elements of the dictionary to make a phenotypic fingerprint for that individual. This is analogous to using features as described above, but now each 'feature' is a distance from an automatically identified behavioural motif from the dictionary. Phenotypic comparisons based on the motif dictionary recapitulated some known genetic relationships and were used to hypothesise connections between previously uncharacterised genes (Brown et al. 2012).

12.4   Outlook

12.4.1   Increasing throughput *and* discrimination

It is clear from the previous sections that the core technologies for high-throughput quantitative phenotyping of *C. elegans* are within reach, but there remain several challenges that must be overcome. A framework for understanding the current state of the art and directions for future improvement is summarised in Fig. 12.4. In essence, it is difficult to achieve both high content and high throughput in a single system. A multi-worm tracker can collect video data from enough individuals to distinguish at least some phenotypes in about 1 hour. This number could be decreased with cameras with more pixels and faster computers, bringing, for example, a genome-wide screen within reach. For groups interested in a particular uncommon phenotype, this could be sufficient.

For example, one of the main phenotypes described by Swierczek et al. was tap habituation. This is the rate at which worms become insensitive to mechanical stimulation provided by repeated taps to their plate. If there are a relatively small number of genes involved, then a large-scale screen could reveal a tractable number of candidate mutants that can be followed up for further functional analysis. If, on the other hand, one is interested in proprioception, a genome-wide search for mutants with a curvature phenotype is likely to yield a large number of hits because there are many ways of disrupting neural connections or signalling pathways that lead to an increase or decrease in curvature. In this case it would be desirable to have a more detailed behavioural fingerprint for clustering worms into related phenotypic classes that may be functionally related as has been demonstrated for a small number of mutants. Then it would be possible to target follow-up experiments to particular phenotypic sub-classes of particular interest.

This consideration of phenotypic content may be especially critical for future large-scale behavioural screens. Multi-worm trackers provide greater throughput, but if they lack the sensitivity to meaningfully distinguish or cluster hundreds or thousands of mutant strains this throughput may be difficult to put to good use. Analysis at this scale has not yet been attempted for worm behaviour so this remains an open question. The development of high-resolution multi-worm trackers may help (Swierczek et al. 2011).

12.4.2   Phenotyping by imaging neuromuscular activity

Behaviour is the result of a potentially complex feedback between sensation, neuromuscular activity, anatomy and the physical properties of the environment. Isolating the contributions from each of these factors is a challenge, but one area where there are likely to be rapid advances is in linking the activity in specific neural circuits and muscles with behaviour and ultimately with genetics. The main technology driving this advance is the optical recording of neural activity with genetically encoded reporters in freely behaving animals. Fluorescence imaging is relatively non-invasive and is perfectly suited to recording from multiple cells simultaneously. Moreover, nematodes are largely transparent, making their neurons easily accessible to optical recording. Indeed, the use of genetically encoded sensors to record neural activity was first demonstrated in *C. elegans* (Kerr et al. 2000).

Currently the most widely used probes are genetically encoded calcium indicators (GECIs) that change their emission in response to changes in Ca^{2+} concentration. This can be either through a change in Förster resonance energy transfer (FRET) efficiency in the case of 'cameleons' (Miyawaki et al. 1997) or intensity in the case of G-CaMP (Nakai et al. 2001) and related sensors. There has also been some recent notable success in genetically encoded sensors designed to directly sense voltage across cell membranes rather than a second messenger like Ca^{2+} (Kralj et al. 2011). Voltage sensors are not limited by the timescale of cellular Ca^{2+} dynamics and do not rely on the proper functioning of endogenous voltage-gated Ca^{2+} channels. Despite their promise, genetically encoded voltage sensors have not yet been developed to the same extent as Ca^{2+} sensors and are not yet widely used in intact animals.

In worms, GECIs have been used to record the activities of a wide range of neurons. In immobilised animals, the responses of individual sensory neurons to mechanical (Suzuki et al. 2003), proprioceptive (Li et al. 2006), chemical (Hilliard et al. 2005, Chalasani et al. 2007) and thermal (Kimura et al. 2004, Biron et al. 2006) stimuli have helped dissect the role of specific neurons in neural circuits (Suzuki et al. 2008, Macosko et al. 2009). It is this ability to correlate a macroscopic behavioural response with the underlying neural circuits that makes neural imaging such an exciting complement to whole-animal behavioural quantification. Demonstrations of neural imaging in unconstrained worms are now emerging. Haspel et al. examined the correlation between motor neuron activity and locomotion but were only able to resolve broad activity differences between forward and backward locomotion (Haspel et al. 2010). This work was extended by Kawano et al. who found that an imbalance in motor neuron activity governs the forward or backward state and identified gap junction genes involved in regulating these states (Kawano et al. 2011). In another study, Faumont et al. showed very fast tracking and high-resolution functional imaging using a quadrant photodiode (rather than image) based tracking system (Faumont et al. 2011).

The combination of neural imaging and worm tracking is in a state analogous to worm tracking itself a decade ago: there have been some interesting demonstrations that have focused on proving feasibility and answering specific questions about locomotion or the action of specific genes. Genetically encoded Ca^{2+} and voltage indicators have ever-improving sensitivity, response time and signal-to-noise ratio (Looger & Griesbeck 2012). At the same time, sensitive cameras are getting faster and cheaper. An inexpensive general-purpose system for both tracking and neural imaging would vastly increase the scope of behavioural phenotyping in worms in part because mutations that lead to subtle defects in particular neural circuits may be masked by compensation in another circuit. This may be a factor that limits the discriminative power in genome-wide screens for behavioural mutants. Even a coarse-grained view of neural circuit dynamics may offer significant improvements in discriminative power (Fig. 12.4).

Any attempt to integrate neural activity into automated behavioural genetics will face an explosion of data, not just in volume, but also dimension. Fortunately, with appropriate analysis it should still be possible to devise an abstraction that makes the data amenable to existing (and future) methods for high-dimensional data. A critical challenge will be to search for a low-dimensional representation that still captures important variation as was done for *C. elegans* locomotion (Stephens et al. 2008). The generality of this approach (Stephens et al. 2011b) bodes well for applications in neural imaging, especially in a relatively tractable nervous system like that of *C. elegans*.

12.4.3   Beyond spontaneous behaviour

The relative simplicity of *C. elegans* has led to speculation that it may be possible to describe quantitatively its entire behavioural repertoire (Stephens et al. 2008). This may be possible in the near future for spontaneous locomotion on a surface, but worms are capable of much more. Exactly how much more is not yet known quantitatively, but finding out will require an extension of behavioural observation beyond spontaneous behaviour on a featureless agar plate.

Caenorhabditis elegans responds to a wide range of stimuli. For example, when gently touched on their anterior half, worms will reverse. When touched on their posterior half, they accelerate forward and these responses are controlled by specific neural circuits (Chalfie et al. 1985). Worms also sense soluble and volatile chemical cues including salt, cAMP, biotin, carbon dioxide and oxygen (Bargmann 2006). Worms actively chemotax up attractant gradients and display escape behaviours around noxious chemicals. Worms also respond to temperature and thermotax toward their most recent cultivation temperature. They will even accurately track isotherms when placed in a temperature gradient (Luo et al. 2006). These behaviours (and the associated genes and neural circuits) have primarily been discovered using manual assays and expert observation, but new technologies are constantly being developed to increase sensitivity and reproducibility and to decrease the subjectivity that can creep into manually scored experiments. Here we review a selection of promising directions to give a sense for what is currently possible and where it might lead.

If a worm crawling on an agar surface is suddenly placed in a drop of liquid, it will rapidly change to a swimming gait characterised by a higher frequency of body bending and a longer body wavelength (Fig. 12.1). This represents quite an extreme change in the worm's mechanical environment and its response seems to be under genetic control (Vidal-Gadea et al. 2011). More subtle transitions are also possible and these have been observed by changing viscosity (Korta et al. 2007, Berri et al. 2009) and more recently with compressive force (Lebois et al. 2012). The advantage of the latter approach is both that the force can easily be changed in real time and also that there is less chance of inhomogeneity, which can complicate interpretation in studies of swimming in viscous fluids (Boyle et al. 2011). What makes these studies so interesting is that an accurate response to viscosity will probably require an integration of senses, for example a sense of the applied force of muscles and of touch. Another possibility would be a sense of applied force combined with sense of posture over time. The role of touch in gait adaptation has been established (Korta et al. 2007), but proprioception has not yet been directly implicated and it is not known whether worms have a direct sense of effort. Further tracking experiments combined with genetic and neural ablations as well as careful theoretical work (Boyle et al. 2008, Fang-Yen et al. 2010, Mailler et al. 2010, Sznitman et al. 2010b, Sauvage et al. 2011) will be required to resolve this case of sensory integration and gait computation.

Microfluidics has opened many opportunities for precisely handling and immobilising worms and for providing specific stimulations. Two-layer devices with pneumatic valves can be used for directing worms to particular chip locations where they can be clamped and assayed (Fig. 12.6). This has been used for studies of laser axotomy and axon regeneration (Yanik et al. 2004), automated (Chung & Lu 2009) and semi-automated (Hulme et al. 2007) neuron ablation and for neural imaging of responses to soluble chemicals (Chronis et al. 2007, Chokshi et al. 2010). See Chronis (2010) for a review of some of the many related applications. Of particular interest for extending behavioural phenotyping are applications that control the environment while allowing

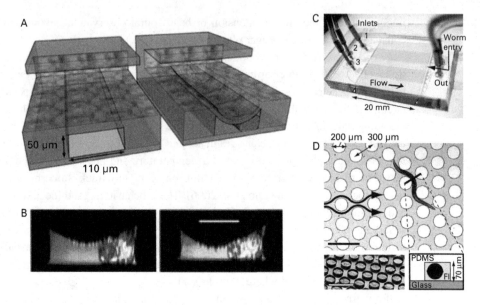

Figure 12.6   (A) Schematic drawing of a pneumatic worm clamping device. (B) Fluorescent image of a clamped worm. (A) and (B) are adapted from Yanik et al. (2004). (C) Microfluidic device for presenting soluble chemicals in sharp steps to worms. (D) Close-up view of a worm in the device shown in (C). (C) and (D) are adapted from Albrecht & Bargmann (2011). A black and white version of this figure will appear in some formats. For the colour version, please refer to the plate section.

relatively free locomotion. Extending the concept of articifcial dirt (Lockery et al. 2008), Albrecht et al. have taken advantage of the laminar flow that exists at low Reynolds number in microfluidic devices to apply precise gradients and sharp boundaries of soluble chemicals to study chemosensation and chemotaxis quantitatively in a potentially high-throughput manner (Albrecht & Bargmann 2011). Another particularly simple and provocative experiment suggested that worms could remember the solution to a simple microfluidic T-maze (Qin & Wheeler 2007).

Another method of applying controlled simuli is provided by optogenetics (Fenno et al. 2011). The basic approach is to express a light-gated ion channel using promoters specific to a subset of neurons. Then, when these channels are activated by light they either depolarise and excite or hyperpolarise and inhibit the neurons where they are expressed. By controlling expression, a simple apparatus can be used to activate or inhibit specific neurons and monitor the effect on behaviour (Nagel et al. 2005). However, because promoters are often not available that are specific to single neurons, it is desirable to be able to target a particular neuron from a population that expresses the light-sensitive channel. This has the added advantage that different neurons can be targeted in the same animal in a single experiment. Following from earlier work that combined activation and imaging in immobilised worms (Guo et al. 2009), two groups simultaneously published reports describing the activation of specific neurons in freely behaving animals using a motorised stage and a digital micromirror device

(Leifer et al. 2011) or a slightly modified projector (Stirman et al. 2011) for local illumination. The obvious application of these systems is for investigating the role of specific neurons and neural circuits in behaviour, but there will also be applications in genetics. In particular, even for mutants with no obvious spontaneous locomotion defects it may be possible to detect changes in responses to neuronal stimulation that may reflect more subtle defects in neural circuit function.

12.5   Conclusions

The promise of automated behavioural phenotyping in *C. elegans* has long been recognised and will soon be realised on a large scale. Quantitative data enable more meaningful comparisons with models of locomotion that are beginning to emerge and can in some cases lead to new insights into the diversity of – as well as constraints on – behaviour. The better we understand behaviour the more able we will be to design useful behavioural metrics and to interpret mutant phenotypes. Already, new algorithms and approaches are showing potential for applications in genetics and with new inexpensive high-throughput methods, open and useable software and new bioinformatic approaches, these will be more widely adopted and applied to a broad range of topics.

The next decade promises substantial advances in our understand of behaviour and the connection between genotype and this fascinating range of phenotypes.

References

Abramoff, M., Magalhaes, P. & Ram, S. (2004), 'Image processing with Image J', *Biophotonics International* **11**(7), 36–42.

Ahringer, J. (2006), 'Reverse genetics', *WormBook*, www.wormbook.org.

Albrecht, D. R. & Bargmann, C. I. (2011), 'High-content behavioral analysis of *Caenorhabditis elegans* in precise spatiotemporal chemical environments', *Nature Methods* **8**(7), 599–605.

Baek, J.-H., Cosman, P., Feng, Z., Silver, J. & Schafer, W. R. (2002), 'Using machine vision to analyze and classify *Caenorhabditis elegans* behavioral phenotypes quantitatively', *Journal of Neuroscience Methods* **118**, 9–21.

Bargmann, C. (2006), 'Chemosensation in *C. elegans*', *WormBook*, www.wormbook.org.

Barriere, A. & Felix, M.-A. (2005), 'High local genetic diversity and low outcrossing rate in *Caenorhabditis elegans* natural populations', *Current Biology* **15**(13), 1176–1184.

Berri, S., Boyle, J. H., Tassieri, M., Hope, I. A. & Cohen, N. (2009), 'Forward locomotion of the nematode *C. elegans* is achieved through modulation of a single gait', *HFSP Journal* **3**(3), 186–193.

Biron, D., Shibuya, M., Gabel, C., Wasserman, S. M., Clark, D. A. et al. (2006), 'A diacylglycerol kinase modulates long-term thermotactic behavioral plasticity in *C. elegans*', *Nature Neuroscience* **9**(12), 1499–1505.

Boyle, J. H., Berri, S., Tassieri, M., Hope, I. A. & Cohen, N. (2011), 'Gait modulation in *C. elegans*: It's not a choice, it's a reflex!', *Frontiers in Behavioral Neuroscience* **5**, 10.

Boyle, J. H., Bryden, J. & Cohen, N. (2008), 'An integrated neuro-mechanical model of *C. elegans* forward locomotion, *in* M. Ishikawa, K. Doya, H. Miyamoto & T. Yamakawa, eds., *Neural Information Processing*, Vol. 4984, Springer, Berlin, pp. 37–47.

Brenner, S. (1974), 'The genetics of *Caenorhabditis elegans*', *Genetics* **77**(1), 71–94.

Brown, A. E. X., Yemini, E. I., Grundy, L. J., Jucikas, T. & Schafer, W. R. (2012), 'A dictionary of behavioral motifs reveals clusters of genes affecting *Caenorhabditis elegans* locomotion', *Proceedings of the National Academy of Sciences of the USA* **110**(2), 791–796.

C. elegans Sequencing Consortium (1998), 'Genome sequence of the nematode *C. elegans*: A platform for investigating biology', *Science* **282**(5396), 2012–2018.

Calixto, A., Chelur, D., Topalidou, I., Chen, X. & Chalfie, M. (2010), 'Enhanced neuronal RNAi in *C. elegans* using SID-1', *Nature Methods* **7**, 554–559.

Chalasani, S. H., Chronis, N., Tsunozaki, M., Gray, J. M., Ramot, D. et al. (2007), 'Dissecting a circuit for olfactory behaviour in *Caenorhabditis elegans*', *Nature* **450**(7166), 63–70.

Chalfie, M., Sulston, J. E., White, J. G., Southgate, E., Thomson, J. N. et al. (1985), 'The neural circuit for touch sensitivity in *Caenorhabditis elegans*', *Journal of Neuroscience* **5**(4), 956–964.

Chokshi, T. V., Bazopoulou, D. & Chronis, N. (2010), 'An automated microfluidic platform for calcium imaging of chemosensory neurons in *Caenorhabditis elegans*', *Lab on a Chip* **10**(20), 2758.

Chronis, N. (2010), 'Worm chips: Microtools for *C. elegans* biology', *Lab on a Chip* **10**(4), 432.

Chronis, N., Zimmer, M. & Bargmann, C. I. (2007), 'Microfluidics for in vivo imaging of neuronal and behavioral activity in *Caenorhabditis elegans*', *Nature Methods* **4**(9), 727–731.

Chung, K. & Lu, H. (2009), 'Automated high-throughput cell microsurgery on-chip', *Lab on a Chip* **9**(19), 2764.

Croll, N. A. (1975a), 'Behavioural analysis of nematode movement', *Advances in Parasitology* **13**, 71–122.

Croll, N. A. (1975b), 'Components and patterns in the behaviour of the nematode *Caenorhabditis elegans*', *Journal of Zoology* **176**, 159–176.

Croll, N. A. & Blair, A. (1973) , 'Inherent movement patterns of larval nematodes, with a stochastic model to simulate movement of infective hookworm larvae', *Parasitology* **67**, 53.

Cronin, C. J., Feng, Z. & Schafer, W. R. (2006), 'Automated imaging of *C. elegans* behavior', *Methods in Molecular Biology* **351**, 241–251.

de Bono, M. & Bargmann, C. I. (1998), 'Natural variation in a neuropeptide Y receptor homolog modifies social behavior and food response in *C. elegans*', *Cell* **94**(5), 679–689.

Dusenbery, D. B. (1985a), 'Using a microcomputer and video camera to simultaneously track 25 animals', *Computers in Biology and Medicine* **15**, 169–175.

Dusenbery, D. B. (1985b), 'Video camera-computer tracking of nematode *Caenorhabditis elegans* to record behavioral responses', *Journal of Chemical Ecology* **11**, 1239–1247.

Fang-Yen, C., Wyart, M., Xie, J., Kawai, R., Kodger, T. et al. (2010), 'Biomechanical analysis of gait adaptation in the nematode *Caenorhabditis elegans*', *Proceedings of the National Academy of Sciences of the USA* **107**(47), 20 323–20 328.

Faumont, S., Rondeau, G., Thiele, T. R., Lawton, K. J., McCormick, K. E. et al. (2011), 'An image-free opto-mechanical system for creating virtual environments and imaging neuronal activity in freely moving *Caenorhabditis elegans*', *PLoS One* **6**(9), e24666.

Feng, Z., Cronin, C. J., Wittig, J. H. J., Sternberg, P. W. & Schafer, W. R. (2004), 'An imaging system for standardized quantitative analysis of *C. elegans* behavior', *BMC Bioinformatics* **5**, 115.

Fenno, L., Yizhar, O. & Deisseroth, K. (2011), 'The development and application of optogenetics', *Annual Review of Neuroscience* **34**(1), 389–412.

Fontaine, E., Burdick, J. & Barr, A. (2006), 'Automated tracking of multiple *C. elegans*', *Proceedings of the Annual International Conference of the IEEE Engineering in Medicine and Biology Society*, pp. 3716–3719.

Fraser, A. G., Kamath, R. S., Zipperlen, P., Martinez-Campos, M., Sohrmann, M. et al. (2000), 'Functional genomic analysis of *C. elegans* chromosome I by systematic RNA interference', *Nature* **408**(6810), 325–330.

Frøkjaer-Jensen, C., Davis, M. W., Hollopeter, G., Taylor, J., Harris, T. W. et al. (2010), 'Targeted gene deletions in *C. elegans* using transposon excision', *Nature Methods* **7**(6), 451–453.

Gaugler, R. & Bilgrami, A. L. (2004), *Nematode Behaviour*, CABI Publications, Wallingford, UK.

Geng, W., Cosman, P., Berry, C. C., Feng, Z. & Schafer, W. R. (2004), 'Automatic tracking, feature extraction and classification of *C. elegans* phenotypes', *IEEE Transactions on Bio-Medical Engineering* **51**(10), 1811–1820.

Guo, Z. V., Hart, A. C. & Ramanathan, S. (2009), 'Optical interrogation of neural circuits in *Caenorhabditis elegans*', *Nature Methods* **6**(12), 891–896.

Hardaker, L. A., Singer, E., Kerr, R., Zhou, G. & Schafer, W. R. (2001), 'Serotonin modulates locomotory behavior and coordinates egg-laying and movement in *Caenorhabditis elegans*', *Journal of Neurobiology* **49**(4), 303–313.

Haspel, G., O'Donovan, M. J. & Hart, A. C. (2010), 'Motoneurons dedicated to either forward or backward locomotion in the nematode *Caenorhabditis elegans*', *Journal of Neuroscience* **30**(33), 11 151–11 156.

Hilliard, M. A., Apicella, A. J., Kerr, R., Suzuki, H., Bazzicalupo, P. et al. (2005), 'In vivo imaging of *C. elegans* ASH neurons: Cellular response and adaptation to chemical repellents', *EMBO Journal* **24**(1), 63–72.

Huang, K.-M., Cosman, P. & Schafer, W. R. (2006), 'Machine vision based detection of omega bends and reversals in *C. elegans*', *Journal of Neuroscience Methods* **158**(2), 323–336.

Huang, K.-M., Cosman, P. & Schafer, W. R. (2008), 'Automated detection and analysis of foraging behavior in *Caenorhabditis elegans*', *Journal of Neuroscience Methods* **171**(1), 153–164.

Hulme, S. E., Shevkoplyas, S. S., Apfeld, J., Fontana, W. & Whitesides, G. M. (2007), 'A microfabricated array of clamps for immobilizing and imaging *C. elegans*', *Lab on a Chip* **7**(11), 1515.

Kamath, R. S., Fraser, A. G., Dong, Y., Poulin, G., Durbin, R. et al. (2003), 'Systematic functional analysis of the *Caenorhabditis elegans* genome using RNAi', *Nature* **421**(6920), 231–237.

Kang, L., Gao, J., Schafer, W. R., Xie, Z. & Xu, X. S. (2010), '*C. elegans* TRP family protein TRP-4 is a pore-forming subunit of a native mechanotransduction channel', *Neuron* **67**(3), 381–391.

Kawano, T., Po, M. D., Gao, S., Leung, G., Ryu, W. S. et al. (2011), 'An imbalancing act: Gap junctions reduce the backward motor circuit activity to bias *C. elegans* for forward locomotion', *Neuron* **72**(4), 572–586.

Kennedy, S., Wang, D. & Ruvkun, G. (2004), 'A conserved siRNA-degrading RNase negatively regulates RNA interference in *C. elegans*', *Nature* **427**(6975), 645–649.

Kerr, R., Lev-Ram, V., Baird, G., Vincent, P., Tsien, R. Y. et al. (2000), 'Optical imaging of calcium transients in neurons and pharyngeal muscle of *C. elegans*', *Neuron* **26**(3), 583–594.

Kimura, K. D., Miyawaki, A., Matsumoto, K. & Mori, I. (2004), 'The *C. elegans* thermosensory neuron AFD responds to warming', *Current Biology* **14**(14), 1291–1295.

Kindt, K. S., Viswanath, V., Macpherson, L., Quast, K., Hu, H. et al. (2007), '*Caenorhabditis elegans* TRPA-1 functions in mechanosensation', *Nature Neuroscience* **10**(5), 568–577.

Korta, J., Clark, D. A., Gabel, C. V., Mahadevan, L. & Samuel, A. D. T. (2007), 'Mechanosensation and mechanical load modulate the locomotory gait of swimming *C. elegans*', *Journal of Experimental Biology* **210**(13), 2383–2389.

Krajacic, P., Shen, X., Purohit, P. K., Arratia, P. & Lamitina, T. (2012), 'Biomechanical profiling of *Caenorhabditis elegans* motility', *Genetics* **191**(3), 1015–1021.

Kralj, J. M., Douglass, A. D., Hochbaum, D. R., Maclaurin, D. & Cohen, A. E. (2011), 'Optical recording of action potentials in mammalian neurons using a microbial rhodopsin', *Nature Methods* **9**(1), 90–95.

Lebois, F., Sauvage, P., Py, C., Cardoso, O., Ladoux, B. et al. (2012), 'Locomotion control of *Caenorhabditis elegans* through confinement', *Biophysical Journal* **102**(12), 2791–2798.

Lee, H., Choi, M.-K., Lee, D., Kim, H.-S., Hwang, H. et al. (2011), 'Nictation, a dispersal behavior of the nematode *Caenorhabditis elegans*, is regulated by IL2 neurons', *Nature Neuroscience* **15**(1), 107–112.

Lehner, B., Calixto, A., Crombie, C., Tischler, J., Fortunato, A. et al. (2006), 'Loss of LIN-35, the *Caenorhabditis elegans* ortholog of the tumor suppressor p105Rb, results in enhanced RNA interference', *Genome Biology* **7**(1), R4.

Leifer, A. M., Fang-Yen, C., Gershow, M., Alkema, M. J. & Samuel, A. D. T. (2011), 'Optogenetic manipulation of neural activity in freely moving *Caenorhabditis elegans*', *Nature Methods* **8**, 147–152.

Li, W., Feng, Z., Sternberg, P. W. & Shawn Xu, X. Z. (2006), 'A *C. elegans* stretch receptor neuron revealed by a mechanosensitive TRP channel homologue', *Nature* **440**(7084), 684–687.

Lockery, S. R., Lawton, K. J., Doll, J. C., Faumont, S., Coulthard, S. M. et al. (2008), 'Artificial dirt: Microfluidic substrates for nematode neurobiology and behavior', *Journal of Neurophysiology* **99**(6), 3136–3143.

Looger, L. L. & Griesbeck, O. (2012), 'Genetically encoded neural activity indicators', *Current Opinion in Neurobiology* **22**(1), 18–23.

Luo, L., Clark, D. A., Biron, D., Mahadevan, L. & Samuel, A. D. T. (2006), 'Sensorimotor control during isothermal tracking in *Caenorhabditis elegans*', *Journal of Experimental Biology* **209**(23), 4652–4662.

Macosko, E. Z., Pokala, N., Feinberg, E. H., Chalasani, S. H., Butcher, R. A. et al. (2009), 'A hub-and-spoke circuit drives pheromone attraction and social behaviour in *C. elegans*', *Nature* **458**(7242), 1171–1175.

Mailler, R., Avery, J., Graves, J. & Willy, N. (2010), 'A biologically accurate 3D model of the locomotion of *Caenorhabditis elegans*', *International Conference on Biosciences (BIOSCIENCESWORLD)*, pp. 84–90.

Miyawaki, A., Llopis, J., Heim, R., McCaffery, J. M., Adams, J. A. et al. (1997), 'Fluorescent indicators for Ca^{2+} based on green fluorescent proteins and calmodulin', *Nature* **388**(6645), 882–887.

Nagel, G., Brauner, M., Liewald, J. F., Adeishvili, N., Bamberg, E. et al. (2005), 'Light activation of channelrhodopsin-2 in excitable cells of *Caenorhabditis elegans* triggers rapid behavioral responses', *Current Biology* **15**(24), 2279–2284.

Nakai, J., Ohkura, M. & Imoto, K. (2001), 'A high signal-to-noise Ca^{2+} probe composed of a single green fluorescent protein', *Nature Biotechnology* **19**(2), 137–141.

Park, E. C. & Horvitz, H. R. (1986), 'Mutations with dominant effects on the behavior and morphology of the nematode *Caenorhabditis elegans*', *Genetics* **113**(4), 821–852.

Pierce-Shimomura, J. T., Morse, T. M. & Lockery, S. R. (1999), 'The fundamental role of pirouettes in *Caenorhabditis elegans* chemotaxis', *Journal of Neuroscience* **19**(21), 9557–9569.

Qin, J. & Wheeler, A. R. (2007), 'Maze exploration and learning in *C. elegans*', *Lab on a Chip* **7**(2), 186–192.

Ramot, D., Johnson, B. E., Berry, T. L., Carnell, L. & Goodman, M. B. (2008), 'The parallel worm tracker: A platform for measuring average speed and drug-induced paralysis in nematodes', *PLoS One* **3**, e2208.

Roussel, N., Morton, C. A., Finger, F. P. & Roysam, B. (2007), 'A computational model for *C. elegans* locomotory behavior: Application to multiworm tracking', *IEEE Transactions on Bio-Medical Engineering* **54**(10), 1786–1797.

Sauvage, P., Argentina, M., Drappier, J., Senden, T., Simon, J. et al. (2011), 'An elastohydrodynamical model of friction for the locomotion of *Caenorhabditis elegans*', *Journal of Biomechanics* **44**(6), 1117–1122.

Simmer, F., Tijsterman, M., Parrish, S., Koushika, S. P., Nonet, M. L. et al. (2002), 'Loss of the putative RNA-directed RNA polymerase RRF-3 makes *C. elegans* hypersensitive to RNAi', *Current Biology* **12**(15), 1317–1319.

Stephens, G. J., Bueno de Mesquita, M., Ryu, W. S. & Bialek, W. (2011a), 'Emergence of long timescales and stereotyped behaviors in *Caenorhabditis elegans*', *Proceedings of the National Academy of Sciences of the USA* **108**(18), 7286–7289.

Stephens, G. J., Johnson-Kerner, B., Bialek, W. & Ryu, W. S. (2008), 'Dimensionality and dynamics in the behavior of *C. elegans*', *PLoS Computational Biology* **4**, e1000028.

Stephens, G. J., Osborne, L. C. & Bialek, W. (2011b), 'Colloquium paper: Searching for simplicity in the analysis of neurons and behavior', *Proceedings of the National Academy of Sciences of the USA* **108**(Supplement 3), 15 565–15 571.

Stirman, J. N., Crane, M. M., Husson, S. J., Wabnig, S., Schultheis, C. et al. (2011), 'Real-time multimodal optical control of neurons and muscles in freely behaving *Caenorhabditis elegans*', *Nature Methods* **8**, 153–158.

Sulston, J. (1977), 'Post-embryonic cell lineages of the nematode, *Caenorhabditis elegans*', *Developmental Biology* **56**(1), 110–156.

Suzuki, H., Kerr, R., Bianchi, L., Frøkjaer-Jensen, C., Slone, D. et al. (2003), 'In vivo imaging of *C. elegans* mechanosensory neurons demonstrates a specific role for the MEC-4 channel in the process of gentle touch sensation', *Neuron* **39**(6), 1005–1017.

Suzuki, H., Thiele, T. R., Faumont, S., Ezcurra, M., Lockery, S. R. et al. (2008), 'Functional asymmetry in *Caenorhabditis elegans* taste neurons and its computational role in chemotaxis', *Nature* **454**(7200), 114–117.

Swierczek, N. A., Giles, A. C., Rankin, C. H. & Kerr, R. A. (2011), 'High-throughput behavioral analysis in *C. elegans*', *Nature Methods* **8**, 592–598.

Sznitman, J., Purohit, P. K., Krajacic, P., Lamitina, T. & Arratia, P. (2010a), 'Material properties of *Caenorhabditis elegans* swimming at low Reynolds number', *Biophysical Journal* **98**(4), 617–626.

Sznitman, J., Shen, X., Purohit, P. K. & Arratia, P. E. (2010b), 'The effects of fluid viscosity on the kinematics and material properties of *C. elegans* swimming at low Reynolds number', *Experimental Mechanics* **50**(9), 1303–1311.

Sznitman, R., Gupta, M., Hager, G. D., Arratia, P. E. & Sznitman, J. (2010c), 'Multi-environment model estimation for motility analysis of *Caenorhabditis elegans*', *PLoS One* **5**, e11631.

Tavernarakis, N., Shreffler, W., Wang, S. & Driscoll, M. (1997), 'unc-8, a DEG/ENaC family member, encodes a subunit of a candidate mechanically gated channel that modulates *C. elegans* locomotion', *Neuron* **18**(1), 107–119.

Tsibidis, G. D. & Tavernarakis, N. (2007), '*Nemo*: A computational tool for analyzing nematode locomotion', *BMC Neuroscience* **8**, 86.

Vidal-Gadea, A., Topper, S., Young, L., Crisp, A., Kressin, L. et al. (2011), '*Caenorhabditis elegans* selects distinct crawling and swimming gaits via dopamine and serotonin', *Proceedings of the National Academy of Sciences of the USA* **108**(42), 17 504–17 509.

Waggoner, L. E., Zhou, G. T., Schafer, R. W. & Schafer, W. R. (1998), 'Control of alternative behavioral states by serotonin in *Caenorhabditis elegans*', *Neuron* **21**(1), 203–214.

Wallace, H. R. (1968), 'The dynamics of nematode movement', *Annual Review of Phytopathology* **6**(1), 91–114.

Wang, W., Sun, Y., Dixon, S. J., Alexander, M. & Roy, P. J. (2009), 'An automated micropositioning system for investigating *C. elegans* locomotive behavior', *Journal of the Association for Laboratory Automation* **14**, 269–276.

White, J. G., Southgate, E., Thomson, J. N. & Brenner, S. (1986), 'The structure of the nervous system of the nematode *Caenorhabditis elegans*', *Philosophical Transactions of the Royal Society B: Biological Sciences* **314**(1165), 1–340.

Williams, P. L. & Dusenbery, D. B. (1990), 'A promising indicator of neurobehavioral toxicity using the nematode *Caenorhabditis elegans* and computer tracking', *Toxicology and Industrial Health* **6**(3–4), 425–440.

Yanik, M. F., Cinar, H., Cinar, H. N., Chisholm, A. D., Jin, Y. et al. (2004), 'Neurosurgery: Functional regeneration after laser axotomy', *Nature* **432**(7019), 822.

Yemini, E., Kerr, R. A. & Schafer, W. R. (2011a), 'Illumination for worm tracking and behavioral imaging', *Cold Spring Harbor Protocols* **2011**(12), pdb.prot067009–pdb.prot067009.

Yemini, E., Kerr, R. A. & Schafer, W. R. (2011b), 'Preparation of samples for single-worm tracking', *Cold Spring Harbor Protocols* **2011**(12), pdb.prot066993–pdb.prot066993.

Yook, K., Harris, T. W., Bieri, T., Cabunoc, A., Chan, J. et al. (2012), 'WormBase 2012: More genomes, more data, new website', *Nucleic Acids Research* **40**, D735–741.

Index

Printed in the United States
By Bookmasters